T0178384

Chemical Admixtures for Concrete

Chemical Admixtures for Concrete

Third edition

Roger Rixom and Noel Mailvaganam

CRC Press
Taylor & Francis Group
Boca Raton London New York

CRC Press is an imprint of the
Taylor & Francis Group, an **informa** business

A TAYLOR & FRANCIS BOOK

First edition published 1978
by Taylor & Francis Ltd

Second edition published 1986
by Taylor & Francis, an imprint of Chapman & Hall

This edition 1999
by Taylor & Francis

CRC Press
Taylor & Francis Group
6000 Broken Sound Parkway NW, Suite 300
Boca Raton, FL 33487-2742

First issued in paperback 2019

ISBN-13: 978-0-419-22520-1 (hbk)
ISBN-13: 978-0-367-44754-0 (pbk)

British Library Cataloguing in Publication Data
A catalogue record for this book is available from the British Library

Library of Congress Cataloging in Publication Data
Rixom, M. R.
    Chemical admixtures for concrete/Roger Rixom and Noel
Mailvaganam. – 3rd ed.
        p.   cm.
    Includes bibliographical references and index.
    1. Concrete–Additives. I. Mailvaganam, N.P. (Noel P., 1938- .
II. Title.
TP884.A3R59 1999
666'.893–dc21   98-43377
                                                                                CIP

Publisher's Note
The publisher has gone to great lengths to ensure the quality of this reprint
but points out that some imperfections in the original may be apparent.

Visit the Taylor & Francis Web site at
http://www.taylorandfrancis.com

and the CRC Press Web site at
http://www.crcpress.com

To my wife Joan and our children Tony, Di, Anne, Sue and Andrew (Roger Rixom)

To my parents Josephine and Arullapa, wife Nalini and sons Dimitri and Stefan (Noel Mailvaganam)

Contents

Disclaimer   xiv
Foreword to third edition   xv
Acknowledgements   xviii

1 Water-reducing agents   1
 1.1 Background and definitions   1
 1.2 The chemistry of water-reducing admixtures   5
     1.2.1 Lignosulfonates   5
     1.2.2 Hydroxycarboxylic acids   9
     1.2.3 Hydroxylated polymers   11
 1.3 The effects of water-reducing admixtures on the
     water–cement system   11
     1.3.1 Rheological considerations   12
     1.3.2 Initial surface effects   14
     1.3.3 Effects on the products and kinetics of hydration   22
     1.3.4 Interpretation in terms of a mode of action   29
 1.4 The effects of water-reducing admixtures on the properties
     of concrete   30
 1.5 The effects of water-reducing admixtures on the properties of
     plastic concrete   30
     1.5.1 Air entrainment   30
     1.5.2 Workability   31
     1.5.3 Workability loss   34
     1.5.4 Water reduction   36
     1.5.5 Setting characteristics of fresh concrete containing
           water-reducing admixtures   40
     1.5.6 The stability of fresh concrete containing
           water-reducing admixtures   41
     1.5.7 Mix design considerations   43
 1.6 The effects of water-reducing admixtures on the properties
     of hardened concrete   45

1.6.1 Structural design parameters   45
1.6.2 Durability aspects   50
1.6.3 Durability guidelines   72
References   73

2 Superplasticizers   77
2.1 Background and definitions   77
2.2 The chemistry of superplasticizers   77
    2.2.1 Sulfonated naphthalene formaldehyde   78
    2.2.2 Sulfonated melamine formaldehyde   79
    2.2.3 Polyacrylates   80
2.3 Effects on the water–cement system   81
    2.3.1 Rheological effects   81
    2.3.2 Zeta potential   82
    2.3.3 Adsorption   83
    2.3.4 Effects on the products and kinetics of hydration   84
    2.3.5 Interpretation in terms of a mode of action   88
2.4 Effects of superplasticizers on the properties of concrete   89
2.5 The effects of superplasticizers on the properties of plastic
    concrete   90
    2.5.1 Air entrainment   90
    2.5.2 Workability   91
    2.5.3 Setting time   93
2.6 The effects of superplasticizers on the properties of hardened
    concrete   93
    2.6.1 Compressive strength   94
    2.6.2 Shrinkage and creep   99
    2.6.3 Freeze–thaw durability   101
    2.6.4 Sulfate resistance   101
References   101

3 Air-entraining agents   104
3.1 Background and definitions   104
    3.1.1 Durability   104
    3.1.2 Cohesion   105
    3.1.3 Density   105
3.2 The chemistry of air-entraining agents   106
    3.2.1 Neutralized wood resins   106
    3.2.2 Fatty-acid salts   107
    3.2.3 Alkyl-aryl sulfonates   107
    3.2.4 Alkyl sulfates   107

        3.2.5 Phenol ethoxylates   108
    3.3 The effects of air-entraining agents on the water–cement
         system   108
        3.3.1 Rheology   109
        3.3.2 Air content and characteristics   110
        3.3.3 Distribution between solid and aqueous phases   115
        3.3.4 Effects on the hydration chemistry of cement   116
        3.3.5 Interpretation as a mechanism of action   119
    3.4 The effect of air-entraining agents on the properties of
         plastic concrete   120
        3.4.1 Volume of air entrained   120
        3.4.2 The stability of the entrained air   128
        3.4.3 Workability   132
        3.4.4 Water reduction   132
        3.4.5 Mix stability   133
        3.4.6 Mix design requirements   133
    3.5 The effects of air-entraining agents on the properties of
         hardened concrete   135
        3.5.1 Structural design parameters   135
        3.5.2 Durability aspects   137
    References   147

4 Concrete dampproofers   149
    4.1 Background and definitions   149
    4.2 The chemistry of concrete dampproofers   150
        4.2.1 Materials which react with cement hydration products   151
        4.2.2 Materials which coalesce on contact with cement
               hydration products   152
        4.2.3 Finely divided hydrophobic materials   152
    4.3 The effects of dampproofers on the water–cement system   153
        4.3.1 Bleeding of cement pastes   153
        4.3.2 Hydration of cement pastes   153
        4.3.3 Effects on the capillary system of hardened paste   153
    4.4 The effects of dampproofers on the properties of plastic
         concrete   156
    4.5 The effects of dampproofers on the properties of hardened
         concrete   157
        4.5.1 Structural design parameters   157
        4.5.2 Durability aspects   158
    References   160

5 Accelerators 162
  5.1 Background and definitions 162
  5.2 The chemistry of accelerators 163
    5.2.1 Calcium chloride 163
    5.2.2 Calcium formate 163
    5.2.3 Triethanolamine 164
  5.3 The effects of accelerators on the water–cement system 164
    5.3.1 Rheological effects 164
    5.3.2 Chemical effects 164
    5.3.3 Effects on cement hydration 167
    5.3.4 Mechanism of action 176
  5.4 The effects of accelerators on the properties of plastic
      concrete 178
    5.4.1 Effect on heat evolution 178
    5.4.2 Effect on setting time 180
  5.5 The effects of accelerators on the properties of hardened
      concrete 181
    5.5.1 Structural design parameters 181
    5.5.2 Durability aspects 182
  References 197

6 Special purpose admixtures 199
  6.1 Introduction 199
  6.2 Alkali–aggregate expansion-reducing admixtures 200
    6.2.1 Alkali–aggregate reaction 200
    6.2.2 Types of admixtures 201
    6.2.3 Mode of action 206
    6.2.4 Effects on the plastic and hardened properties of
          mortar and concrete 207
  6.3 Antifreeze admixtures 208
    6.3.1 Chemical composition and mode of action 209
  6.4 Antiwashout admixtures 212
    6.4.1 Categories 212
    6.4.2 Formulating non-dispersible underwater concrete 214
    6.4.3 Effects produced on plastic and hardened concrete 215
    6.4.4 Factors affecting the performance of antiwashout
          admixtures 216
    6.4.5 Mixture and storage 217
    6.4.6 Applications 219
  6.5 Corrosion-inhibiting admixtures 219

6.5.1 Material parameters   220
6.5.2 Types of corrosion inhibitors   220
6.5.3 Research on other corrosion inhibitors   226
6.6 Calcium-sulfoaluminate-based expanding admixtures   227
6.6.1 Chemical composition   228
6.6.2 Mode of action   228
6.6.3 Mix proportioning, mixing and curing   230
6.6.4 Factors influencing the reaction   231
6.6.5 Effects on the plastic and hardened properties of
      mortar and concrete   234
6.6.6 Applications   235
6.7 Polymer-based admixtures   235
6.7.1 Categories   236
6.7.2 Material parameters influencing performance   238
6.7.3 Modification of the cementitious matrix   239
6.7.4 Mix proportioning   241
6.7.5 Mixing, placing and curing   242
6.7.6 Properties of latex-modified mortar and concrete   243
6.7.7 Applications   248
6.7.8 Standards and specifications   249
6.8 Admixtures for recycling concrete waste   249
6.8.1 Chemical composition and mechanism of action   252
6.8.2 Effects on hardened properties of concrete   252
6.9 Shotcrete admixtures   252
6.9.1 Types of admixtures and mode of action   255
6.9.2 Factors influencing the effects of the admixture   260
6.9.3 Effects on the plastic and hardened properties of
      concrete and mortars   261
6.9.4 Guidelines for use   263
6.10 Shrinkage-reducing admixtures   265
6.10.1 Effects on the fresh concrete properties   268
6.10.2 Effects on the hardened properties of concrete   268
6.10.3 Factors affecting the performance of
       shrinkage-reducing admixtures   269
References   271

7 Applications of admixtures   276
7.1 Introduction   276
7.1.1 Reasons for use of admixtures   278
7.2 Air-entraining admixtures   278

7.2.1	Control measures used to ensure proper air entrainment	281
7.2.2	Methods of placing	283
7.2.3	Air-entraining admixture/superplasticizer compatibility	284
7.2.4	Composite air-entraining–water-reducing admixtures	285
7.3	Normal-setting water-reducing admixtures	288
7.3.1	Ready-mixed concrete	289
7.3.2	High-strength/high-performance concrete	292
7.3.3	High-workability mixes	297
7.3.4	Pumping	297
7.3.5	'Watertight' concrete	299
7.3.6	Piling	300
7.4	Set-retarding and water-reducing admixtures	300
7.4.1	Retarded concrete for large pours	301
7.4.2	Slip-forming	304
7.4.3	Marine structures	309
7.4.4	Tilt-up construction	311
7.5	Accelerating admixtures	312
7.5.1	Purpose and advantages resulting from the use of accelerators	312
7.5.2	Non-chloride admixtures	313
7.5.3	Accelerators for use in blended cement (fly ash or slag) mixtures	313
7.6	Superplasticizers (high-range water reducers)	314
7.6.1	Flowing concrete	315
7.6.2	High-range water-reduced concrete	325
7.6.3	High-performance concrete and mortar	329
7.7	Viscosity-enhancing admixtures	340
7.7.1	Grouting applications	341
7.7.2	Underwater concrete	345
7.7.3	Formulation of construction products	348
7.8	Damp-proofing admixtures	349
7.9	Recycling of cementitious wastes	350
7.10	Hot-weather concreting	351
7.11	Cold-weather concreting	359
7.11.1	Acceleration of hydration and depression of freezing point of the water mix	362
7.11.2	Reduction of freezable water	365
7.11.3	Case studies	366
7.12	Economic aspects of admixture use	367
7.12.1	Economies in mix proportioning	368

7.12.2 Economies from improved durability   369
7.12.3 Economies from improved placing characteristics
        and construction methods   370
7.12.4 Precast concrete   372
7.12.5 Economic benefits of cold-weather admixtures   375
7.12.6 Economic benefits from the recycling of plastic
        concrete and wash water   377
7.13 Guidelines for the use of admixtures   377
7.13.1 Evaluation and selection   378
7.13.2 Admixture uniformity   380
7.13.3 Precautions in the use of admixtures   380
7.13.4 Mix proportioning using computers   382
7.13.5 Safety and hygienic aspects in the handling of
        admixtures   383
7.13.6 Admixture problems – limitations and incompatibility 383
7.14 Batching and dispensing of admixtures   402
7.14.1 Manufacture   403
7.14.2 Packaging and delivery   404
7.14.3 Labels   406
7.14.4 Storage   407
7.14.5 Dispensing of admixtures   409
7.14.6 Dispensing equipment   412
7.14.7 Calibration and maintenance of batching systems   421
7.14.8 Computer batching   422
References   424

Index   431

Disclaimer

The information and recommendations presented in this book are represented in good faith and believed to correct as of the date of publication. However the publisher, the authors and the organizations to which the authors belong make no representation or warranties, either express or implicit as to the completeness or accuracy thereof. Information is presented upon the condition that the persons receiving the same will make their own determination as to its suitability for their purposes prior to use. In no event will the publisher, the authors and the organizations to which the authors belong be responsible for damages whatsoever resulting from the use or reliance on the information contained in this book.

Foreword to the third edition

It is now 20 years and 13 years since the first and second editions respectively of *Chemical Admixtures for Concrete* were published. A first glance at the international admixture business could lead to the impression that not a lot had changed in 20 years; certainly not enough to justify the complete revision of the second edition of the book. While it is true that products based on lignosulfonates, sodium glucoheptonate and corn syrup are still provided to the market in thousands of tonnes every year, there have been several significant changes:

1. A change in the supply situation of Vinsol™ resin with a commensurate price increase as well as the more widespread use of supplementary materials led almost all admixture companies to reformulate their air-entraining agents using synthetic or alternative natural products such as tall oil or rosins. In a period of about 2 years, a product that had been in dominant use for 40 years was relegated to a lesser role in the industry.
2. A whole new category of water-reducing agents, the mid-range products, have emerged as a major benefit to the concrete industry, allowing easier-to-place high-slump concrete to be produced at minimal cost and with little effect on setting time. In the USA, this is the fastest growing sector of the admixture industry.
3. Whilst the use of superplasticizers has not grown as fast as had been anticipated, and still accounts for less than 10% of concrete produced, there have been technical developments in this area. New chemical types of these products have been developed based on polyacrylates and have begun to be used in formulations. With the advent of high-performance concrete, such developments have been timely. Since the new acrylate-based materials maintain higher slumps for longer periods, and can therefore be added at a central mixing facility, ready-mixed high-perfomance concrete is now a reality.
4. Other types of chemical admixtures that were supplied in very small quantities 15 or 20 years ago, such as corrosion inhibitors etc., have become high-volume products and have helped to double the concrete admixture market since the last edition.

Even the companies supplying the international market have seen changes: Master Builders Technologies has changed ownership to the German company SKW GmbH (a member of the VIAG Group), which has resulted in the largest construction chemicals business in the world with sales approaching $2 billion per year and having a clear commitment to the industry and a strong track record in research. Further consolidation is evident by the acquisition by W.R. Grace & Company of Cormix Construction Chemicals, which itself had absorbed the Gifford Hill and American Admixtures business some years previously. Again, this company is committed to the advancement of the industry through research and development. The vacuum left by the acquisition of smaller companies by the two larger ones is predictably being filled by the spawning of new companies, usually on a strong regional basis, such as GRT in Minnesota and Arrmacrete Construction Chemicals in Florida. These companies are bringing their own changes to the industry in terms of innovation and service.

In academia too, workers have advanced our knowledge of how water-reducing agents and superplasticizers function and have made our earlier theories look simplistic and pedantic.

Other authors have also been active; when *Chemical Admixtures for Concrete* was first published, it was almost the only book available on the subject. Now there are several others, and two of them in particular, by Ramachandran and Dodson, belong on the shelf of anyone with a serious interest in chemical admixtures.

Against this background therefore, the authors felt that there was still a need for a comprehensive treatise that would fill the needs of all those interested in the subject, whether they be a student, a user, a manufacturer, or a specifier, and that not just a general update, but a deep revision was justified. Specifically, the following changes have been made:

1. A general update of references and developments in all chapters including a major new chapter on 'superplasticizers', or high-range water-reducing agents.
2. Addition of a new section on miscellaneous admixtures including shotcrete admixtures, corrosion inhibitors, and admixtures for recycling wash water and plastic concrete.
3. Expansion of the chapter on applications, including an additional section on troubleshooting.

The book is a survey of current work and thinking and contains both the latest understanding of the complex process of admixture–cement interactions and the techniques that enable improvements in the performance of concrete. We have attempted to present the subject in a manner that can be readily applied and is therefore structured for the practicing engineer, providing a balance between theory and a hands-on approach of admixture use on the job site.

Reluctantly, we have dropped the chapter on specifications and standards on the grounds of space for the new expanded areas; also the timing of the new unified European Standards did not meet the deadline for publication.

Both authors have been resident and active in the construction chemicals field in North America for some years now and despite having previously been active on three other continents for a combined total of over 60 years, the parochialism does show through and we apologize for this.

As always, we welcome suggestions and constructive criticism so that in the unlikely event of a fourth addition, we can do a better job.

Acknowledgements

The authors of the third edition of *Chemical Admixtures for Concrete* acknowledge with appreciation the help and cooperation of the following companies and individuals for providing:

The Canadian National Research Council's Institute for Research in Construction, Ottawa, Canada ● Master Builders Technologies, Canada and USA ● Sternson Ltd, Brantford, Canada ● Euclid Chemicals Canada and USA ● Marcotte Instruments, Quebec, Canada ● Cormix Inc., UK ● Conchem Research Associates, Orleans, Canada ● Structural Preservation Systems, Baltimore, USA ● Fritz Admixtures, Texas, USA ● Sika Corporation, Canada and Switzerland ● Halsall Associates, Ottawa, Canada ● W.R. Grace, Canada and USA ● Arr-Maz Products Company, Florida, USA ● Dr D.R. Morgan, Prof. Paulo, Helene (University of Sao Paulo) and Prof. L. Pudencio (University of St. Catarina) of Brazil ● Prof. N. Banthia (Univ. of British Columbia, Canada) ● Claude Beddard, Prof. C. Jollicoeur (Univ. of Sherbrooke, Canada) ● Dr C. Nmai, Dr A. Vaysburd, Prof. N. Swamy (Univ. of Sheffield, UK) ● Prof. Bassheer (Univ. of Ulster, N. Ireland) ● Dr V.S. Ramanchandran, Prof. K. Hover (Cornell University, USA) ● Prof. G. Razaqpur (Carleton University, Ottawa) ● Mr C. Chun (Project Manager Sehoe Bridge Project, Pago Korea) ● Mr M.Y. Shim (Dongnam Industries, Seoul, Korea).

Special recognition is given to the following individuals who helped at various stages of the preparation of the manuscript: Mrs Nalini Mailvaganam, Dimitri and Rex Mailvaganam, Dr W. Repette, O. Maadani, Mrs L. Dessereault. The information presented in the sections on polymer-based admixtures and cold and hot water concreting has been largely derived from the publications of Prof. Y. Ohama (Univ. of Kyoto, Japan) Prof. I. Soroka (Technion, Israel) and Dr C. Nmai (Master Builders Technologies). Their cooperation in giving approval for the use of a number of figures is gratefully acknowledged.

Chapter 1

Water-reducing agents

1.1 Background and definitions

The water-reducing admixtures are the group of products which possess as their primary function the ability to produce concrete of a given workability, as measured by slump or compacting factor, at a lower water–cement ratio than that of a control concrete containing no admixture.

The earliest known published reference to the use of small amounts of organic materials to increase the fluidity of cement containing compositions, was made in 1932 [1] where polymerized naphthalene formaldehyde sulfonate salts were claimed as useful in this role. This was followed during the mid 1930s to early 1940s by numerous disclosures regarding the use of lignosulfonates and improved compositions [2–9].

The lignosulfonates formed the basis of almost all the available water-reducing admixtures until the 1950s when the hydroxycarboxylic acid salts were developed which have grown to occupy a significant but, nevertheless, still a minority position in this product group. Materials such as glucose and hydroxylated polymers obtained by the partial hydrolysis of polysaccharides have been widely used in North America. The polymers usually have a low molecular weight and contain glycoside units ranging from 3–25. In addition. other chemical and admixture types have been included into the water-reducing admixtures formulations to produce five types within this category.

The **normal** water-reducing admixtures allow a reduction in the water–cement ratio at a given workability without significantly affecting the setting characteristics of the concrete. In practice, this effect can be utilized in three ways:

1. By the addition of the admixture with a reduction in the water–cement ratio, a concrete having the same workability as the control concrete can be obtained, with unconfined compressive strengths at all ages which exceed those of the control.
2. If the admixture is added directly to a concrete as part of the gauging water with no other changes to the mix proportions, a concrete

possessing similar strength development characteristics is obtained, yet having a greater workability than the control concrete.

3. A concrete with similar workability and strength development characteristics can be obtained at lower cement contents than a control concrete without adversely affecting the durability or engineering properties of the concrete.

In all three ways of use, this type of admixture can be regarded as a cement saver, as illustrated in Fig. 1.1.

Corresponding mixes are, therefore, concrete mixes having the same workability and 28-day strength characteristics, but the mix containing the water-reducing admixture will have a lower cement content than the other mix. In practice, of course, the parameters of workability and strength are dictated by the requirements of the particular situation; in areas of high steel content, a high workability will be required, whilst in the production of extruded prestressed lintels, a very low workability is needed. In both cases the strength requirements will be dictated by the load-bearing characteristics of the application. Thus in comparing any properties of admixture-containing concrete, the results of corresponding mixes should be studied, whether the investigation be related to strength, durability factors or statistical considerations, such as standard deviation.

Although the pictorial comparison shown in Fig. 1.1 and discussed above is true at low and average cement contents up to about 350 kg m^{-3} it is more difficult to obtain higher strengths and workability by further increasing the cement content. It is in this area that the hydroxycarboxylic acid water-

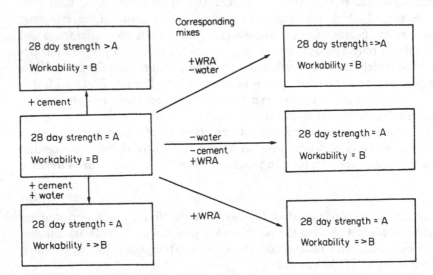

Fig. 1.1 The concept of corresponding mixes.

reducing admixtures are particularly beneficial, enabling considerable increases in strength to be obtained without the expense and undesirable side effects of large cement increments.

The other members of the water-reducing admixture group possess some other function which could not be obtained by mix design considerations.

The **accelerating** water-reducing admixtures, whilst possessing the water-reducing capability of the 'normal' category, give higher strengths during the earlier hydration period and faster setting times which allow finishing operations to be carried out in a timely manner, particularly at lower temperatures.

A special type of the accelerating water-reducing admixture is finding increased application and is known in North America as a 'mid-range water-reducing admixture'. This product type is formulated using water-reducing ingredients which have minimal set-retarding effects (such as low-sugar processed lignosulfonates, or blends of lignosulfonates with superplasticizers), high proportions of accelerators (chloride or non-chloride), and often a non-ionic surfactant. The mid-range water-reducing agent can be used at higher dosages and hence give greater workability or water reduction without extending set times to unacceptable levels.

The **retarding** water-reducing admixtures again behave in a similar manner to the 'normal' materials and are often of similar chemical composition used at a higher dosage level, but extend the period of time when the concrete is in the plastic state. This means that the time available for transport, handling, placing and finishing is lengthened. In fact, although a few materials are available which exert only a retarding influence on concrete and have little or no water-reducing capacity, the vast majority (about 95%) of materials called 'retarders' are actually retarding water-reducing admixtures and are considered as such in this book.

The way in which the four types of water-reducing admixtures discussed so far affect the strength gain characteristics of concrete containing them is shown in Fig. 1.2. The four concrete mixes have been designed to have approximately the same 28-day compressive strength, i.e. the admixture-containing mixes would contain approximately 10% less cement than the control mixes.

The **air-entraining** water-reducing agents possess the ability to entrain microscopic air bubbles into the cement paste whilst allowing a reduction in the water–cement ratio greater than that which would be obtained by the air entrainment itself. They are available in the normal and retarding form and also fall into two types depending on the level of air entrainment; the first type entrains only about 1–2% additional air and is normally used to increase the internal surface of the concrete to redress any deficiencies in fine aggregate gradings. The second type results in concrete containing 3–6% air and is used to enhance the durability of the concrete to freeze–thaw conditions.

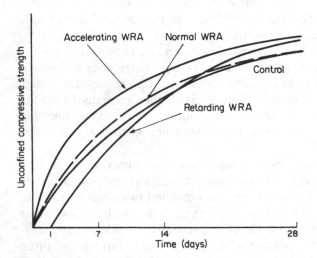

Fig. 1.2 Compressive strength development of concrete containing various types of water-reducing admixture.

The advantages of using this type of material rather than a straight air-entraining agent are based mainly on minimizing the deleterious effect that air entrainment has on compressive strength, as shown in Fig. 1.3. Thus in a typical concrete mix, up to 3% air can be entrained without any alteration to the mix design or reduction in compressive strength when an air-entraining water-reducing admixture is used.

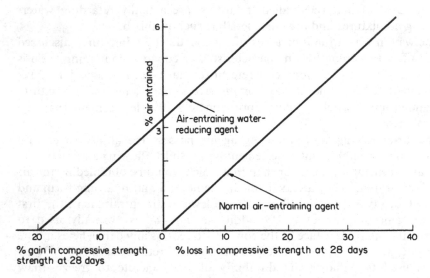

Fig. 1.3 The effect of air entrainment on compressive strength of concrete containing a water-reducing air-entraining agent and a normal air-entraining agent.

1.2 The chemistry of the water-reducing admixtures

Although the variety of admixtures that are commercially available are marketed under a multitude of benefit-orientated classifications, namely waterproofers, densifiers, workability aids, etc., it is possible to categorize the basic chemicals used as shown in Table 1.1.

It can be seen, therefore, that only three chemical materials form the basis of all the water-reducing admixtures, i.e. lignosulfonate, hydroxycarboxylic acid, and hydroxylated polymers.

1.2.1 Lignosulfonates

Lignin is a complex material which makes up approximately 20% of the composition of wood. During the process for the production of paper-making pulp from wood, a waste liquor is formed as a by-product containing a complex mixture of substances, including decomposition products of lignin and cellulose, sulfonation products of lignin, various carbohydrates (sugars) and free sulfurous acid or sulfates. Subsequent neutralization, precipitation and fermentation processes [10] produce a range of lignosulfonates of varying purity and composition depending on a number of factors, such as the neutralizing alkali, the pulping process used, the degree of fermentation and even the type and age of the wood used as pulp feedstock [11].

Table 1.1 Formulations of water-reducing admixtures

	Type of water-reducing admixture		
Normal	Accelerating	Retarding	Air-entraining
Purified lignosulfonate	Lignosulfonate + CaCl$_2$	High sugar lignosulfonate	Impure lignosulfonate
Lignosulfonate + air-detraining agent	Lignosulfonate + triethanolamine	Hydroxycarboxylic acid	Lignosulfonate + surfactant
Hydroxy-carboxylic acid at low dosage	Lignosulfonate + Ca formate	Hydroxylated polymer	Hydroxycarboxylic acid + surfactant
Hydroxylated polymer at low dosage	Hydroxycarboxylic acid + CaCl$_2$ Lignosulfonate + Sod. thiocyanate		

Commercial lignosulfonates used in admixture formulations are predominately calcium or sodium based with sugar contents of 1–30%. Typical analyses of two commercially available lignosulfonate water-reducing admixtures are shown in Table 1.2 [12].

The lignosulfonate molecule is a substituted phenyl propane unit containing hydroxyl, carboxyl, methoxy and sulfonic acid groups [13–15]. A possible representation is a polymer containing the repeating unit shown in Fig. 1.4. The polymer could typically have an average molecular weight of about 20 to 30 000 with a range varying from a few hundred to 100 000 [12, 16]. The range of molecular weight present in the lignosulfonate is dependent on the manner and conditions under which refinement is done. Three such methods are used, namely, ultra-filtration, heat treatment at a specified pH and fermentation.

Figure 1.5 is a typical molecular weight distribution curve for a lignosulfonate obtained by means of gel permeation chromatography (a sophisticated analytical method where molecules are sieved according to their molecular size).

It has been found [17] that the lignosulfonate polymer is not a simple linear flexible coiled 'thread', as found in many high molecular weight materials, but forms spherical microgels of the type shown in Fig. 1.6. Thus the charges are predominately on the outside of the spheroid with the internal carboxyl groups and sulfonate group being non-ionized. Conductivity studies have confirmed that lignosulfonates are only 20–30% ionized [16].

Table 1.2 Typical analyses of lignosulfonate-based water-reducing admixtures (after Edmeades)

Type	Spruce wood sulfite lye calcium lignosulfonate (%) [12]	Sodium lignosulfonate (%)
Solid content	54	30
Ash	6.6	—
Sulfated ash	10.9	—
CaO (in ash)	44	0.1
Reducing sugars (as glucose)	5.8	0.9
Total sulfur	3.2	2.6

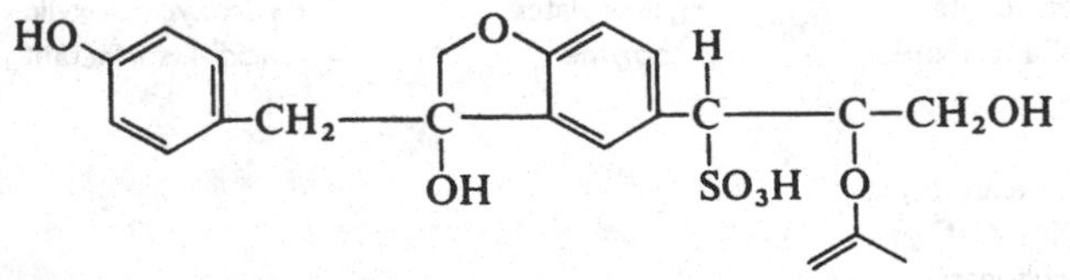

Fig. 1.4 Repeating unit of a lignosulfonate molecule.

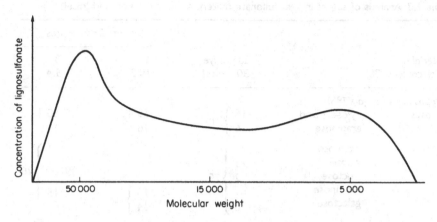

Fig. 1.5 Molecular weight distribution of a typical lignosulfonate.

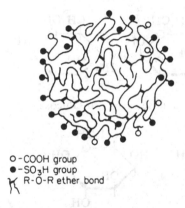

O -COOH group
● -SO$_3$H group
ⵊ R-O-R ether bond

Fig. 1.6 Schematic representation of a lignosulfonate polyelectrolyte microgel unit.

The sugars contained in lignosulfonates vary both in type and concentration, depending on the source, type and degree of refining that has taken place. In the fermentation process the micro-organisms used preferentially consume the hexoses rather than the pentoses so that the residual sugars present in the refined lignosulfonates are mainly pentoses. The types of sugars found are shown in Fig. 1.7, and Table 1.3 gives a breakdown of sugars found in untreated sulfite lye [15] and two commercial water-reducing admixtures [11].

Although several different salts of lignosulfonates are commercially available, the calcium and sodium derivatives are the most widely used in admixture formulation. The sodium salt tends to maintain its solubility at

Table 1.3 Analysis of sugars in lignosulfonate materials (after Mouton and Joisel)

		Sulfite lye	Fermented admixtures	
			A	B
Material				
Sugar content (%)		30 typical	10.2	5.4
Composition of sugars (%)				
Pentoses	xylose + acid	15 }21	55 }70	60 }74
	arabinose	6	16	14
Hexoses	mannose	48		11
	gluose	15		}26
	fructose	2 }75		
	rhamnose	—	16 }30	15
	galactose	10	14	
Others		4		

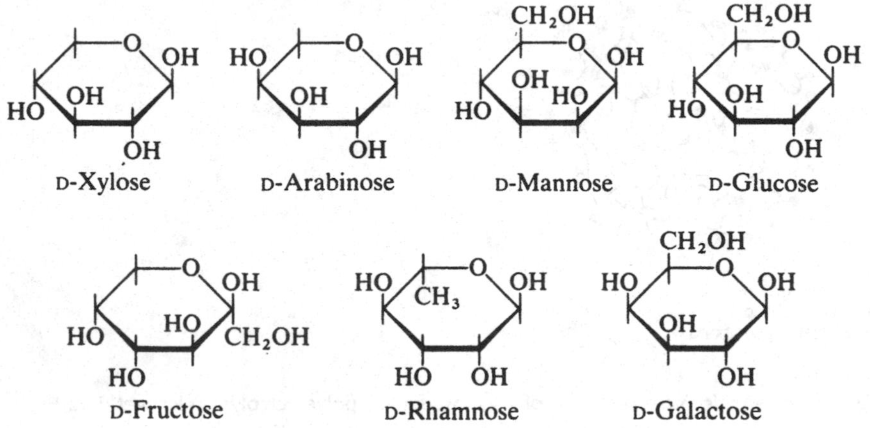

Fig. 1.7 Formulae of sugars found in untreated and purified lignosulfonate materials.

low temperatures, thus avoiding sedimentation in winter conditions. In addition, the sodium salt has a higher degree of ionization in solution [16] than the calcium salt. This is reflected in the observation that solutions of higher concentrations of the calcium salt are required to obtain the same reduction in water–cement ratio obtained by using the same dosage of a sodium-salt-based water-reducing admixture. However, calcium lignosulfonate raw materials are invariably cheaper than sodium lignosulfonates so that the higher concentrations can be offered on an approximately equal cost-effectiveness basis.

In the formulation of admixtures from lignosulfonate (Table 1.1), the following comments are relevant:

1. Many lignosulfonates and, in particular, the less pure types and those produced from hardwood lignins entrain a small proportion of air into the concrete. This can be desirable where an air-entraining material is required to enhance durability or cohesion, but is often an unwanted side effect. Thus in the production of normal water-reducing admixtures, a small quantity of an air-detraining agent can be added. The usual material is tributylphosphate at a level of less than 1% of the lignosulfonate, although dibutyl phthalate, water-insoluble alcohols, borate esters and silicone derivatives find some application [18].

2. The lignosulfonate molecule itself and, of course, the sugars present in the lignosulfonate materials do have a retarding influence on the hydration of the cement. In the case of the higher-sugar-content materials, this is utilized to produce the retarding water-reducing admixtures which allow longer transport or placing times. However, for normal water-reducing admixtures, this is an undesirable effect and, therefore, additions of triethanolamine are occasionally made at a level of about 15% of the lignosulfonate content of the admixture [19]. At this level of addition the triethanolamine acts as an accelerator and compensates for the retarding influence of the lignosulfonate and its impurities. This has been shown to have certain deleterious effects on some properties of the resultant concrete.

3. The accelerating water-reducing admixtures are simple blends of either calcium chloride, nitrate, thiocyanate or formate with a lignosulfonate or a hydroxycarboxylic acid salt. In some cases it may not possible to obtain a completely sediment-free solution and agitation of store tanks may be necessary. Typically, a mixture of approximately 33% calcium chloride and 4% calcium lignosulfonate by weight in water would be used.

4. Air-entraining water-reducing admixtures containing lignosulfonates can be based on impure lignosulfonate raw materials, as stated earlier, where only 2–3% additional air is required. However, this air may not be of the amount, type, and stability required, therefore additions of surfactants are made. Several different types can be used but in the majority of cases they are based on alkyl-aryl sulfonates (e.g. sodium dodecyl benzene sulfonate) or fatty-acid soaps (e.g. the sodium salt of tall-oil fatty acids). Additions of these types will allow incorporation of sufficient stable air of the correct bubble size to meet durability requirements under freeze–thaw conditions.

1.2.2 Hydroxycarboxylic acids

As the name implies, these are organic chemicals which have both hydroxyl and carboxyl groups in their molecules. Generally, the sodium salt is used, although occasionally the materials are found as salts of ammonia or triethanolamine. They are produced from pure raw material feedstocks, by either chemical or biochemical means and, therefore, are of high and consistent purity. Indeed, the primary use of the materials is often in foodstuffs or pharmaceuticals.

In the form of sodium salts all are very soluble and have low freezing points, so that solidification in winter conditions is unlikely. Figure 1.8 shows the types and formulae of materials which have been reported to find application in the formulation of this type of water-reducing admixture. However, the only materials finding widescale application in formulations are the salts of gluconic and heptonic acids.

		Citric acid [23, 25]	Tartaric acid [23, 25]	Mucic acid [10, 14, 25]
Functionality	OH groups	1	2	4
	COOH groups	3	2	2
Molecular weight		192	150	210

Formula

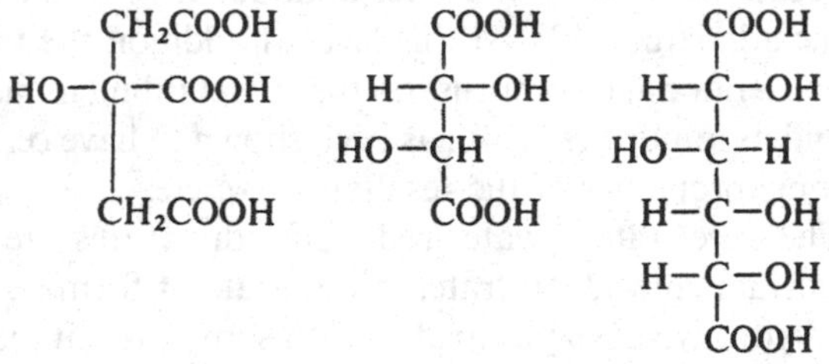

$$CH_2COOH$$
$$HO-C-COOH$$
$$CH_2COOH$$

$$COOH$$
$$H-C-OH$$
$$HO-CH$$
$$COOH$$

$$COOH$$
$$H-C-OH$$
$$HO-C-H$$
$$H-C-OH$$
$$H-C-OH$$
$$COOH$$

		Gluconic acid [10, 20, 21, 26]	Salicylic acid [24]	Heptonic acid [22]	Malic acid [25]
Functionality	OH groups	5	1	6	1
	COOH groups	1	1	1	2
Molecular weight		196	138	230	134

Formula

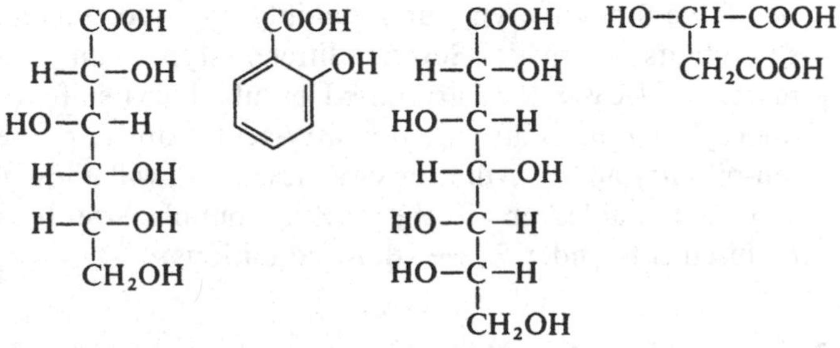

$$COOH$$
$$H-C-OH$$
$$HO-C-H$$
$$H-C-OH$$
$$H-C-OH$$
$$CH_2OH$$

$$COOH$$ OH (benzene ring)

$$COOH$$
$$H-C-OH$$
$$HO-C-H$$
$$H-C-OH$$
$$HO-C-H$$
$$HO-C-H$$
$$CH_2OH$$

$$HO-CH-COOH$$
$$CH_2COOH$$

Fig. 1.8 Hydroxycarboxylic acids used in admixtures.

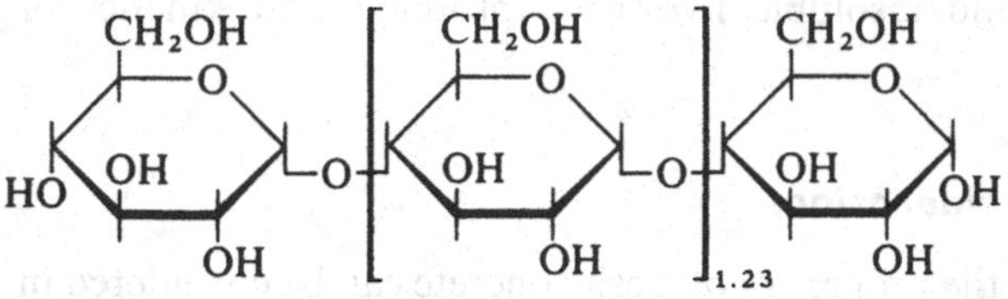

Fig. 1.9

Normally, approximately 30% solutions of the salts would be used with additions of other chemical types, depending on the proposed function in concrete. Thus the salts may be present alone to produce normal water-reducing admixtures at low dosages and retarding water-reducing admixtures at higher dosages. Small amounts can be blended with calcium chloride to produce accelerating water-reducing admixtures which are almost colorless, sediment-free solutions and, in a similar manner to lignosulfonates (see earlier), air-entraining agents can be added to form the air-entraining water-reducing admixtures which may or may not be retarding, depending on the amount of hydroxycarboxylic acid salt present in the formulation.

1.2.3 Hydroxylated polymers

The hydroxylated polymers are derived from naturally occurring poly-saccharides, such as corn starch, by partial hydrolysis to form lower-molecular-weight polymers containing from 3 to 25 glycoside units (Fig. 1.9) [27]. Unlike the monosaccharide glucose, these materials are stable under the alkaline conditions of a cement-containing composition and behave as efficient water-reducing agents. They do impart a retardation to the concrete in which they are incorporated, which can be overcome by the addition of small quantities of calcium chloride or triethanolamine [27].

The three categories of major ingredients discussed above for the formulation of water-reducing admixtures account for the majority of commercially available products, but there may be limited use of insitol [28], polyacrylamide [29], polyacrylic acids [30] and polyglycerol [31].

1.3 The effects of water-reducing admixtures on the water–cement system

In order to understand more fully the effect that water-reducing admixtures have on the plastic properties of fresh concrete, and to gain an insight into the mechanism of action of this category of materials, it is useful to study the effect on the water–cement system. The topic can be considered from the following points of view: (a) rheological and initial surface effects, and (b) effects on soluble and insoluble hydration products and kinetics of hydration.

1.3.1 Rheological considerations

It is known that some of the properties of fresh concrete can be considered in terms of the rheological properties of the cement paste contained in the concrete. Thus a high water–cement ratio concrete will contain a paste content which is more fluid than that of a low water–cement ratio concrete.

The 'fluidity' of the cement paste can be measured in rheological terms by the torque transmitted to a stationary 'bob' inside a revolving outer cylinder placed in a water–cement system as shown in Fig. 1.10. The shear stress measured at the stationary bob is plotted against the rate of applied shear when, for pastes of varying water–cement ratios, the results shown in Fig. 1.11 are obtained for readings taken of the shear stress as the shearing rate is increased (the up curve).

Systems which give linear shear-stress–shear-rate relationships with an intercept on the shear stress axis are said to exhibit plastic flow, and the intercept value is known as the 'yield stress'. The viscosity of the system is given by the slope of the line over its linear portion.

The properties of an overall concrete system in the plastic state will be a function of many parameters such as aggregate types and shapes, cement

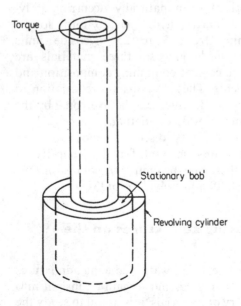

Fig. 1.10 A concentric cylinder viscometer.

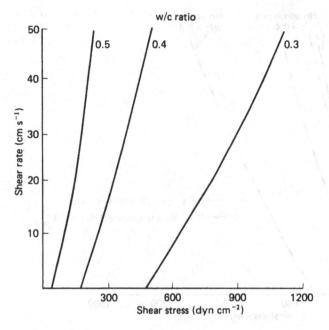

Fig. 1.11 Shear-stress–shear-rate relationships for cement pastes at various water–cement ratios.

contents and characteristics, etc., but it is useful to isolate the effect of the paste rheology where it can be stated that:

- The **consistency** or **fluidity** of the concrete will be a function of the viscosity of the cement paste.
- The **cohesion** of the concrete will be a function of the yield stress of the cement paste.

Turning now to the effect that water-reducing admixtures have on the rheology of cement pastes, it can be seen from Fig. 1.12 that the addition of these types of materials does not appear to alter the shape of the shear-stress–shear-rate relationship but merely moves it to a lower level (the lignosulfonate material has had tributylphosphate added to it so that the effect of air entrainment is eliminated). In view of the relationship shown in Fig. 1.12 it is useful to use paste viscosities as a means of assessing and studying water-reducing admixtures, and Figs 1.13 and 1.14 show respectively the effect of varying addition levels of two water-reducing admixtures on the paste viscosity and the effect of two materials at typical dosages at various water–cement ratios. These data would indicate that the water

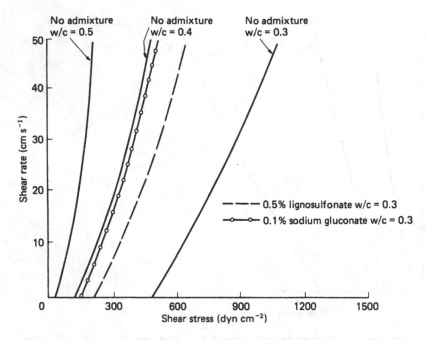

Fig. 1.12 Shear-stress–shear-rate relationships for cement pastes containing various water-reducing agents.

reductions possible using the two types of water-reducing agents are different and depend on the water–cement ratio of the system. Typical results are given in Table 1.4 and these values are somewhat higher than those obtained in concrete mixes, emphasizing the importance of other mix parameters. The lignosulfonates also entrain some air into concrete which could increase the water reduction obtained.

1.3.2 Initial surface effects

The rheological characteristics of cement pastes are related to the nature of the attractive and repulsive forces which exist between cement and cement hydration product particles and can be categorized as follows:

1. Van der Waals forces of attraction, which are large in magnitude but only at interparticle distances of up to 5–7 mm.
2. Electrical repulsion due to the cationic nature of the free valences at the surface of the cement particles due to the Ca, Al and Si atoms [32]. This repulsion is smaller in magnitude than the van der Waals forces but,

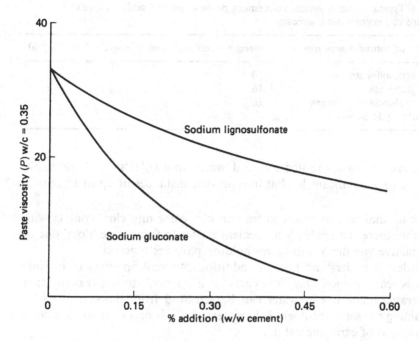

Fig. 1.13 The effect of water-reducing admixtures on paste visosity at different addition levels.

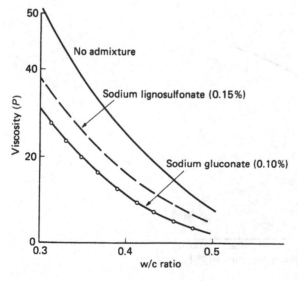

Fig. 1.14 The effect of water-reducing admixtures on paste viscosity at various water – cement ratios.

Table 1.4 Typical water reductions of cement paste at normal addition levels of admixture to maintain paste viscosity

Admixture at normal dosage rate	Average water reduction over $w/c = 0.3–0.5$ (%)
Sodium lignosulfonate	10
Sodium gluconate	16
Sodium naphthalene sulfonate formaldehyde polymer	26

because of an associated ion and water molecule 'sheath', probably exists to a significant level at interparticle distances of up to 15 nm.

Thus in normal cement pastes where particles come into close contact with each other there is a tendency for cement pastes to form large 'flocs' due to the attractive van der Waals forces holding particles together.

In order to understand how the addition of small amounts of organic materials can reduce this interparticle attraction to an extent that considerable quantities of water can be removed from the system whilst maintaining the same rheological characteristics, it is necessary to consider a large amount of experimental data:

1. The addition of a water-reducing agent to the aqueous phase is followed by a rapid reduction in the quantity of admixture in solution and an

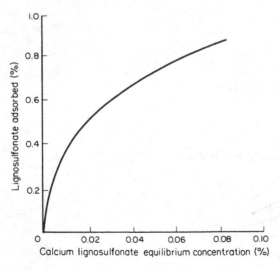

Fig. 1.15 Adsorption isotherm of calcium lignosulfonate on ordinary Portland cement (Ernsberger).

increase in the amount adsorbed at the cement particle/water interface [27, 33–39]. This effect is illustrated by an adsorption isotherm where various addition levels of admixtures are added to the water–cement systems and shaken for a period of time when the amount left in solution is estimated. The amount of material on the surface is calculated by difference. A typical curve for calcium lignosulfonate on ordinary Portland cement is given in Fig. 1.15 [33].

2. Water-reducing admixtures are not adsorbed equally by the various anhydrous and hydrated cement constituents and in studies with calcium lignosulfonate, the approximate maximum adsorption figures shown in Table 1.5 have been obtained [38, 39]. In addition, adsorption isotherms have been studied at various ages of C_3A hydration [36] and it has been shown that it is the initial hydration products (less than 15 min) which have the highest adsorptive capacity for calcium lignosulfonate, suggesting that it may be the precursor to the hexagonal phases which is responsible for the initial high adsorption, i.e. C_4AH_{19}. Figure 1.16 indicates this effect. Data for salicylic acid (a hydroxycarboxylic acid) yield similar information, although adsorption levels are usually lower.

3. The various types of water-reducing admixtures possess different but characteristic adsorption isotherms which qualitatively reflect their effect on cement hydration kinetics, as shown in Fig. 1.17.

4. In the absence of knowledge of the surface area of cement hydrates available for adsorption at the time of addition, it is difficult to estimate how many layers of water-reducing admixture molecules are adsorbed, but attempts have been made [40] indicating that over 100 layers may be formed with calcium lignosulfonate and salicylic acid at normal levels of addition. However, these calculations were based on specific surface areas of $0.3–1.0 \text{ m}^2\text{g}^{-1}$, whereas other studies [27, 38, 39] have indicated

Table 1.5 Adsorption of calcium lignosulfate by cement constituents (after Feldman and Ramachandran)

Cement constitutent*	Adsorption (%)
C_3A (anhydrous)	0
C_3A hydrates – C_2AH_8 $\Big\}$ hexagonal C_2AH_{13}	2.2
C_2AH_6 cubic	0
CH	12
C_3S	0
C_3S hydrates	7

* The usual abbreviations are used here and throughout this book: C = CaO, A = Al_2O_3, S = SiO_2, H = H_2O.

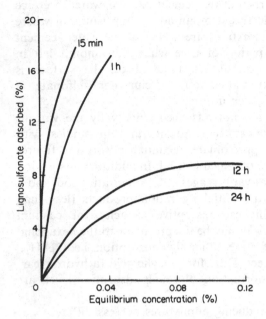

Fig. 1.16 Effect of hydration time on the adsorption of calcium lignosulfonate on C_3A hydrates (Rossington).

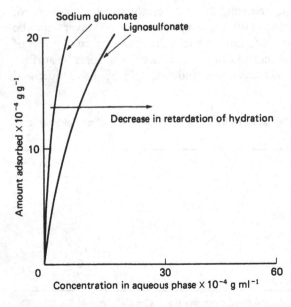

Fig. 1.17 Adsorption isotherms for various water-reducing admixtures.

Table 1.6 Surface area of cement constituent hydrates (Feldman and Ramachandran)

Phase	Surface area $(m^2 g^{-1})$	Adsorbed gas
C$_3$A hydrate (hexagonal)	11–14	N$_2$
Calcium hydroxide	16	—
C$_3$S hydrate	136	H$_2$O
C$_3$S hydrate	70	N$_2$

considerably higher surface areas as shown in Table 1.6. The earlier C$_3$A hydrates would be expected to have an even higher surface area, although all the surface available to nitrogen molecules would not be available to the larger admixture molecules. For two materials, calcium lignosulfonate and sodium gluconate, it is possible to compile the data shown in Table 1.7 from which it can be seen that a surface area available to the admixtures of about 3–7 m^2 g^{-1} would be required for formation of a monolayer of admixture molecules. The values given in Table 1.6 suggest that even after a few minutes hydration, this level of available surface may be present. At levels much higher than the addition required for the minimum paste viscosity, it is likely that multilayers are formed, possibly of insoluble aluminum salts. It has been pointed out [41] that if several tens of molecular layers are built up, then steric hindrance could be as important a factor as ionic repulsion in reducing paste viscosity.

5. In the case of those materials known to possess an adsorptive capacity towards cement constituents and the ability to reduce the paste viscosity, there is a linear correlation between the amount adsorbed on the surface and the reduction in paste viscosity at addition levels in the region of those normally used. This is shown in Fig. 1.18 [33]. These data can be used to obtain a viscosity reducing index for various materials by measuring the slope of the line. Some values are shown in

Table 1.7 Adsorption data for calcium lignosulfonate and sodium gluconate

Material	Molecular area $(Å^2)$	Surface covered by 1 mg material (m^2)	Addition level for minimum paste viscosity (%)	Amount adsorbed (from adsorption isotherm) (%)	Area for monolayer completion at minimum paste viscosity $(m^2 g^{-1})$
Calcium lignosulfonate	250 000	0.6	1.0	0.6	3.6
Sodium gluconate	50	1.5	0.5	0.45	6.8

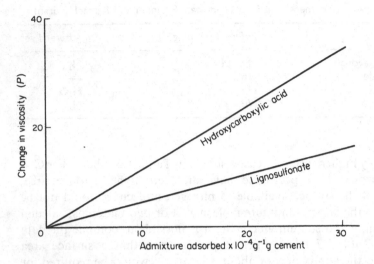

Fig. 1.18 Reduction in cement paste viscosity by various water-reducing admixtures as a function of amount adsorbed.

Table 1.8 although it should be emphasized that this technique is only used by the author as a means of comparing series of experimental products with known standards since different values are obtained for different batches of cement.

6. The nature of the bond between the molecules of the water-reducing admixture and the surface of the cement constituent hydrates has been described as 'ionic group outwards' in many references [33, 42,], mainly based on work [33, 43] showing migration of cement particles under the influence of an electric current when lignosulfonate molecules are adsorbed on the surface. Similar results have been reported for hydroxycarboxylic acids [44]. Other relevant data are summarized below:

(a) All classes of water-reducing admixtures have been found to be irreversibly adsorbed on to hydrating Portland cement. This is shown experimentally by the addition of further quantities of solvent during the determination of the adsorption isotherms; in

Table 1.8 Viscosity reduction of cement pastes by various water-reducing admixtures

Material	Viscosity reduction index ($P\ m^{-1}g^{-1}$) (water–cement ratio = 0.30)
Calcium lignosulfonate	5
Polysaccharide	9
Sodium gluconate	12

true physical adsorption the original isotherms should be followed, but, as shown for calcium lignosulfonate on the C_3A hexagonal hydrates (Fig. 1.19), there is almost no tendency for desorption [38]. The adsorption has also been shown to be irreversible on hydrated C_3S phase for calcium lignosulfonate materials [39].

(b) The amount of calcium lignosulfonate adsorbed on to hydrating cement is almost independent of initial water–cement ratio within the range 0.4 to 1.5 [34].

(c) In model studies with a hydroxycarboxylic acid (salicylic acid), a coordination complex has been isolated from the reaction with the C_3A hydrates [45], as shown in Fig. 1.20.

(d) Differential thermal analysis has indicated the formation of a complex between C_3S hydrates and calcium lignosulfonates [39].

(e) The value and sign of the surface potential of Portland cement in the presence of varying amounts of ammonium lignosulfonates are shown in Table 1.9 [46]. Comparing these data with a typical adsorption isotherm would support a progressive surface coverage up to a level of 0.25–0.50% lignosulfonate where an imperfect monomolecular layer would be formed.

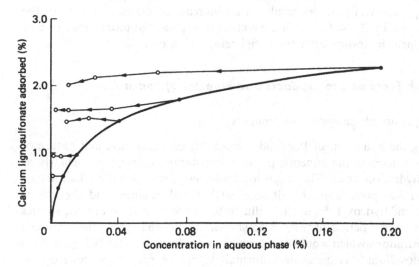

Fig. 1.19 Adsorption–desorption isotherms of calcium lignosulfonate on C_3A hydrate (hexagonal phase) (Ramachandran).

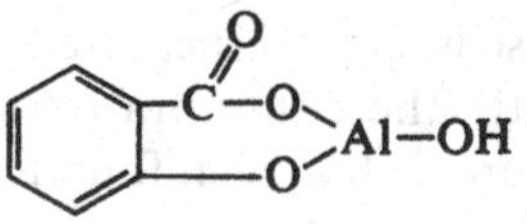

Fig. 1.20 Chemical compound formed from salicylic acid and C_3A hydrates.

Table 1.9 Reduction of surface potential by ammonium lignosulfonate (after Zhuravhev)

Ammonium lignosulfonate on cement weight (%)	0	0.01	0.05	0.1	0.25	0.8	1.0
Surface potential (mV)	+7.2	+6.0	+5.8	+4.1	+3.5	+3.3	+1.1

(f) The lignosulfonate is absorbed more by the hydrated C_3A than the hydrated C_3S [44], presumably because the C_3A is hydrating faster. There is also some indication that in the case of lignosulfonates, the intermediate adsorption isotherms indicate that there may be some reaction involving hydroxyl groups.

7. The addition of a water-reducing admixture to a cement suspension can be shown to disperse the agglomerates of cement particles into smaller particles [33, 38, 47] and can be seen clearly in photomicrographs as shown in Fig. 1.21. Maximum dispersion occurs at a level of 0.3–0.5% by weight of calcium lignosulfonate [33, 34] which would indicate the presence at the surface of about 0.2–0.4% calcium lignosulfonate. The separation of particles results in an increase in the surface area of the system by 30–40% [33, 38], which may explain the more rapid rate of cement hydration after the initial retardation period.

1.3.3 Effects on the products and kinetics of hydration

(a) Aqueous phase hydration products

During the hydration of Portland cement the concentration and nature of dissolved ions in the aqueous phase is constantly altering, particularly at early hydration times. The major ionic species present are the alkali metals sodium and potassium, the alkaline earth metal calcium, and the anions sulfate and hydroxyl. Figure 1.22 illustrates the level of these anionic species present over a period of time. Also shown are the changes in these anionic concentrations which would occur where an addition of a fairly high dosage of lignosulfonate is made or a normal dosage of sodium gluconate type water-reducing admixture [34, 37, 48–50]. The following observations can be made:

1. The calcium ion concentration of the solution phase is slightly increased in the early stages of hydration, but subsequently the concentration approaches that of a system containing no admixture. It has been found that the greater the period of set retardation, the larger is the difference in calcium concentration, and it takes longer to reach the same level as a

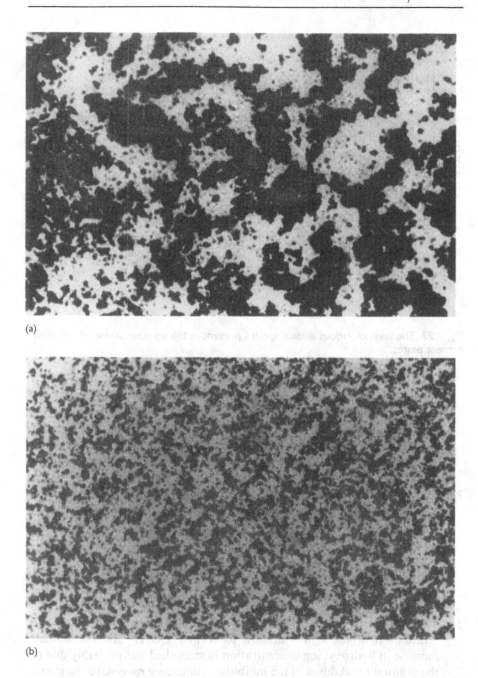

(a)

(b)

Fig. 1.21 Dispersion of cement particles by a water-reducing admixture, (a) before addition, (b) after addition.

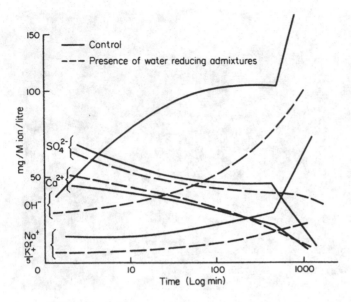

Fig. 1.22 The level of various anionic species present in the aqueous phase of hydrating cement pastes.

control system. The retarded hydration of the C_3S phase would be expected to be reflected in a lower concentration, but it is believed that the observed increase is due to a delay in a drop of the calcium and sulfate ion concentrations from the supersaturated to the saturated level. As can be seen this drop occurs very early in the control water–cement system.

2. The sulfate ion concentration normally diminishes rapidly, whereas in the presence of a water-reducing admixture, the high concentration is maintained due to the reasons given in (1) which, in turn, can be related to a delay in the reaction of the gypsum by the C_3A phase to form ettringite. Indeed, if the cement constituents are given a prehydration period prior to the addition of the admixture, results intermediate between those shown in Fig. 1.22 are obtained because part of the $C_3A + CaSO_4$ reaction has been permitted to proceed [49].

3. The hydroxyl ion concentration is initially reduced due to the retardation of the C_3S hydration to form $Ca(OH)_2$ and the sudden increase in hydroxyl ion concentration is smoothed out probably due to the gradual breakdown of the inhibiting admixture monolayer to give a faster diffusion of hydration products.

4. The alkali metal ions Na^+ and K^+ behave in a similar manner to the hydroxyl ion concentration, again due to the delay in C_3S hydration,

since the majority of the soluble alkaline metal ions originates from the C_3S crystalline phase.

These observations on the aqueous phase are consistent with the concept of the adsorption of water-reducing admixtures on to the initial hydrates of C_3A and C_3S with a corresponding modification of the normal process of reaction of C_3A with gypsum to form ettringite and a delay in hydration of the C_3S phase.

(b) Effect on solid hydration products

The hydration process of Portland cement is extremely complex but can be largely considered in terms of the hydration of the major phases present and reactions with calcium sulfate. A number of studies have been made of the various phases and it is generally considered that the behavior of a typical Portland cement is largely that of the sum of its components [51]. The changes on addition of water-reducing admixtures can be considered in terms of the chemistry, morphology and reaction kinetics. Figure 1.23 shows a typified heat evolution carried out under isothermal conditions using a sample of a normal Portland cement containing C_3A, C_2S, C_3S and gypsum [52].

The chemistry associated with peaks 1, 2 and 3 can be summarized as follows.

PEAK I

1. $C_3A + H_2O = C_4AH_{19}$ (quickly stopped by formation of a protective coating of ettringite).
2. $C_3A + CaSO_4 = C_3A.3CaSO_4.32H_2O$.

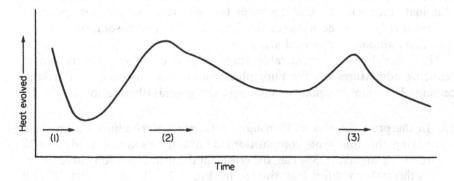

Fig. 1.23 Isothermal calorimetric curve of hydrating Portland cement.

3. $C_3S + H_2O = C-S-H$ initial C–S hydrate layer about 100 Å (10 nm) thick and low in Ca.
4. Heat of solution of free alkali.

The C_3A phase is subjected to competitive reaction between water and gypsum to form a small amount of the initial C_3A hydrates covered by a protective coating of ettringite, and also a proportion of the C_3S hydrates to form a calcium silicate hydrate of low calcium content in the region of 100 nm thick.

PEAK 2

During this phase the initial self-retarding calcium silicate hydrate layer appears to break down, allowing further hydration products from the C_3S and C_2S phases to be formed. These appear to be produced in three morphological forms, plate-like, needles and crumpled foils. During this reaction considerable quantities of hexagonal prisms of precipitated lime are also formed.

$$
\left.\begin{matrix} C_2S \\ C_3S \end{matrix}\right\} + H_2O \quad \begin{array}{c} C-S-H \text{ (gel)} \\ \text{approx. } C_3S_2H_{25} \\ + \\ Ca(OH)_2 \\ \text{hexagonal prisms} \end{array} \left\{\begin{array}{c} \text{plates } C:S = 75 - 1.00:1.00 \\ 80 \text{ nm} \times 20 \text{ nm in twos or threes} \\ \text{needles } C:S = 2:1 \\ \text{(probably from } C_3S) \\ \text{crumpled foils } C:S = \text{variable} \\ 1 - 20 \text{ nm thick} \end{array}\right.
$$

PEAK 3

In this final peak, the remaining tricalcium aluminate reacts with both gypsum and water to form ettringite and subsequently the tricalcium aluminate monosulfate. C_3A hydrates formed are of varying composition but eventually form the stable cubic phase C_3AH_6 in solid solution with the tricalcium aluminate monosulfate.

There has been a considerable study of the effect of various water-reducing admixtures on the pure phases and also on ordinary Portland cement. The following points summarize the general observations.

1. In the presence of a sodium lignosulfonate water-reducing admixture having the following composition: 57.6% lignosulfonic acid, 11.6% reducing sugars, 18.5% ash, the effect on the heat evolution curve under isothermal conditions is shown in Fig. 1.24. These results [53, 54] indicate that the second peak, i.e. the hydration of the C_2S and C_3S

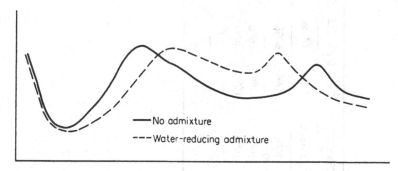

Fig. 1.24 **The effect of a water-reducing admixture on the heat evolution of cement.**

phases, is retarded, whilst that of the third peak is accelerated. Some results have been obtained for sodium gluconate [55] at least up to the end of the second peak. It seems, therefore, that the C_2S and C_3S phases are retarded and that the presence of a water-reducing admixture interferes with the ettringite reaction, particularly in the conversion of the ettringite to the monosulfate.

2. The point in (1) is borne out by the analysis of mixtures of C_3A and water at various times in the presence or absence of gypsum and calcium lignosulfonate (CLS). Table 1.10 illustrates the results obtained [56]. These results reinforce the suggestion that calcium lignosulfonate merely delays the conversion of C_3A to its stable hydrated form C_3AH_6 in a similar manner to gypsum. In addition, the rate of formation and conversion of ettringite to monosulfate is also delayed. This work also suggested that the hexagonal plates of C_4AH_x hydrates were thinner in the presence of calcium lignosulfonate, which may be responsible for an increase in strength observed.

3. As far as the final hydration products of ordinary Portland cement are concerned, there is an indication from isothermal calorimetry [57] that there is very little difference in the presence or absence of a calcium lignosulfonate water-reducing admixture. In this work, the heat evolved per unit of water incorporated into the hydrate has been determined for two cements, with the results shown in Fig. 1.25. It can be seen that the relationship between the amount of heat evolved and the amount of water combined with the cement is maintained whether the admixture is present or not. This work also indicated that the retardation in the early stages is compensated for at later times by an acceleration.

4. In the presence of calcium lignosulfonate [58], the calcium silicate hydrate gel from the C_3S and C_2S phases tends to have a greater proportion of the crumpled foil morphology type than the corresponding system without the admixture. This observation tends to be made

Table 1.10 Effect of calcium lignosulfonate on C_3A reaction (after Chatterji)

	$C_3A + H_2O$			$C_3A + gypsum$			$C_3A + gypsum + 0.2\%$ CLS			$C_3A + 0.2\%$ CLS		
	1 day	14 day	3 month	1 day	14 day	3 month	1 day	14 day	3 month	1 day	14 day	3 month
C_3A	80%	20%	0	60%	30%	20%	70%	60%	30%	80%	60%	10%
Gypsum	—	—	—	20%	0	0	20%	10%	0	—	—	—
C_2AH_8	10%	trace	trace	—	trace	10%	0	0	10%	10%	10%	trace
C_4AH_x	10%	20%	20%	—	10%	10%	0	0	10%	10%	20%	30%
C_3AH_6	trace	60%	80%	—	—	—	—	—	—	—	10%	60%
Ettringite	—	—	—	20%	0	0	10%	20%	0	—	—	—
Monosulfate	—	—	—	trace	60%	60%	trace	10%	50%	—	—	—

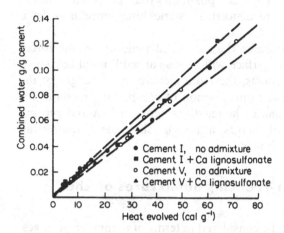

Fig. 1.25 The heat evolved from hydrating cement with and without the addition of calcium lignosulfonate as a function of combined water (Khalil).

only at high concentrations of calcium lignosulfonate in the region of two to four times the normal dosage. This morphological type is believed to have greater drying potential than the other hydrate types of C–S–H gel and may be the reason why occasionally high drying shrinkages are observed in the presence of water-reducing admixtures.

5.   The addition of calcium lignosulfonate or a hydroxycarboxylic acid water-reducing admixture [50] to tricalcium silicate systems can alter the morphology of the resultant calcium hydroxide to produce irregular crystals differing dramatically from the normal hexagonal form. However, this is not always observed and in one study only one calcium lignosulfonate out of four altered the morphology of the calcium hydroxide crystals. In addition, there seems to be an effect on the number of crystals per unit volume, although again this can be either increased or decreased depending on the type of material used.

1.3.4 Interpretation in terms of a mode of action

It is considered that water-reducing admixtures operate by strong adsorption on to the early hydration products of the C_3A phase and to some extent with the C_3S initial hydration products to form a layer of admixture at the water–cement interface. This layer will have the highly charged SO_3^- and COO^- groups pointed outwards into the solution phase to give an increased charge at the surface of the cement particles. This results in a repulsion of the particles by electrostatic means and the interparticle friction in the system is reduced so that the energy required to induce flow into the system is also reduced. There is a possibility that, particularly where multiple layers of molecules are concerned, steric hinderence may be a factor.

The presence of the admixture at the surface, depending on the forces between the admixture and the surface will impose an additional barrier to the diffusion of hydration products, therefore increasing the length of the dormancy period. Because of the introduction of competitive material for sites, particularly on the C_3A phase, the reactions between C_3A and gypsum are slightly modified in the early stages, although subsequently the overall products of reaction are very similar.

1.4 The effects of water-reducing admixtures on the properties of concrete

The properties of concrete can be considered in terms of a number of stages:

1. The **initial plastic** state of the fresh concrete subsequent to the mixing process, where properties such as the air content, density and workability are normally measured by relevant standard tests, and utilized as a means of control of production. The magnitude of these properties is affected by the addition of water-reducing admixtures, either intentionally or as a side effect, which could result not only in a change in the characteristics in the plastic state, but could also be reflected in changed properties in the hardened state.
2. The **later plastic** state when the concrete may be transported, handled and placed and where changes in properties such as workability and the ability of the mix to resist segregation and bleeding may affect these operations.
3. The **hardened** state at a relatively early date, usually 28 days, when the mechanical properties such as compressive and flexural strength and stiffness are used as a basis of structural design.
4. The **subsequent hardened** state during the life of the structure where the concrete material must fulfill its structural or aesthetic role without deterioration. It is important that the durability of the concrete should not be adversely affected by the presence of a water-reducing admixture.

1.5 The effects of water-reducing admixtures on the properties of plastic concrete

1.5.1 Air entrainment

During the mixing of concrete, the 'folding' action of the mixing sequence causes air voids to be formed in the system, which in normal concrete would be reduced by the mechanical forces used in placing the concrete, leaving perhaps up to 1.5% air by volume trapped under aggregate particles. In UK practice it is generally considered undesirable to allow air contents to rise much above this level for structural concrete, because of the effect on compressive strength. In North America, where air-entrained concrete is more widely used, the use of those water-reducing admixtures which have a tendency to increase air contents will necessitate the reduction of the dosage of the air-entraining agent, often by as much as 50%.

The presence of a water-reducing admixture can alter the air content of concrete, either as a deliberate measure (the air-entraining water-reducing admixtures) or as a side effect of the material in lowering the surface tension of the aqueous phase.

The amount of air entrainment obtained will obviously vary according to the type and quantity of admixture used, as well as mix design parameters, but in general at normal dosage levels, in a 50 mm slump sand/gravel mix of 300 $kg\,m^{-3}$ cement content the changes in air content shown in Table 1.11 will be observed. Where the water-reducing admixture has been added to produce a concrete of high workability, for those materials which result in an increase in the air content, approximately 1% more air will result.

The presence of entrained air will, of course, be reflected in a reduced density in the plastic and hardened stage and its effect on subsequent properties of the hardened concrete will be discussed later.

1.5.2 Workability

The ease with which concrete can be deformed by an applied stress is known as the workability of the concrete and is measured by standard tests such as

Table 1.11 Air entrainment by water-reducing admixtures

Category of water-reducing admixture	Chemical type	Additional air content (% by volume)	Reference
Normal	Lignosulfonate	0.4–2.7	[63–65]
	Lignosulfonate + tributyl phosphate	0.3–0.6	[59]
	Hydroxycarboxylic acid	−0.2–0.3	[24, 64]
Accelerating	Lignosulfonate + CaCl$_2$ or formate	0.3–0.5	[64]
	Hydroxycarboxylic acid + CaCl$_2$	0.8–1.6	[18]
Retarding	High sugar lignosulfonate	1–2	
	Hydroxycarboxylic acid	0	[63]
	Hydroxylated polymer	−0.2–0	[24]
Air-entraining	Lignosulfonate + surfactant	0.9–2.6	[60, 61]
	Hydroxycarboxylic acid + surfactant	3–5	[62]

compacting factor, VeBe or slump under arbitrarily chosen conditions of sample preparation and magnitude of applied stress. The amount of deformation obtained under standard conditions would depend on the volume fraction of the aggregate and the shear resistance or viscosity of the cement paste. The effect that water-reducing admixtures have on the cement paste viscosity has been described earlier, but other factors can alter their effect on concrete such as the lubrication of aggregate particles in higher aggregate–cement ratio mixes, etc. [67].

When a normal, accelerating, or retarding water-reducing admixture is utilized to increase the workability of a concrete mix by direct addition, it would be reasonable to assume that the extent of the effect would be markedly affected by changes in mix design parameters such as cement content, aggregate size, shape and grading, and the water–cement ratio. A study of many hundreds of results, however, indicates that this is not the case and Fig. 1.26 illustrates the relationship between initial and final slump for water-reducing admixtures at normal dosage levels. The hydroxycarboxylic acid type appears to be generally superior to the lignosulfonates in increasing the value of slump, and this difference is maintained over the initial slump of 0–100 mm. This non-dependence of mix design parameters on the effect of water-reducing admixtures is perhaps less surprising when it is considered that factors such as wetting and adsorption of aggregates, attrition between aggregate particles, and sufficient excess water to achieve the required slump, have already been taken into consideration during the developments of the initial mix design to produce the relevant workability. Therefore the effect of water-reducing admixtures is above and beyond these requirements and leads to approximately the same increase in slump across the initial slump range.

This independence of efficiency in relation to mix design parameters is only true with regard to workability increases; where a concurrent change in water–cement ratio is made, a number of variables must be considered and this will be discussed later.

The increase in workability obtained is, or course, a function of the dosage of admixture used and this is illustrated in Fig. 1.27 for ligno-sulfonates and the hydroxycarboxylic acid material. It will be appreciated that considerable retardation would be obtained at the higher dosage levels.

(c) Workability control

The relationship between water–cement ratio and workability for mixes containing water-reducing admixtures in comparison to mixes not containing them can be studied by consideration of Fig 1.28, which is for a mortar containing an undisclosed normal water-reducing admixture [67]. This figure illustrates that a given range of workability can be obtained over a smaller range of water–cement ratios for an admixture-containing mix. In practice

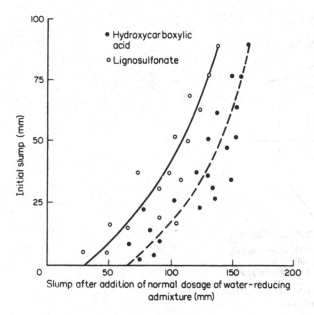

Fig. 1.26 The relationship between initial slump and the slump after the addition of water-reducing admixtures.

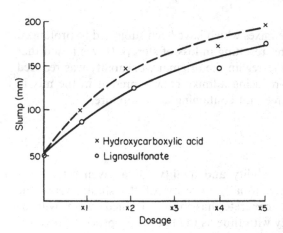

Fig. 1.27 The effect on slump of varying addition levels of water-reducing admixtures.

this can mean that the normal variabilities in water added to the mix produce a wider range of slump values, which is not conducive to accurate control of workability. However, this effect can be considered as beneficial in allowing regain of workability by addition of further water with the minimum effect on concrete quality in terms of strength. This effect has been studied [68] as a

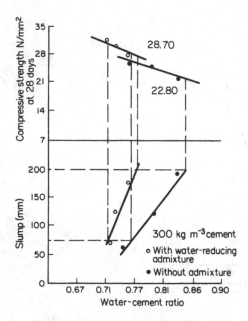

Fig. 1.28 The relationship between the slump and the water–cement ratio for mixes with and without a water-reducing admixture (Howard).

means of retempering concrete mixes which have been subjected to prolonged mixing at elevated temperatures resulting in loss of slump. It was found that the amount of water required to regain the original workability was reduced by up to 20% when a water-reducing admixture was present in the mix in comparison to a control concrete not containing an admixture.

1.5.3 Workability loss

Concrete is judged for its suitability and quality for a given set of mix proportions by its workability, usually in terms of the slump. Once the required workability of the concrete has been attained there will be progressive loss of workability with time as the hydration process proceeds. This process continues through the mixing, discharging, handling, placing, vibrating and finishing and any changes in the rate at which workability is lost can affect any or all of these steps. The loss of workability generally appears to be more pronounced with mixes containing water-reducing admixtures and is illustrated in Fig. 1.29 [68]. All mixes were designed to initial slump (ASTM) of 10 cm and had a cement content of 300 kg m^{-3}. An increase in the dosage apparently reduces the slump loss as shown in Fig. 1.30 [68].

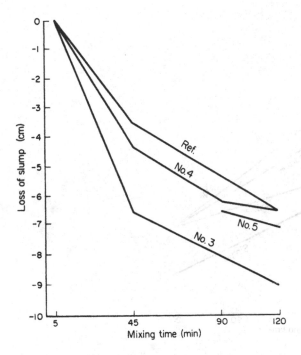

Fig. 1.29 The loss of slump with time (Ravina).

Both Figs 1.29 and 1.30 illustrate the loss of slump from those mixes designed to initial slump equivalent to a mix containing no admixture. However, when the water-reducing admixture has been used to increase the workability by a straight addition, although the rate of slump loss is still greater in the case of the admixture-containing mixes, the high workability is maintained for a longer time as shown in Fig. 1.31 [68].

Similar results are obtained for hydroxycarboxylic-acid-based retarding water-reducing admixture and are shown in terms of loss of workability measured by BS 1881 slump test and by the VeBe in Fig. 1.32. The general conclusion can be reached that the use of retarding water-reducing admixtures to increase the initial workability, so that the initial rate of the slump loss is compensated for, will prolong the time available for the transporting, handling and placing of concrete. Even when these types of materials are used to produce concrete of normal workability, it is generally found that the increased slump loss would cause no problems in normal concrete production unless particular circumstances such as hot weather or long hauls are involved. In these cases the amount of water required to correct the loss of slump is reduced in the presence of a water-reducing admixture. This statement applies to the majority of cases, but there have been instances of

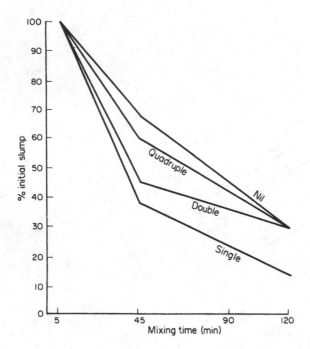

Fig. 1.30 The effect of different dosage levels on loss of slump (Ravina).

severe loss of slump, which have hampered concreting operations and it has been suggested [69] that this is more likely to occur in high-alkali cements. The problem is minimized by the addition of the admixture after the mixing ingredients have been given an initial mixing cycle of 2 min.

1.5.4 Water reduction

The most widely used application of water-reducing admixtures is to allow reductions in the water–cement ratio whilst maintaining the initial workability in comparison to a similar concrete containing no admixture. This, in turn, allows the attainment of a required strength at lower cement content to effect economies in mix design.

The amounts of water reduction possible depend on numerous factors and these are summarized below.

(a) The aggregate–cement ratio

The efficiency of water-reducing admixtures, and their relative usefulness are dependent on the aggregate–cement ratio. Hydroxylated polymer and

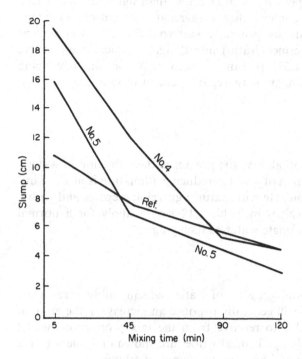

Fig. 1.31 The loss of slump with time when straight addition of a water-reducing agent is made (Ravina).

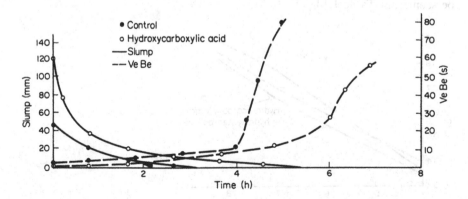

Fig. 1.32 Changes in slump and VeBe values for concrete containing straight addition of a hydroxycarboxylic-acid-based water-reducing agent.

hydroxycarboxylic acid types are more effective than lignosulfonate-based materials at higher cement contents (lower aggregate–cement ratios), whilst the lignosulfonate materials are generally preferred for the lower cement contents (high aggregate–cement ratio) mix designs. Typical comparative data are shown in Fig. 1.33. It can be seen that the water-reducing admixtures are most effective at an aggregate–cement ratio in the region of 6.5–7.0 in these mixes.

(b) Designed workability

The higher the required workability, the greater is the reduction in water–cement ratio when an addition of a water-reducing admixture is made. Thus for a typical 300 kg m^{-3} concrete with natural gravel aggregates and with a zone 3 sand, the typical values in Table 1.12 would apply for a normal addition level of a lignosulfonate water-reducing agent.

(c) Addition level

It is possible to vary the addition level of water-reducing admixtures when an increase in dosage level will generally produce an increase in the amount of water which it is possible to remove from the mix proportions whilst maintaining the required slump. Typical values are shown in Table 1.13 for an aggregate–cement ratio of 5.85 : 1 and a slump of 50 mm.

The amount of water reduction possible is also a function of the way in which an admixture is added to the concrete; if a period between mixing with water is allowed prior to the addition of the admixture, greater adsorption of the admixture on to the initial hydrates is obtained and a higher workability or alternatively a greater reduction in water–cement ratio is obtained, as can be seen from Table 1.14 [73].

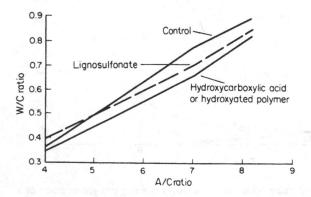

Fig. 1.33 Reductions in water–cement ratio as a function of aggregate–cement ratio for lignosulfonate and hydroxycarboxylic-acid-based water-reducing agents.

Table 1.12 Water reduction by water-reducing agents as a function of workability

Designed slump (BS 1881) (mm)	Reduction in water–cement ratio (%)
50	5–8
75	8–10
100	10–12
150	12–15

Table 1.13 Effect of addition level of water-reducing admixtures on the water reduction

Water-reducing admixture type	Addition level	Water–cement ratio
None		0.55
Lignosulfonate	Normal	0.51
	2 × normal	0.49
	5 × normal	0.47
None		0.55
Hydroxycarboxylic acid	2 × normal	0.48
	5 × normal	0.46

Table 1.14 Effect of varying the point of addition on workability and/or water reduction (after Dodson)

Method of addition of retarder (0.225% calcium lignosulfonate by wt cement)	Water–cement ratio	Slump (mm)	Water reduction (%)
No retarder added	0.59	100	—
Added with mix water	0.55	88	6.8
Addition delayed 2 min	0.55	163	6.8
Addition delayed 2 min	0.51	81	13.6

(d) Cement characteristics

In the case of lignosulfonate water-reducing agents, the effectiveness in reducing the water–cement ratio diminishes with an increase in either the the C_3A or alkali content. In a comparative experiment with three cements varying in C_3A content from 9.44 to 14.7% in comparable mixes, the percentage water reduction for a calcium-lignosulfonate-based material varied from 4 to 10% to achieve a similar level of workability. It has also been shown that the $C_3A/CaSO_4$ ratio is important in determining the effect of water-reducing agents on the fluidity and subsequent stiffening of the cement paste [71–74].

(e) Type of water-reducing agent

Products based on hydroxycarboxylic acid salts are more effective than lignosulfonates in reducing the water–cement ratio as illustrated in Table 1.15 [75].

1.5.5 Setting characteristics of fresh concrete containing water-reducing admixtures

The effect of mix and environmental factors on the setting characteristics of concrete in the presence of admixtures which exert a retarding influence is discussed later. As a generalization, Table 1.16 can be used as a guide for a $300 \ kg \ m^{-3}$ ordinary Portland cement concrete mix having a slump in the range 50–100 mm and the initial setting time (measured by Proctor needle, ASTM C403) in the region of 7–8 h.

In view of the differences in initial setting time, most calcium lignosulfonates and hydroxycarboxylic-acid-based materials would extend

Table 1.15 Effect of water-reducing admixtures (0.1% by weight of cement) on the water reduction at a given workability (cement content = $300 \ kg \ m^{-3}$) (Maniscalco and Collepardi)

Water reducer	Slump (mm)	Water–cement ratio	Water reduction (%)
Nil	95	0.68	—
Sodium gluconate	100	0.61	10.3
Glucose	95	0.63	7.3
Sugar-free Sodium lignosulfonate	100	0.65	4.4

Table 1.16 Extension of initial setting by various water-reducing admixtures

Admixture type	Dosage	Extension of initial setting time (h) at 20°C
High-grade calcium lignosulfonate (normal water-reducing)	1 × normal	4
	2 × normal	10
	3 × normal	16
Hydroxycarboxylic acid type (retarding water-reducing)	1 × normal	6
	2 × normal	12
	3 × normal	17
Lignosulfonate/calcium chloride (accelerating water-reducing)	1 × normal	−1
Sodium lignosulfonate (normal water-reducing)	1 × normal	0.5
	2 × normal	2.0
	3 × normal	3.5

the time available for transport, handling, placing and finishing of concrete, whereas the materials based on sodium lignosulfonate will have only a marginal effect. Some lignosulfonates contain a small amount of triethanolamine to overcome the retardation effect of the major component but, as described later, this can be undesirable where possible volume deformation changes are important in the resultant concrete.

1.5.6 The stability of fresh concrete containing water-reducing admixtures

The stability of the concrete mix can be considered in terms of its 'cohesion', which is a subjective term used to describe its ability to maintain a homogeneous appearance when subjected to applied stress. Lack of cohesion leads to segregation of the mix components into layers relevant to their densities. A further term associated with mix stability is that of 'bleeding', which is the movement of water to the surface of the fresh concrete. This phenomenon can occur either in isolation or as a manifestation of segregation. Bleeding in excess is normally considered to be undesirable because of the dangers of water runs at the shutter/concrete interface and cracking due to plastic settlements, and there is also the possibility of adverse effect on the concrete–reinforcement bond due to the collection of water beneath the steel.

(a) Cohesion

There is little published data on the cohesion of mixes containing water-reducing agents, presumably because of the absence of a truly quantitative method of measurement. A general observation is that when a water-reducing admixture is used to produce a higher-strength concrete at a reduced water–cement ratio, the concrete appears to be more cohesive. In addition, it is often noted that increases in workability produced by the addition of a water-reducing admixture can be made without the loss of cohesion associated with redesigning the mix at higher water content. The quantitative data useful in this context in the published information [76] are the rheological characteristics of concrete containing a lignosulfonate water-reducing admixture. Figure 1.34 shows the rheological results; the slope of the line is a function of the workability of the concrete, whilst the intercept on the energy input axis should be an indication of the cohesion or the inherent structure of the concrete. Figure 1.35 relates the workability to the cohesion in these terms.

These limited results give some indication that it is possible to achieve high-workability material without a consequential loss in cohesion by the use of water-reducing admixtures of the lignosulfonate.

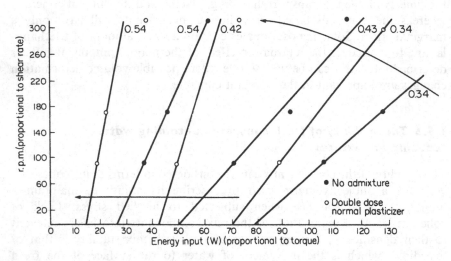

Fig. 1.34 Rheology of concrete containing a normal plasticizer (300 kg m^{-3}) (Hewlett).

(b) Bleeding

The movement of clear water to the surface of the concrete can lead to aesthetic problems of surface finish and plastic settlement and interfere with the reinforcement–concrete bond. On the other hand, promotion of bleeding can, in certain circumstances, be beneficial; in hot and windy weather conditions, plastic cracking can occur when the rate of evaporation exceeds the bleeding rate. Also concrete which has shown severe bleeding is . often stronger [77] because of the reduction in water–cement ratio. In general it can be said that the addition of any water-reducing admixture which does not significantly increase the air content will lead to an increase in the rate and capacity of bleeding of the plastic concrete. Lignosulfonate-based materials usually lead to an increase in air content which can result in an overall decrease in bleeding, whilst normal or retarding water-reducing admixtures containing hydroxycarboxylic acid materials invariably give an increase in bleeding rate.

However, when all categories are used to reduce the water–cement ratio to the same workability as a concrete containing no admixture, this, in turn, reduces bleeding so that the net effect is an increase or decrease. It is reasonable to assume that an air-entraining water-reducing admixture based on any raw material type will not increase the bleeding rate whether used as a means of increasing workability or decreasing the water–cement ratio. Limited data on water-reducing admixtures containing calcium chloride do not indicate any increase in bleeding rates for the accelerating water-reducing admixtures.

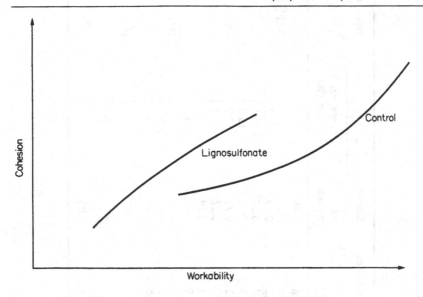

Fig. 1.35 Concretes containing water-reducing admixtures tend to have more structure than a plain concrete at a given workability (after Hewlett).

Table 1.17 summarizes published data for a variety of water-reducing admixtures in a range of mix designs.

1.5.7 Mix design considerations

It is clearly not the purpose of this book to give a guide to the principles of good mix design, which is already catered for by a number of excellent books and reviews [79–82] but rather to set down a few points which are relevant to the use of water-reducing admixtures.

1. When water-reducing admixtures are used at normal dosage levels to obtain a higher workability for a given concrete mix, there is no necessity to make any alteration to the mix design from that produced for the concrete of the initial lower slump. There is generally no loss of cohesion or excess bleeding even when the hydroxycarboxylic acid materials are used.
2. If this class of product is used to decrease the water–cement ratio, again no change in mix design will be required, although small alterations in plastic and hardened density will be apparent and should be used in any yield calculations. One interesting area is in the production of concrete of high elastic modulus. Most high-modulus aggregates are obtained as crushed rocks [83] but this potential advantage is lost because of the high water demand to obtain the required workability. The use of water-

Table 1.17 Effect of water-reducing admixtures on bleeding rates and capacities

Product type	Mix details		Air content	Bleeding ml m⁻¹ cm⁻² surface	Total mixing water (%)	Reference
Lignosulfonate	Crushed limestone 310 k m⁻³ cement Approx 100 mm slump	Control	1.6	0.48	14.2 }	[80] [77]
		Admixture	4.8	0.19	6.3 }	
Lignosulfonate	300 kg m⁻³ cement Natural gravel 50 mm control No reduction in w/c ratio with admixture	Control	0.4	0.50	14.1	Own data
		Na ligno	0.7	0.92	15.2	
		Ca ligno 1	1.2	0.22	7.8	
		Ca ligno 2	1.7	0.22	5.7	
		Cal ligno 3	1.4	0.19	4.4	
Hydroxycarboxylic acid	310 k m⁻³ Rounded aggregate— increased workability	Control	1.5	—	4.3 }	[77]
		Admixture	1.3	—	5.5 }	
	310 kg m⁻³ cement Angular aggregate— reduction in w/c ratio with admixture	Control	1.0	—	7.1 }	[77]
		Admixture	1.1	—	8.6 }	
Lignosulfonate/CaCl₂	300 kg m⁻³ cement Slump 7.5 ± 0.5 cm	Control	1.4	0.13	3.4 }	[78]
		Admixture	4.4	0.08	1.5 }	

reducing admixtures can overcome this problem by reducing the paste water–cement ratio.

3. Economies in mix design are effected by reducing the cement content whilst maintaining the same water–cement ratio. In view of the reduction of paste volume, conflicting recommendation are made of how this should be compensated for. The following is a guide [84]:

   (a) If the mix appears to be 'sticky', the paste volume being removed should be balanced by an increase in the sand content.

   (b) If strength is the major consideration, the paste volume should be balanced by an increase in the coarse aggregate content.

   (c) Most usually the paste volume is replaced by an increase in the total aggregate without a change in the coarse to fine proportioning. This technique has been used in many hundreds of applications, and only occasionally has it been found necessary to revert to (a) or (b).

4. It was shown earlier that aggregate types do not materially affect the performance of water-reducing admixtures. This is not true for cement and mixes containing special cements require particular care. Examples here are increased retardation with low C_3A cement (for example, sulfate-resistant cement) and even an almost complete reduction in expansive properties with expansive cements in the presence of water-reducing admixtures. However, pozzolans such as fly ash appear to behave normally with water-reducing admixtures.

1.6 The effects of water-reducing admixtures on the properties of hardened concrete

The major physical attributes of concrete as a construction material are a high compressive strength and stiffness, an ability to protect and restrain steel and, most important of all, to retain these properties over a considerable period of time. The effects that water-reducing admixtures have on these properties can be considered from the point of view of design parameters, i.e. those properties of concrete at a relatively early age (usually 28 days) which are used for structural calculations, and longer-term aspects or durability.

1.6.1 Structural design parameters

The three most important properties of concrete used in calculations for load-bearing applications are the compressive strength, the tensile strength and the modulus. However, for certain applications, e.g. water-retaining structures, the permeability or porosity of the concrete will be a relevant design criterion and this is also considered here.

(a) Compressive strength

The compressive strength at 28 days of concrete containing water-reducing admixtures of the lignosulfonate, and hydroxycarboxylic acid types is a function of the water–cement ratio and conforms to Abram's rule in the manner of concrete or cement paste [85] which does not contain an admixture. It is often claimed that materials of these types produce higher 28-day compressive strength for a given water–cement ratio, but the author has not found this in his own work. Typical data for British cements and aggregates are shown in Fig 1.36 and span a range of aggregate and mix design types for lignosulfonates and hydroxycarboxylic acid water-reducing agents. Therefore, for materials of these types, no special consideration has to be taken into account for design purposes as far as 28-day compressive strength is concerned.

Air-entraining water-reducing admixtures require special consideration; the presence of entrained air leads to a reduction in compressive strength, whilst the water reduction results in a compensatory increase in strength. The effect can be quantified, however, by considering the amount of entrained air in terms of an equivalent volume of water to calculate the (air and water)–cement ratio. This new factor can be used to estimate the expected strength from Fig. 1.37.

(b) Tensile strength

The tensile strength can be measured in two ways: (1) direct tensile strength from 'dumbbell' specimens; (2) splitting tensile strength from cylinders. Alternatively

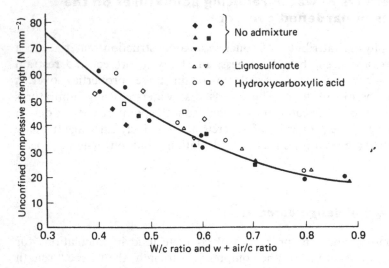

Fig. 1.36 The relationship between water–cement ratio (water and air–cement ratio) and compressive strength of concrete containing lignosulfonate and hydroxycarboxylic-acid-based water-reducing agents.

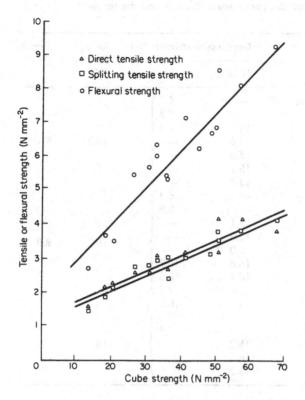

Fig. 1.37 The relationship between cube strength and tensile strength (Kromloš).

the flexural strength can be measured using rectangular prisms. Methods (1) and (2) give similar values, whilst flexural strength, where the applied and resultant forces are not entirely tensile in nature, give somewhat higher values. Typical values are shown in Fig. 1.37 [86] which shows the relationship between tensile and compressive strength. Only limited data are available to illustrate the effects of water-reducing admixtures on the relationship between compressive strength and tensile strength. However, Table 1.18 summarizes the tensile flexural and compressive strengths for some published results and also includes some comparative figures for control concretes.

It can be concluded that water-reducing admixtures of the lignosulfonate and hydroxycarboxylic acid types will not alter the relationship between the compressive strength and the tensile and flexural strengths.

(c) Modulus of elasticity

There is a paucity of recorded comparative data on the elastic modulus of concretes containing water-reducing admixtures. The one Investigation of

Table 1.18 Relationship between the compressive strength and the tensile and flexural strengths

Reference	Admixture type	Compressive strength (%)		Average flexural (%)	Tensile (%)
		Flexural	*Tensile*		
[87]	Hydroxycarboxylic acid	—	6.9		
		—	9.3		
		14.7	—		
[88]		14.6	—	15.2	8.1
		17.8	—		
[89]		15.7	—		
		13.4	—		
[87]	None	—	6.3		
		—	8.9		
		15.1	—		
[88]		13.8	—	16.2	8.8
[89]		16.0	—		
		16.8	10.7		
[86]		18.2	8.5		
		17.0	7.6		
[90]			10.6		
[87]	Lignosulfonate		7.1		
			7.8		
		15.2		14.9	7.5
		16.3			
		13.2			

significance studied a lignosulfonate based material in corresponding mixes using five different cements [65] and the results are given in Table 1.19 as a ratio of the admixture-containing mix to the non-admixture-containing mix of similar workability and 28-day compressive strength parameters. There are strong indications that after 28 to 35 days curing, there is little or no difference in the modulus of elasticity between the corresponding mixes, and at earlier ages the trend is towards a higher modulus.

Work [91] on a hydroxycarboxylic-acid-based material revealed the data given in Table 1.20.

(d) Permeability or porosity

The permeability of concrete is a guide to its durability (Section 1.5.2) but it can also be relevant to the design of structures which are intended to withstand a hydraulic head of water or other liquid. Extreme porosity is usually due to continuous passages in the concrete, due to poor compaction or cracks which can be minimized by the use of water-reducing admixtures to give increased workability whilst maintaining a low water–cement ratio.

Table 1.19 Elastic modulus of concrete containing a lignosulfonate-based water-reducing agent as a ratio of a plain mix (Tam)

Age (days)	Ratio of dynamic modulus Cement					Average
	1	2	3	4	5	
1	1.05	1.10	1.00	1.05	1.25	1.10
3	1.15	1.10	1.05	1.00	1.15	1.10
7	1.15	1.10	1.05	1.05	1.10	1.10
14	1.05	—	1.05	1.05	1.05	1.05
21	1.05	1.05	1.00	1.00	1.00	1.00
28	1.05	1.00	1.05	1.05	1.00	1.05
35	1.05	1.00	1.05	1.05	1.00	1.05
63	1.00	1.00	1.05	1.05	1.00	1.00
91	1.00	1.00	1.05	1.05	1.00	1.00
119	1.05	1.00	1.00	1.00	1.00	1.00
147	1.05	1.00	1.05	1.05	1.00	1.05
182	1.00	1.00	1.00	1.00	1.00	1.00

In the absence of cracks and large channels in the concrete, the permeability is a function of the paste water–cement ratio.

The graphs given in Fig. 1.38 show the logarithmic relationship between the water–cement ratio and the permeability coefficient of hardened cement paste. Thus concrete with a paste water–cement ratio of 0.4 will be almost impermeable. Water-reducing agents can be used to reduce the water–cement ratio, so ensuring that the permeability is kept to a minimum.

Table 1.20 Elastic modulus of concretes containing a hydroxycarboxylic acid water-reducing agent (Brookes)

Concrete mix number	Aggregate type	Water–cement ratio	Admixture	28-day strength $(N\ mm^{-2})$	Modulus of elasticity at 28 days $(N\ mm^{-2})$
1.1	Quartz	0.65	No	30.0	29.6
1.2		0.65	Yes	29.3	29.2
1.3		0.60	Yes	41.8	30.5
2.1		0.45	No	38.2	33.8
2.2		0.45	Yes	40.6	35.2
2.3		0.40	Yes	46.5	39.2
3.1	Limestone	0.65	No	29.2	30.5
3.2		0.65	Yes	26.7	32.9
3.3		0.58	Yes	41.2	35.9
4.1		0.43	No	47.3	40.5
4.2		0.43	Yes	46.9	37.2
4.3		0.38	Yes	52.1	42.1

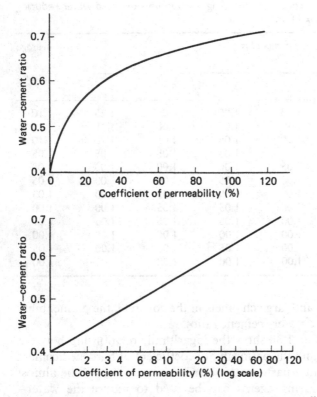

Fig. 1.38 The relationship between the permeability coefficient of concrete and its water–cement ratio.

Typically a concrete of paste water–cement ratio of 0.55 could be reduced to 0.50 resulting in a permeability less than half the original value.

The overall conclusion from the available data is that when a concrete mix is designed incorporating a water-reducing admixture of the normal or retarding type, then the properties of the resultant concrete at 28 days will conform to the normal relationships used for concrete not containing an admixture at the same water–cement ratio.

1.6.2 Durability aspects

The durability of concrete is the ability of the material to maintain its structural integrity, protective capacity, and aesthetic qualities over a prolonged period of time. It is important that the benefits conferred to concrete in the plastic and early hardened state by water-reducing admixtures are not negated by any adverse effect on the long term durability.

Concrete durability can be considered in terms of the following properties:

1. The resistance to attack by aggressive liquids which would commonly be chlorides from marine environments or de-icing salts and sulfates from ground water.
2. The resistance to freeze–thaw cycling which may be experienced during the winter months in many countries. This will not be a function of the average wintertime temperatures of the various countries because, in fact, the very cold environments will have only a small number of freeze–thaw cycles. In countries such as Great Britain, the winter daytime temperatures are often above 0°C and the night-time temperature below. In view of this, more freeze–thaw cycles would be experienced than in countries such as Scandinavia or North America where daytime temperatures in the winter tend to remain below 0°C.
3. The protection of steel reinforcements. Concrete produces a layer of passivity at the steel/concrete interface and any breakdown of this can increase the chance of reinforcement corrosion. In addition, it is important that concrete be maintained in a state of low permeability to minimize the passage of moisture and air to the steel.
4. The majority of load calculations for concrete structures are based on 28-day compressive strengths of concrete; this is based on the knowledge that concrete continues to gain in strength over the subsequent years. Any significant change in this gain of strength would obviously be deleterious for the integrity of the structure and certainly any sudden change in strength characteristics could be disastrous.
5. Concrete undergoes volume changes, particularly under drying conditions. In an unloaded state, the volume change is called shrinkage, whilst the additional volume change under an applied load is known as creep. Alterations in the rate at which a concrete shrinks and creeps due to added materials can be problematical, particularly where concretes of different volume deformation characteristics are in contact with each other, or where joints have been designed for a given rate of movement.

(a) Resistance to aggressive liquids

The deterioration of concrete under the action of materials which aggressively attack the cement matrix will be a function of the permeability or porosity of the concrete [92] and can be measured indirectly by means of the ISA test [93]. It has been shown [94] that for concrete mixes containing $255–300 \, \text{kg m}^{-3}$ cement designed to the same workability and 28-day compressive strength, there is no significant difference between those mixes containing no admixture and those containing a lignosulfonate water-reducing agent as far as the initial surface adsorption is concerned. The same concretes were also subjected to a pore size distribution assessment by means of a mercury porosimeter and these results are shown in Figs 1.39 and 1.40.

Direct measurement of the effect of aggressive reagents on concrete durability appears to be confined to sea water and sulfate attack, where in both areas it is recognized that the lower the water–cement ratio, the greater will be the resistance to attack and the use of a water-reducing admixture will be obviously helpful. This is confirmed by work carried out in Holland [95] and Japan [84] and a general conclusion is that a reduction in the water–cement ratio from 0.5 to 0.40, would allow a reduction in thickness of cover of the reinforcement by about 50%.

Sulfate resistance

In this area, the effects of various types of lignosulfonate and hydroxy-carboxylic acid water-reducing admixture have been studied from the point of view of the effect on concrete having the same mix design but with a lower water–cement ratio in the case of the admixture-containing mixes and also a small amount of work on corresponding mixes containing lower cement contents and the same water–cement ratio, and hence 28-day strength, in the case of the admixture-containing mixes. Table 1.21 shows a set of results for various types of water-reducing admixture using a test method where the concrete is given periodic exposure to sulfate-containing solutions and the number of cycles to achieve a given expansion is noted together with the reduction in Young's modulus, E.

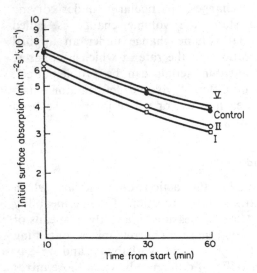

Fig. 1.39 Initial surface absorption of oven-dried concretes containing lignosulfonate water-reducing agents (see Fig. 1.40 for key) (Hewlett).

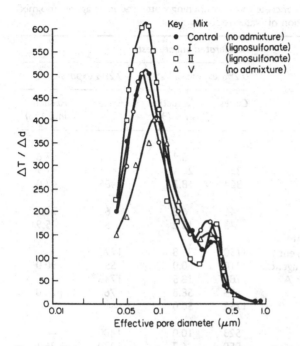

Fig. 1.40 **Frequency curve of pore size distribution (Hewlett).**

It can be seen that generally the lignosulfonate and hydroxycarboxylic acid type materials lead to an improvement in sulfate resistance [96]. This is borne out [97] by Russian work shown in Table 1.22 where again a direct addition of an unspecified water-reducing agent was made and the test method used here was measuring the resonance frequency of concrete specimens after different periods of immersion in a 5% sodium sulfate solution. A formula was developed to give a durability factor Ck. This work showed that by reducing the cement content in the presence of the water-reducing agent, the durability in the presence of sulfate solutions was adversely affected. In view of this, it is concluded that both lignosulfonate and hydroxycarboxylic acid water-reducing agents can be used to reduce the water–cement ratio of concrete mixes, which would be reflected in an enhancement of the durability to sulfate attack. However, when cement reductions are made to the same workability and strength characteristics, testing should be carried out prior to use in any sulfate-sensitive applications.

In view of the known deleterious effect of admixtures containing calcium chloride and the possibility of the same effect being found with calcium formate, it is suggested that accelerating water-reducing admixtures should not be used in those areas where sulfate resistance is of importance.

Table 1.21 Sulfate resistance of concrete mixes containing water-reducing agents (no mix design changes other than addition of water-reducing admixture)

Mix no.	Remarks	Accelerated sulfate test			
		0.2% expansion		0.2% expansion	
		Cycles	Reduction in E (%)	Cycles	Reduction in E (%)
	KIRWIN DAM				
80	No agent, control	226	18.4	332	57.6
81	0.25% agent L	324	20.4	449	61.9
82	0.2% agent A	354	18.7	485	47.0
	GROSS DAM				
83	No agent, control	46	6.8	65	21.6
84	0.2% agent 2	43	10.2	57	25.9
	MONTICELLO DAM				
75	No pozzolan, no agent	1150	31.8	1725*	—
76	30% pozzolan, no agent	310	60.0	554	+90.0
77	No pozzolan, agent A	780	18.5	1746*	—
78	Pozzolan, agent A	495	58.8	760	+76.0
79	Fly ash, agent A	1752	50.0	—	—
	PPT SERIES				
8	No agent, control	523	10.0	1480	—
5	0.3% agent G	550	6.7	1270	26.1
6	0.6% agent C	550	6.5	1120	20.7
7	0.25% agent D	660	8.4	1430*	—
16	No agent, control	690	6.1	1385*	—
12	0.2% agent D	662	4.3	1385*	—
13	0.3% agent D	600	7.2	1385*	—
14	0.4% agent D	895	7.1	1385*	—
15	0.5% agent D	720	14.3	1385*	—

*Cycles to time of report without expanding to that indicated by the column heading.
Agents: A: ammonium lignosulfonate solution. D: hydroxycarboxylic acid solution.
      G: calcium lignosulfonate solution. L: calcium lignosulfonate solution.

(b) Resistance to freeze–thaw cycling

Concrete is damaged by exposure to freeze–thaw conditions due to the expansion of water in the capillaries on freezing to form ice. The expansion results in micro-cracking with a consequential loss of strength and modulus of elasticity. In addition, such concrete would become aesthetically unacceptable because of spalling at the surface, and the possibility of the ingress of water and air through the micro-cracks could lead to reinforcement corrosion. In view of this, any reduction in the water–cement ratio would be beneficial in enhancing the durability under these conditions. This is illustrated in Fig. 1.41 [98].

Table 1.22 Sulfate resistance of reduced water–cement ratio and corresponding mix

Specimens and type of admixture	Water–cement ratio	Durability factor Ck		
		10 months	20 months	28 months
Without admixture, control	0.50	1.00	0.73	0.45
With plasticizer	0.47	1.02	0.94	0.87
With plasticizer, reduced cement content	0.50	0.94	0.45	0.33

There is a considerable amount of recorded data on this aspect of durability, and this is summarized in Table 1.23.

LIGNOSULFONATES

For concrete used in dam construction, the results shown in Table 1.23 [99] have been obtained. From these results, it will be seen that a reduction in the water–cement ratios was obtained and in the large majority of cases (80% of the specimens) an improvement in freeze–thaw resistance was obtained. In fact, the average resistance of admixture-containing concrete was 39% greater than the control specimens. The ability of the lignosulfonates to

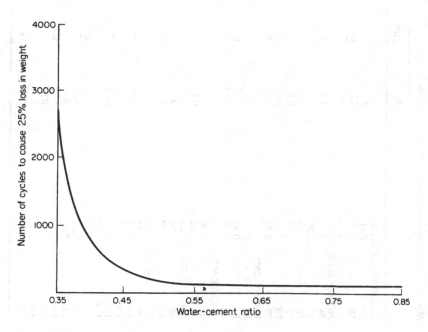

Fig. 1.41 The relationship between water–cement ratio and freeze–thaw resistance.

Table 1.23 Freeze–thaw resistance of concretes used in dam construction

Mix no.	Water–cement ratio	Agent %	Type	Pozzolan (%)*	Cycles of freezing and thawing to 25% weight loss — 28-day fog cure	Cycles of freezing and thawing to 25% weight loss — 14-day fog plus 76-day 50% RH
AINSWORTH CANAL						
59	0.51	0		0	570	780
60	0.53	0.2	G	0	750	720
61	0.52	0.4	G	0	930	850
MONTICELLO DAM						
74	0.57	0.2	B	30	620	580
75	0.50			0	1180	2900
76	0.53			30	650	530
77	0.44	0.1	A	0	1540	2830
78	0.50	0.1	A	30	960	620
79	0.43	0.1	A	30	1490	2050
KIRWIN DAM						
80	0.58	0		0	270	1050
81	0.52	0.25	L	0	430	
82	0.50	0.20	A	0	350	490
GROSS DAM						
83	0.66	0		0	210	1070
84	0.62	0.2	A	0	430	1640
FLAMING GORGE DAM						
96	0.54	0		33.3	630	410
97	0.50	0.37	G	33.3	670	440
GLEN CANYON DAM						
98	0.51	0		42.8	550	320
99	0.46	0.37	G	42.8	1020	410
100	0.56	0		33.3	800	400
101	0.49	0.37	G	33.3	860	540
104	0.64	0		0	600	900
105	0.60	0.27	G	0	660	800
106	0.56	0.54	G	0	450	450
107	0.66			20		590
108	0.63	0.27	G	20	530	710
109	0.61	0.54	G	20	590	810
110	0.67	0		20	360	230
111	0.61	0.37	G	20	460	250

*Fly ash used as pozzolan.
Agents: A: ammonium lignosulfonate. B: ammonium lignosulfonate. G: calcium lignosulfonate. L: calcium lignosulfonate.

provide an improved resistance to freeze–thaw was particularly important in the pozzolan-containing concretes because the substitution of cement by pozzolan generally in this work resulted in a reduction in the freeze–thaw resistance. The use of water-reducing admixtures permitted the inclusion of pozzolan for other beneficial side effects such as reduction of sulfate attack or to enhance resistance to marine environments. Indeed where this latter environment, i.e. sea water, is important, freeze–thaw cycling of concretes with or without sulfite liquor (a crude lignosulfonate) has been carried out [100] in sea water of $34\ g l^{-1}$ salt, with the graphical results shown in Fig. 1.42. It can be seen that there is a considerable enhancement in the durability to such conditions in the presence of the lignosulfonate-containing material.

When lignosulfonates are used to reduce the cement content whilst maintaining the workability and strength characteristics, it has been found [94] that there is still a considerable enhancement of durability of those mixes containing less cement and a lignosulfonate water-reducing agent, in comparison to a control. This is illustrated in Fig. 1.43 where progressive reductions in cement content have been made using a lignosulfonate water-reducing agent to maintain the 28-day strength.

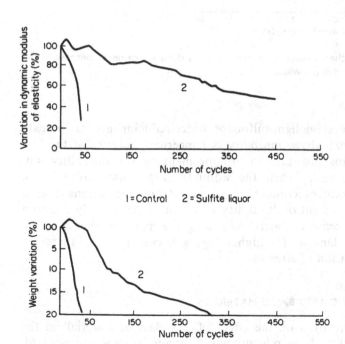

Fig. 1.42 The freeze–thaw resistance of concrete containing lignosulfonates (sulfite liquor) under saline conditions.

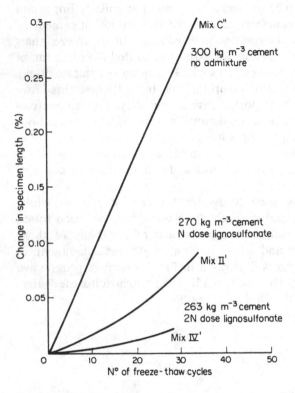

Fig. 1.43 The freeze–thaw resistance of concretes of different cement contents in the presence of lignosulfonates (Hewlett).

The recorded data on lignosulfonate water-reducing agents indicate that, as far as freeze–thaw durability is concerned, because of the low water–cement ratios possible, an enhancement to the durability will invariably be obtained. When the admixtures are used to effect a reduction in the cement content, there are strong indications that a considerable enhancement of durability is obtained, presumably due to a reduction in the cement matrix which is the part of the concrete susceptible to frost damage. The higher aggregate content would therefore allow easier dissipation of stresses.

HYDROXYCARBOXYLIC-ACID-BASED MATERIALS

There is little information on the effect of this class of material on the freeze–thaw durability of concrete into which the admixture is incorporated. However, one set of data is shown in Table 1.24 [101].

Table 1.24 Freeze–thaw resistance of concretes containing a hydroxycarboxylic acid water-reducing agent

Admixture	Dose (%)	Cement content of mix (gravel aggregate, slump = 75 to 100 mm) (kg m^{-3})	Water–cement ratio	Air content	Relative durability to freeze–thaw cycling– number of cycles to reduce dynamic modulus by 50%
None	—	313	0.61	2.7	26
Hydroxycarboxylic acid	0.16	308	0.59	3.5	56

AIR ENTRAINING WATER-REDUCING ADMIXTURES

Mixes containing water-reducing admixtures based on both lignosulfonate and hydroxycarboxylic acids have been incorporated into air-entrained concrete and compared for freeze–thaw durability in comparison to straight air-entrained mixes. All concretes were designed to have between 5 and 7% of air by volume in the plastic state and the freeze–thaw cycling was carried out under water using a cabinet complying with the requirements of ASTM C666-92. Beams were tested up to 300 cycles of freezing and changes in fundamental longitudinal frequency and weight loss were determined. The changes in frequency were used to determine the durability factor and weight loss was used as a measure of surface deterioration [102]. The results shown in Table 1.25 were obtained. It can be seen that the durability to

Table 1.25 Freeze–thaw resistance of air-trained concrete containing water-reducing admixtures

Water-reducing admixture	Dosage	Air-entraining agent type	Water–cement ratio	Freeze–thaw data Durability factor (%)	Freeze–thaw data Weight loss (%)
None (control)	—	Vinsol resin	0.53	86	0.48
None (control)	—	Vinsol resin	0.53	84	1.27
(A) Salt of	1 × normal	Vinsol resin	0.50	99	1.79
hydroxycarboxylic	2 × normal	Vinsol resin	0.48	98	0.52
acid	3 × normal	Vinsol resin	0.45	97	2.05
(B) Calcium	1 × normal	Vinsol resin	0.47	95	0.92
lignosulfonate	2 × normal	Vinsol resin	0.42	95	1.60
	4 × normal	Vinsol resin	0.39	93	0.88
(C) Calcium	1 × normal	Vinsol resin	0.50	92	0.59
lignosulfonate	2 × normal	Vinsol resin	0.47	96	1.64
	4 × normal	Vinsol resin	0.45	97	0.20

freeze–thaw cycling measured by the durability factor is enhanced by the reduction in water–cement ratio due to the presence of the water-reducing admixtures of both types. The means of measuring this durability factor, i.e. the change in resonance frequency, would indicate that the internal integrity of the material is maintained at a higher level in the presence of the water-reducing admixtures. However, the weight loss from the surface data are somewhat inconclusive and it does appear that no similar enhancement of resistance to surface spalling is obtained. Freeze-thaw data have been published [103] comparing concrete containing no admixture with concrete air-entrained to the UK Department of the Environment paving quality concrete specification using neutralized wood resins and an air-entraining water-reducing agent based on hydroxycarboxylic acid. The results are shown in Table 1.26, where the freeze–thaw dilation tests according to the draft British Standard using temperature cycling under water and also a Canadian test using cycling under salt water and measuring surface scaling indicate improvements over both the control concrete and the straight air-entrained concrete.

In the area of air-entraining water-reducing admixtures, it is difficult to find data relevent to corresponding mixes where properties of concrete of similar workability and strength characteristics subjected to freeze–thaw cycling are compared, but the data given in Table 1.27 [104] show that when the strength is at least similar to that of the control containing more cement but no admixture, there appears to be no detrimental effect on the durability to freeze–thaw cycling.

It can be concluded from the assessment of the data in this section that inclusion into a concrete mix of a water-reducing admixture of the lignosulfonate, hydroxycarboxylic acid and air-entraining type should not lead to any deterioration in the durability of that concrete to freeze–thaw cycling. Indeed there are strong indications that, when used either as a means of reducing the water–cement ratio or, alternatively, of reducing the cement content, more durable concrete may result.

Table 1.26 Freeze–thaw resistance of concrete containing an air-training water-reducing

Mix no.	1	2	3
Concrete containing	No admixture	Neutralized wood resin	Air-entraining water-reducing agent
Air content (%)	1.2	4.5 ± 0.3	4.5 ± 0.3
Freeze–thaw dilation test (% dilation at 50 cycles)	0.035	−0.033	−0.031
DHO test (number of cycles to cause 0.5 mg mm^{-2} scaling)	7.5	15.0	25.0

Table 1.27 Freeze-thaw resistance of corresponding mixes containing water-reducing admixtures and air-entraining agents (after Mielenz)

Series	Mix	Air-entraining agent	Water-reducing admixture	Cement content (kg m^{-2})	Water-cement ratio	28-day compressive strength (N mm^{-2})	Air content (%)	Slump (mm)	Durability factor ASTM C290 1967
I	F	Vinsol resin	None	261	0.62	28.1	5.6	102	28
	G		Calcium lignosulfonate	227	0.63	26.8	5.2	89	25
II	F	Vinsol resin	None	314	0.53	33.8	5.1	159	34
	G		Calcium lignosulfonate	311	0.45	36.7	6.0	165	51
III	F	Vinsol resin	None	307	0.53	32.9	5.8	159	34
	G		Calcium lignosulfonate	307	0.47	35.8	5.6	171	42

(c) Protection of steel reinforcement

One of the prime functions of concrete in load-bearing structures is to protect the steel incorporated to increase the tensile strength of the cementitious material. If exposed to air and water the steel will rust (an electrochemical reaction), the reaction products having a lower density than the original steel. The expansive forces caused result in stresses being applied internally so that the concrete fails in tension causing large pieces to be broken away at the surface. In view of this, water–cement ratios are kept as low as possible to prevent the ingress of water, and minimum cover of concrete over the reinforcement is often specified.

In order to study the effect that water-reducing admixtures may have on the role that concrete plays in protecting steel reinforcement, it is necessary to consider the following aspects.

CONCRETE PERMEABILITY

This has already been dealt with in previous parts of this section where it can be seen that a reduction in the water–cement ratio by use of admixtures is beneficial in reducing the permeability of the concrete. Even when cement contents are reduced, whilst maintaining the workability and strength characteristics of corresponding mixes containing water-reducing admixtures, there is no deleterious effect on the permeability.

CHEMICAL ATTACK ON THE REINFORCEMENT

The action of an admixture in relation to attack on reinforcement can be considered either in direct chemical reaction with the steel or, alternatively, a breakdown of the passive layer imparted by concrete which normally prevents corrosion at the cement/steel interface. In this respect, any accelerating water-reducing admixtures containing calcium chloride can be considered hazardous as far as raising susceptibility of steel reinforcement to corrosion is concerned. It is particularly so at calcium chloride contents in the concrete at or above 1.5% by weight of cement as discussed in the section on accelerators. The use of such materials has been controlled by relevant codes of practice where embedded metal is present in the concrete.

Some data are available [105, 106] concerning lignosulfonate-based materials, and Figs 1.44 and 1.45 show, respectively, the potential–time curves for steel embedded in cement paste containing sulfite black lye (an impure lignosulfonate) at various levels, and potential–time curves for steel embedded in cement paste containing 2% calcium chloride and different concentrations of black lye. It is clear that the lignosulfonate-containing admixture not only enhances the passivation of the embedded steel, but also can counteract the effect of calcium chloride on it.

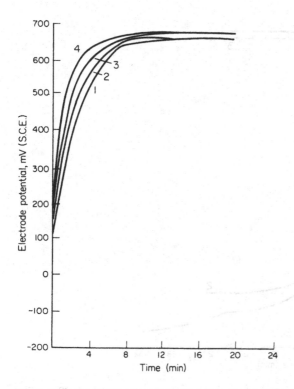

Fig. 1.44 Potential–time curve for steel embedded in cement paste containing black lye. Curve 1–0.0% black lye; curve 2–0.2%; curve 3–0.5%; curve 4–1%.

REINFORCEMENT BOND

The bond between the concrete and reinforcement is an important engineering aspect of the composite material and any slippage could possibly create an easier ingress of air and moisture, leading to corrosion. Information has been obtained on the effect that calcium lignosulfonate has on the bond of concrete with steel and, in one particular investigation [105], the effect before and after the application of anodic-connected direct current. These results are shown in Table 1.28 where it can be seen that the presence of calcium lignosulfonate increases the bond strength in comparison to the control over the applied voltage range.

Similar data have been obtained for other lignosulfonate-based water-reducing admixtures where cement and workability were kept constant with a reduction in the water–cement ratio and the reinforcement bond was measured by ASTM C234:91a, with the result shown in Table 1.29 [107].

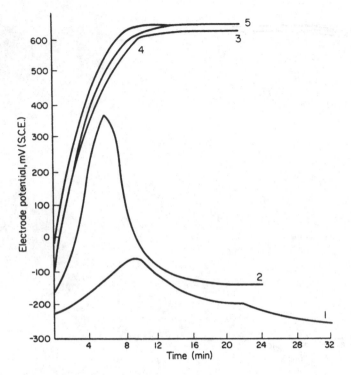

Fig. 1.45 Potential–time curve for steel embedded in cement paste containing 2% CaCl$_2$ and different concentrations of black lye. Curve 1–0.0% black lye; curve 2–0.2%; curve 3–1.5%; curve 4–2.0%; curve 5–3.0%.

The data presented in this section illustrate that, with the exception of those accelerating water-reducing admixtures containing calcium chloride, there is an abundance of evidence to support the conclusion that water-reducing admixtures of lignosulfonate chemical form certainly will not accelerate any kind of corrosion with reinforcement and, when used to reduce the water–cement ratio, will form a more permeable and durable protective cover for the reinforcement. In view of the chemical nature of the other types of materials such as the hydroxycarboxylic acids and hydroxylated polymers, it seems most likely that these materials too would have no deleterious effect in this respect.

(d) Gain in compressive strength

As has been clearly shown in the preceding section, it is possible to produce concrete containing an admixture having a 28-day strength equal to a

Table 1.28 Effect of a lignosulfonate-based water-reducing agent on reinforcement bond under an applied external voltage (after Kondo)

	Applied voltage (V)					
	0		10		20	
Admixture	No	Yes	No	Yes	No	Yes
Bond (N mm^{-2})	3.2	3.4	3.3	3.5	3.7	3.8

Table 1.29 Effect of a lignosulfonate-based water-reducing agent on reinforcing bond

Admixture	Water–cement ratio	Reinforcement bond	
		Horizontal	Vertical
None	0.62	100	100
Lignosulfonate	0.52	121	116

control concrete but being designed at a significantly lower cement content. However, the choice of 28-day compressive strengths is purely arbitrary and, as far as durability is concerned, it is important that the concrete continues to gain in strength over subsequent years. There are some data available in this area where concretes containing various types of water-reducing admixtures have been tested for strength at ages up to 13 years and Fig. 1.46 shows a graphical summary of this information. The filled points indicates the type of results obtained for up to 50 years on many hundreds of concrete specimens not containing an admixture and it can be seen that by and large the available data fit in the filled points at least as often as results for tests on control concretes carried out during the admixture evaluations.

In view of this it seems reasonable to conclude that of the types of water-reducing admixture reported, there is no indication of any subsequent loss of strength or of any significant decrease in the gain of strength.

(e) Volume deformations

The volume deformations of concrete are shrinkage, which occurs under drying conditions, and creep, which is the additional deformation obtained under an applied stress. Creep does occur under saturated conditions (basic creep) but increases considerably under conditions of moisture loss. The picture is rather complicated in that creep is made up of a recoverable and irrecoverable portion on removal of the applied stress.

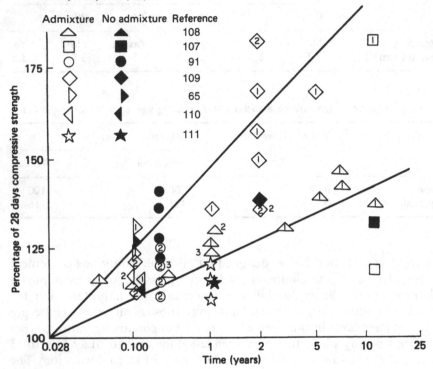

Fig. 1.46 The increase in compressive strength of concretes containing various water-reducing agents.

Both deformations are believed to proceed by the same mechanism which, although not fully understood, is associated with the bound water in the tobermorite gel. These volume deformations are recognized as important design considerations and are compensated for in codes of practice for structural design. It will be appreciated that any change in the known behavior of concrete by the addition of minor ingredients could result in durability problems in concrete structures; increased deflection in beams and slabs, the transfer of load to reinforcement, loss of prestress in prestressed concrete, changes in relative movements between components, and changes in stability of slender columns and walls [112] could all lead to minor or major failures.

Volume deformations are largely a function of the nature and quantity of the cement paste in the concrete and it has been shown [113] that studies on cement paste properties correlate well with those of the concrete into which it is incorporated, certainly as far as comparison of the effect of added materials.

It is difficult to prepare a definitive synopsis of the effect that the various classes of admixtures will have on the volume deformations of concrete because of the conflicting results between workers and different types of test methods used, but the points given below should enable some guidelines to be set.

VOLUME DEFORMATIONS UNDER SATURATED CONDITIONS

Shrinkage/swelling Under saturating conditions where there is no weight loss due to the evaporation of water, there is little volume change in cement paste specimens for admixtures based on lignosulfonate alone, and accelerating water-reducing admixtures based on lignosulfonate with the addition of either calcium chloride or triethanolamine [114] (Fig. 1.47). This has been studied in greater detail [91] for a hydroxycarboxylic-acid-based material containing no additional accelerating materials where it can be seen in Fig. 1.48 that any shrinkage/swelling properties are not significantly changed by the addition of the admixture and are largely a function of the thermal expansion of the resultant concrete.

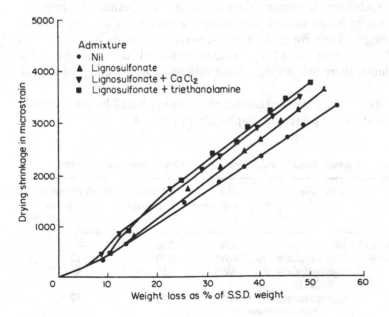

Fig. 1.47 Volume changes of cement pastes under near saturated conditions (Morgan).

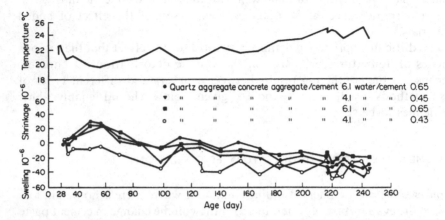

Fig. 1.48 Volume changes of concretes containing a hydroxycarboxylic acid water-reducing agent under saturated conditions (Neville).

Basic creep   The creep of concrete under near-saturated conditions (98% RH) has been studied using a lignosulfonate water-reducing admixture, a lignosulfonate with the addition of calcium chloride, and a lignosulfonate with the addition of triethanolamine [115]. The mix designs used in this work were identical for each mix, therefore the water-reducing admixtures were used as direct additions to increase the workability. The results are shown in Table 1.30, where it can be seen that there are only marginal increases in creep over the plain mix for each of the admixture types studied.

Statistical evaluation of all the results indicated that, at the 90% confidence limit, there was no significant difference in the creep of any of the mixes.

Some results are available for the hydroxycarboxylic-acid-based materials [91] and these data are presented graphically in Fig. 1.49.

Table 1.30 Effect of lignosulfonate-based admixtures on basic creep of concrete

Mix details	Admixture	Mix designation	Observed creep microstrain	Increase in creep over plain mix (%)
Direct addition to plain mix 98% RH 56 days under load	Nil	AW	240	—
	Lignosulfonate	AW1	270	+12
	Lignosulfonate + CaCl₂	AW2	250	+4
	Lignosulfonate + triethanolamine	AW3	270	+12

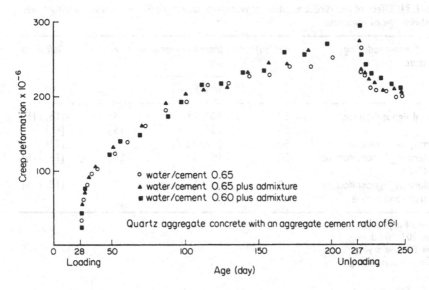

Fig. 1.49 The creep and creep recovery of concrete containing a hydroxycarboxylic acid water-reducing agent under saturated conditions (Neville).

Volume deformations under drying conditions   It is under conditions where moisture is lost from concrete that volume deformations under loaded or unloaded conditions occur to any magnitude. It is difficult to say what degree of relative humidity structural concrete will be subjected to in actual practice, but certainly for thin sections or near the surface of large sections, considerable interchange of water due to changing climatic conditions will occur.

Although there are some anomalies in the literature, it is generally agreed that both types of volume deformation are a function of the same fundamental mechanism and that the influence of other factors such as admixtures will affect both shrinkage and creep in a similar manner. As outlined earlier, water-reducing admixtures can be used to obtain different effects on the plastic/hardened concrete and it is this factor, together with the admixture type, that is important in determining the effect on the volume deformations of concrete.

Direct addition of water-reducing admixture   This increases the workability of the concrete. The effect of all types of water-reducing admixture under these conditions is invariably to increase the shrinkage and creep of the concrete. Some typical values are shown in Table 1.31. The considerable increases in both shrinkage and creep in the presence of admixtures containing calcium chloride and triethanolamine are clearly illustrated.

Table 1.31 Effect of the direct addition of water-reducing admixtures on the drying creep and shrinkage of concrete

Type of water-reducing admixture	RH(%)	Increase over plain mix (%)		Reference
		Shrinkage	Creep	
Normal lignosulfonate	50	10* 1†	34‡	[18, 115]
	50	15§	15§	[65]
Hydroxycarboxylic acid	50	(−4) to (2)		[116]
Accelerating lignosulfonate + CaCl₂	50	19* 4†	54‡	[18, 115]
Accelerating lignosulfonate + triethanolamine	50	35* 13†	79‡	[18, 115]

* After 14 days drying.
† After 203 days drying.
‡ After 84 days drying.
§ After 175 days drying.

Admixture addition with simultaneous reduction in cement content This produces corresponding mixes having similar 28-day strengths and workability characteristics to a plain concrete control. It has been studied with regard to volume deformations and, although again there are some conflicting results, it does seem that lignosulfonate materials used in this context produce little if any change in the creep and shrinkage characteristics of the concrete. However, where triethanolamine and calcium chloride have been included in the formulations to produce accelerating materials, it is possible to obtain increases in the creep and shrinkage, and a limited amount of work on hydroxycarboxylic acids suggests that even in corresponding mixes there may be an increase in volume deformations (Table 1.32). It will be appreciated from the spread of results and the fact that many tests were carried out under different conditions, that it is difficult to make comparisons, and only the trends can be observed.

Addition of admixture to obtain higher strengths In this situation the small amount of reported work indicates that where the water–cement ratio is reduced, the shrinkage and creep of the concrete is also reduced. Figure 1.50 shows the effect of a hydroxycarboxylic acid plasticizer on the creep of the concrete where the material has been used to effect a reduction in the water–cement ratio without any other changes in the mix design. Thus it seems that the reduction of the water–cement ratio will compensate for the increases in creep observed in the data above.

This is a difficult field for interpretation, where techniques and procedures vary from worker to worker and where conflicting results are apparent.

Table 1.32 Effect of water-reducing admixtures on the drying creep and shrinkage of corresponding mixes

Type of water-reducing admixture	RH (%)	Increase over plain mix (%)		Reference
		Shrinkage	Creep	
Normal lignosulfonate	50	−4 to +22	−33 to +17	[65]
	35	—	+2 to +17	[117]
	95	—	−2 to +19	[115]
	50	+1	—	[18]
	50	+1 to +4	—	[116]
Retarding hydroxycaroxylic acid	95	—	−32 to +13	[115]
	94	−52 to −6	−5.8 to +49	[118]
Accelerating lignosulfonate + CaCl₂	50	+3	—	[18]
	50	+11	—	[18]
Accelerating lignosulfonate + TEA	35	—	−8 to +35	[117]

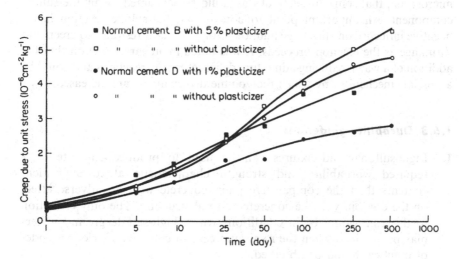

Fig. 1.50 The effect of a hydroxycarboxylic acid water-reducing agent on the creep of concrete when used to lower the water–cement ratio (Rodrigues).

However, it is felt that the intrinsic property of most water-reducing admixtures is to increase the creep and shrinkage of the concretes into which they are incorporated. This is minimized by utilizing the effect of the admixture to reduce the cement and, therefore, the paste content of the mix. For corresponding mixes with lignosulfonate water-reducing agents, the FIP/CEB recommendation for creep computation can be used with confidence because of the incorporation of a cement content/water–cement ratio function in the calculation.

When materials other than a lignosulfonate water-reducing agent are used, in conditions where significant drying of concrete can take place, caution must be exercised where straight additions are used and where differential creep rates could be obtained in comparison to adjacent concrete containing no admixtures.

It is interesting to speculate on the reasons for such varied effect on creep and it is felt that although current theories have been based on the effect on the tobermorite gel layering [119, 120] some thought should be given to the possibility that it is the known interference (see earlier) on the ettringite reaction of water-reducing admixtures. Indeed the sulfate component of cement has been shown to influence the shrinkage of cement paste [113] according to the following equation:

$$\text{Paste shrinkage (microstrain)} = (120 + 260\ C_3A + 101\ C_4AF - 770\ SO_4$$
$$+ 1200\ \text{alkali} + 0.4\ \text{fineness}).$$

It can be seen, therefore, that on typical paste shrinkage in the region of 4000 microstrain, that some 30–40% of this could be accounted for by the sulfate component. An important point to note is that the sulfate reaction has a negative influence on shrinkage and, therefore, acts as a restraint to creep and shrinkage as the reaction proceeds. It was noted in an earlier section that the addition of a water-reducing admixture delays ettringite reactions and could be a possible mechanism by which the volume deformations are increased.

1.6.3 Durability guidelines

1. Lignosulfonate admixtures can be used to produce concrete of a required workability and strength characteristic at lower cement contents than the comparative plain concrete with no adverse effect on the durability of the concrete or total structure. The only exception to this rule would be in conditions where high-sulfate ground waters may be involved when the minimum cement contents of relevant codes of practice should be observed.
2. Water-reducing admixtures containing calcium chloride should not be used in concrete containing embedded metal or where volume deformations are important.
3. In shrinkage- and creep-sensitive situations, sections of a concrete structure should not be treated with any type of water-reducing admixture either as a means of obtaining extra workability or for the retempering of slightly matured mixes where the differential volume changes of adjacent parts of the structure could lead to cracking. This is also true, of course, when cement contents are increased for the same reasons.
4. Lignosulfonate admixtures containing triethanolamine should not be used in situations sensitive to increased volume deformations.

References

1   US Patent (1932). 643 740.
2   Roş, M. (1934). *Eigenschoften des Betons*, Report No.79. EMPA Zurich.
3   US Patent (1937). 2 081 642.
4   US Patent (1937). 2 081 643.
5   US Patent (1938). 2 127 451.
6   US Patent (1938). 2 141 570.
7   US Patent (1939). 2 169 980.
8   US Patent (1939). 2 174 051.
9   US Patent (1941). 2 229 311.
10 Cook, H.K. (1967). *Proceedings of the International Symposium on Admixtures for Mortar and Concrete*, Brussels, 135–6.
11 Mouton, Y. (1972). *Bulletin of the Liaison Laboratory*, **58**, Ref. 1185.
12 Edmeades, R.M. and Hewlett, P.C. (1975). *Proceedings of the First International Congress on Polymer Concretes*. London, Section 7, Paper 10.
13 Hansen, W.C. (1959). *ASTM Special Publication* No. 266, 20–2.
14 Chaiken, B. (1961). *Public Roads*, **31**, 126–35.
15 Joisel, A. (1973). *Physico Chemistry of Admixtures for Cement and Concrete*, 40.
16 Ernsberger, F.M.Z. (1948). *Journal of Physics and Colloid Chemistry*, **52**, 267–76.
17 Rezanowich, A. (1960). *Journal of Colloid Science*, **15**, 452–71.
18 Foster, D.E. (1963). *ACI Journal* Title **60–64**, 1481–523.
19 Morgan, D.R. (1974). *Proceedings of the First Australian Conference on Engineering Materials*, University of New South Wales, 97–108.
20 Hewlett, P.C. (1975). Private communication.
21 Anon. (1967). *Admixtures for Concrete*. Concrete Society TRCS, 1.
22 Anon. (1974). *Heptonates as Additives for Concrete and Cement*, Croda Ltd., data sheet and Sandberg reports L337 and M13.
23 Bruere, G.M. (1964). *Constructional Review, Australia*, **37**, 16–21.
24 Diamond, S. (1971). *Journal of the American Ceramic Society*, **54**, 2734.
25 Mielenz, R.C. (1968). *Proceedings of the Fifth International Symposium on the Chemistry of Cement*, Tokyo, 32.
26 Danielson, U. (1967). *Proceedings of the International Symposium on Admixtures for Mortar and Concrete*, Brussels, 58–67.
27 British Patent (1967). 1 068 886.
28 Japanese Patent (1972). 41 897.
29 Japanese Patent (1975). 36 517.
30 Japanese Patent (1973). 117 519.
31 Japanese Patent (1974). 71 416.
32 Hanson, W.C. (1972). *Journal of Materials JMSLA*, **5**, 842–55.
33 Ernsberger, F.M. (1945). *Industrial and Engineering Chemistry*, **37**, 598–600.
34 Manabe, T. (1959). *Japan Cement Engineering Association, Review*, 40–6.
35 Blank, B. *et al.* (1963). *Journal of the American Ceramics Society*, **46**, 395–9.
36 Rossington, D.R. (1968). *Journal of the American Ceramics Society*, **51**, 46–50
37 Mielenz, R.C. (1968). *Proceedings of the Fifth International Symposium on the Chemistry of Cement*, Tokyo, 15–19.
38 Ramachandran, V.S. (1971). *Cement Technology*, **2**, 21 9.

39 Ramachandran, V.S. (1972). *Cement and Concrete Research*, **2**, 179–94.
40 Young, J.F. (1972). *Cement and Concrete Research*, **2**, 415–33.
41 Banfil, B.F.G. (1979) *Cement and Concrete Research*, **9**, 795–6.
42 Flateau, A.S. (1974). *Concrete*, **8**, 45–7.
43 Uchikawa, H. and Hanehara, S. (1997). *Proceedings of the Fifth Canmet/ACI International Conference on Superplasticizers*, Rome, Italy.
44 Mielenz, R.C. (1968). *Proceedings of the Fifth International Symposium on the Chemistry of Cement*, Tokyo, **4**, 1–29.
45 Diamond, S. (1972). *Journal of the American Ceramics Society*, **55**, 405–8.
46 Zhuravhev, U.F. (1952). *Journal of Applied Chemistry (USSR)*, **25**, 1317–24.
47 Prior, M.E. (1959). *ASTM Special Publication No. 266*, 173–4.
48 Singh, W.B. (1975). *Cement and Concrete Research*, **5**, 545–50.
49 Roberts, M.H. (1967). *Proceedings of the International Symposium on Admixtures for Mortar and Concrete*, Brussels, 7–27.
50 Dodson, V.H. (1967). *Proceedings of the International Symposium on Admixtures for Mortar and Concrete*, Brussels, 59–64.
51 Taylor, H.F.W. (1966). *The Chemistry of Cements*. RIC Lecture Series. No.2.
52 Marianpolski, N.A. (1974). *Neft. Khoz*, **10**, 27–30.
53 Stein, H.W. (1961). *Journal of Applied Chemistry*, **42**, 474–82.
54 Forrester, J. (1967). Private communication.
55 Singh, W.B. (1975). *Cement and Concrete Research*, **5**, 548.
56 Chatterji, S. (1967). *Indian Concrete Journal*, **4**, 151–60.
57 Khalil, S.M. (1973). *Cement and Concrete Research*, **3**, 677–88.
58 Ciach, T.D. (1971). *Cement and Concrete Research*, **1**, 159–76.
59 Edmeades, R.M. (1975). *Proceedings of the Conference on Ready-Mixed Concrete*, Dundee, Theme 3.
60 Hodkinson, K. (1974). Private communication.
61 Hodkinson, K. (1974). Private communication.
62 Rixom, M.R. (1975). *Chemistry and Industry*, **7**, 162–5.
63 Anon. (1967). *Admixtures for Concrete*, Concrete Society, Report 1.
64 Fletcher, K.E. (1971). *Concrete*, **5**, 175–9.
65 Tam, C.T. (1974). *Proceedings of the First Australian Conference on Engineering Materials*, University of New South Wales, 73–96.
66 Warris, B. (1967). *Proceedings of the International Symposium on Admixtures for Mortar and Concrete*, Brussels, 30–3.
67 Howard, F.L. *et al.* (1959). *ASTM Special Publication No. 266*, 149–54.
68 Ravina, D. (1975). *ACI Journal*, **72**, 291–5.
69 Hersey, A.T. (1975). *ACI Journal*, **72**, 52–8.
70 Dodson, V.H. (1976). *ACI Journal*, **73**, 1739–53.
71 Luke, K. and Aitcin, P.C. (1991). *Ceramic Transactions. Advances in Cement Materials*, **16**, 147–166.
72 Basile, S. *et al.* (1997) *Cement and Concrete Research*, **17**, 715–22.
73 Locher, F.W. *et al.* (1976). *Zement-Kalk-Gips*, **29**(10), 435–42.
74 Meyer, L.M. and Perenchio, W.F. (1980). *PCA R&D Bulletin* RD069.01T.
75 Maniscalco, V. and Collepardi, M. Unpublished data given to V.S. Ramachandran
76 Anon. (1976). *Superplasticising Admixtures in Concrete*, Cement and Concrete Association, p. 17.
77 Vollick, C.A. (1959). *ASTM Special Publication No. 266*, 19–5.

78 Okada, K. (1967). *Proceedings of the International Symposium on Admixtures for Mortar and Concrete*, Brussels, 115–29.
79 Shacklock, B.W. (1975). *Concrete Constituents and Mix Proportions*, Cement and Concrete Association.
80 Anon. (1974). *ACI Publication SP46*, 97–108.
81 Cordon, W.A. (1975). *ACI Journal*, **72**, 46–9.
82 Anon. (1997) *ACI Manual of Concrete Practice* Pt1 (ACI 211.1–91).
83 Pomeroy, C.D. (1972). *C & CA Technical Report* UDC 666 972, 12; 666 970, 17.
84 Anon. (1970). *Materiaux et Construction*, **3**, 130.
85 Lawrence, C.D. (1974). *Proceedings of the International Symposium on Pore Structure and the Properties of Materials*, RILEM/IUPAC, **5**, 167–76.
86 Kromloš, K. (1970). *Magazine of Concrete Research*, **22**, 232–8.
87 Wallace, G.B. (1959). *ASTM Special Publication No. 266*, 4–7.
88 Kreiger, P.C. (1967). *Proceedings of the International Symposium for Mortar and Concrete*, Brussels, 28–9.
89 Larson, T.D. *et al.* (1963). *ACI Journal*, **60**, 1739–53.
90 Authors' data (1975).
91 Brookes, J.J. (1975). *Concrete*, **9**, 33–5.
92 Anon. (1971). *Supplement to the Consulting Engineer*, **27**, 9.
93 Levitt, M. (1971). *Journal of Non-destructive testing*, **12**, 106–12.
94 Hewlett, P.C. (1975). *Proceedings of the Workshop on the use of Chemical Admixtures in Concrete*, University of New South Wales, 43–73.
95 Anon. (1970). *Materiaux et Construction*, **3**, 125.
96 Ore, E.L. (1959). *ASTM Special Publication No. 266*, 86.
97 Stolnikov, U.V. (1967). *Proceedings of the International Symposium on Admixtures for Mortar and Concrete*, Brussels, 37.
98 Anon. (1960). *Conf. Laboratory Report No. C. 810*. US Bureau of Reclamation.
99 Ore, E.L. (1959). *ASTM Special Publication No. 266*, 84.
100 Batarakov, V.G. (1967). *Proceedings of the International Symposium on Admixtures for Mortar and Concrete*, Brussels, 76.
101 Fischer, H.C. (1959). *ASTM Special Publication No. 266*, 214.
102 Larson, T.D. *et al.* (1963). *ACI Journal*, **60**, 1739–53.
103 Rixom, M.R. (1975). *Proceedings of the Workshop on the Use of Chemical Admixtures in Concrete*, University of New South Wales, 149–76.
104 Mielenz, R.C. (1968). *Proceedings of the Fifth International Symposium on the Chemistry of Cement*, Tokyo, 11.
105 Kondo, Y. *et al.* (1959). *ACI Journal*, **56**, 299–312.
106 Gouda, U.K. *et al.* (1973). *Journal of Colloid and Interface Science*, **43**, 294–302.
107 Anon. (1976). *Superplasticizing Admixtures in Concrete*, Cement and Concrete Association, 28.
108 Browne, R.D. (1976). Private communication.
109 Wallace, J.J. (1959). *ASTM Special Publication No. 266*, 38–96.
110 Hope, B.B. (1970). *ACI Journal*, **67**, 673–8.
111 Kobayashi, M. (1967). *Proceedings of the International Symposium on Admixtures for Mortar and Concrete*, Brussels, 79–96.
112 Browne, R.D. (1971). Private communication

113   Roper, H. (1974). *Proceedings of the First Australian Conference on Engineering Materials*, University of New South Wales, 45–71.
114   Morgan, D.R. (1974). *Materiaux et Construction*, **7**, 283–9.
115   Morgan, D.R. (1973). Ph.D. Thesis, University of New South Wales.
116   Bruere, G.M. *et al.* (1971). *CSIRO Report PB 203.263*, Australia.
117   Hope, B.B. (1971). *ACI Journal*, **68**, 361–5.
118   Hope, B.B. (1967). *Proceedings of the International Symposium on Admixtures for Mortar and Concrete*, Brussels, 19–32.
119   Morgan, D.R. (1974). *Civil Engineering Transactions*, **56**, 7–10.
120   Jessop, E.L. *et al.* (1968). *Proceedings of the Fifth International Symposium on the Chemistry of Cement*, Tokyo, Supplement IV, V, 36–41.

Chapter 2

Superplasticizers

2.1 Background and definitions

The superplasticizers are a special category of water-reducing agents in that they are formulated from materials that allow much greater water reductions, or alternatively extreme workability of concrete in which they are incorporated. This is achieved without undesirable side effects such as excessive air entrainment or set retardation.

The materials originally developed as the basis for superplasticizers in the 1960s were sulfonated naphthalene formaldehyde (SNF) [1–8] and sulfonated melamine formaldehyde (SMF) [9] in Japan and Germany respectively, which have found increasing application world-wide over the intervening years. In the early 1980s, work began on designing polyacrylate-based polymers [10] for superplasticizer formulations and after some difficulties with severe retardation, and in some cases excessive air entrainment, products began to appear in the marketplace, initially in Germany, and then in Japan and the United States [11–15]. The polyacrylate-based products are based on three different types of polymer and are being heralded as the next generation of super-plasticizers.

2.2 The chemistry of superplasticizers

The three major types of raw materials used in superplasticizers, SNF, SMF, and polyacrylates are shown in Fig. 2.1, which also illustrates the three different types of polyacrylates. Minor amounts of other materials are often added such as triethanolamine (to counteract retardation), tributyl phosphate (to cut down excessive air entrainment) and hydro-xycarboxylic acid salts or lignosulfonates (to increase retardation). In addition proprietary superplasticizers can be blends of two of the main ingredients

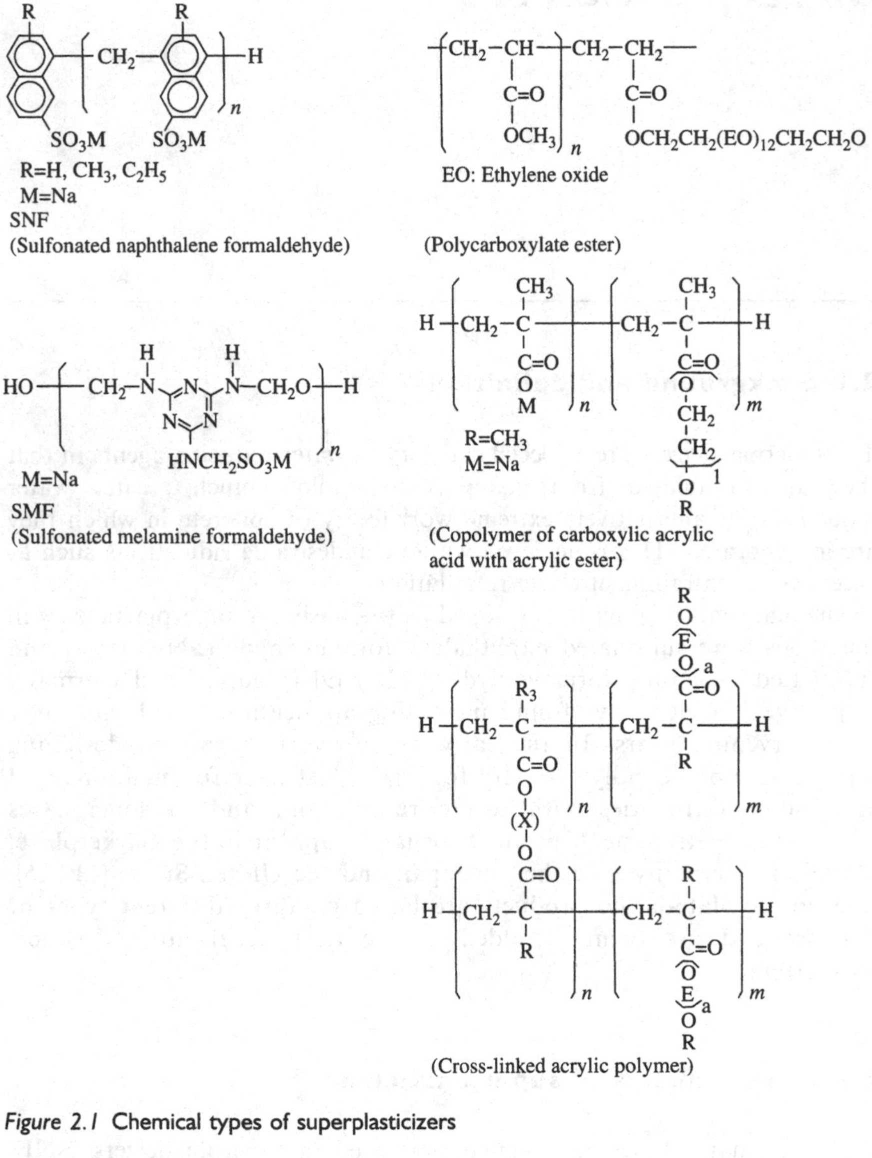

R=H, CH$_3$, C$_2$H$_5$
M=Na
SNF
(Sulfonated naphthalene formaldehyde)

EO: Ethylene oxide

(Polycarboxylate ester)

M=Na
SMF
(Sulfonated melamine formaldehyde)

R=CH$_3$
M=Na

(Copolymer of carboxylic acrylic acid with acrylic ester)

(Cross-linked acrylic polymer)

Figure 2.1 Chemical types of superplasticizers

2.2.1 Sulfonated naphthalene formaldehyde

This raw material was one of the first materials referred to in the literature as a water-reducing agent [16] yet only since 1970 has it found extensive application in admixture formulations.

The material is produced from naphthalene by oleum or sulfur trioxide sulfonation under conditions conducive to the formation of the β sulfonate. Subsequent reaction with formaldehyde leads to polymerization and the sulfonic acid is neutralized with sodium hydroxide [17] or lime. The process is illustrated in Fig. 2.2. The value of n is typically low but conditions are chosen to get a proportion of higher-molecular-weight product as it is believed to be more effective [18]. The quantity of sodium sulfate by-product formed by the neutralization of excess sulfonating reagent will vary depending on the process used, but can be reduced by a subsequent precipitation process using lime [19].

Commercial products vary from 25 to 45% w/w solids contents and addition levels required to produce concrete of almost self-compacting properties would be 1.0–3.0% by weight of cement.

2.2.2 Sulfonated melamine formaldehyde

This type of chemical product was originally developed [20] in the 1950s as a dispersant for a wide variety of industries, but it was not until some 10 years later that the possibilities for its use in concrete were recognized.

It is manufactured by normal resinification techniques according to the process [21] shown in Fig. 2.3. The normal production procedure results in a product having the characteristics given in Table 2.1 [22]. The length of the polymerization time will influence the molecular weight of the product, the most useful average molecular weight being about 30 000. This material is normally used alone or in combination with SNF. When used alone it is typified by having minimal effect on air entrainment or setting time. Hydroxycarboxylic acids [23] have been incorporated into some formulations in order to cut down slump loss.

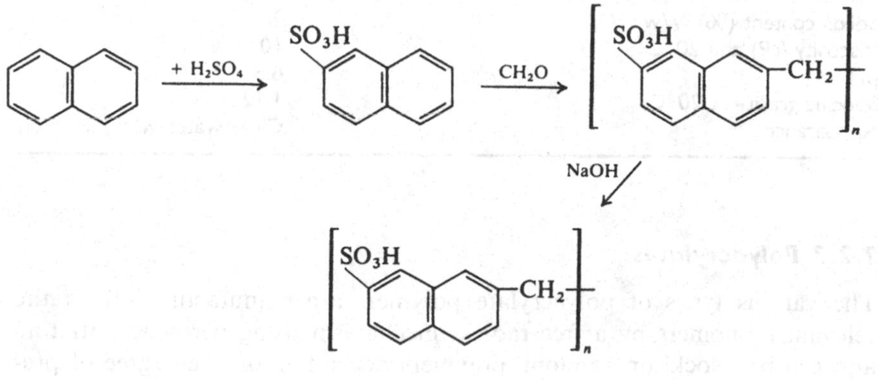

Figure 2.2 SNF manufacturing process.

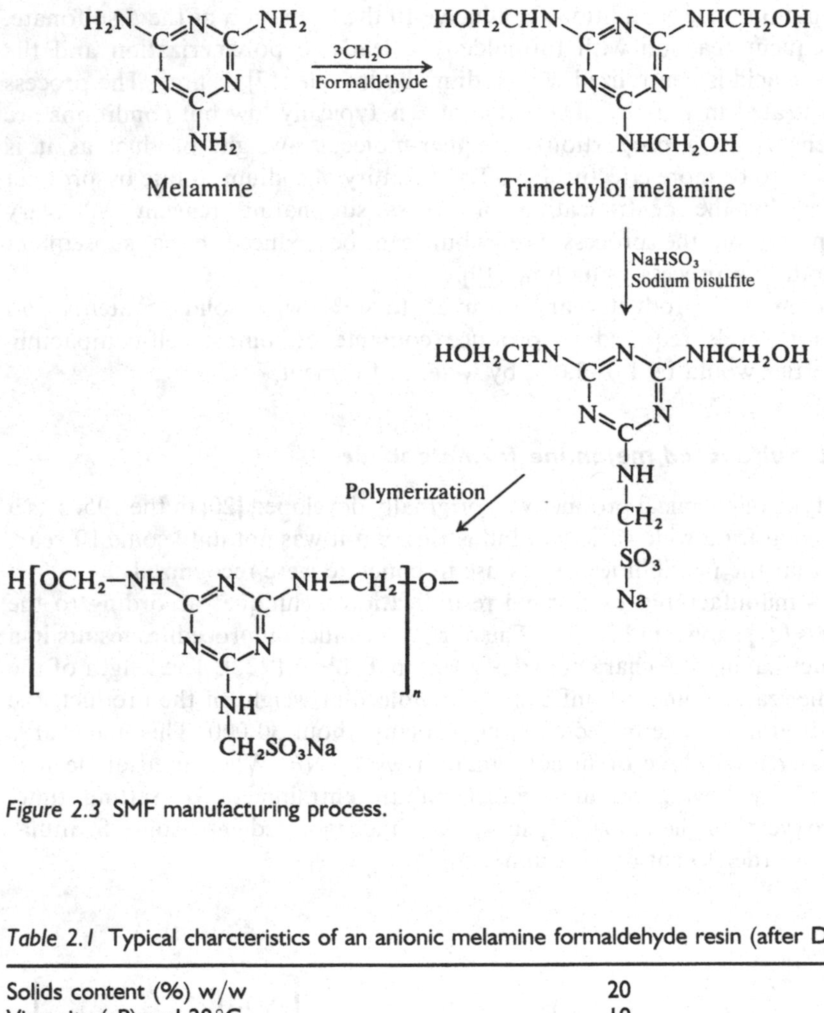

Figure 2.3 SMF manufacturing process.

Table 2.1 Typical charcteristics of an anionic melamine formaldehyde resin (after Davis)

Solids content (%) w/w	20
Viscosity (cP) and 20°C	10
pH	8.5
Specific gravity at 20°C	1.12
Appearance	Clear water white solution

2.2.3 Polyacrylates

The various types of polyacrylate polymers are manufactured from the relevant monomers by a free radical mechanism using peroxide initiators and can be 'block' or 'random' polymers depending on the degree of pre-polymerization of the monomers used.

2.3 Effects on the water–cement system

In order to understand the way in which superplasticizers affect the properties of concrete, studies have often been made on cement pastes in view of the convenience of such investigations.

2.3.1 Rheological effects

A viscometer can be used to study the yield stress and viscosity of cement pastes (Section 1.3.1). This is carried out by plotting the shear rate against shear stress as shown in Fig. 2.4 for cement pastes of various water cement ratios. These cement pastes are generally considered to exhibit Bingham plastic behavior where the yield value is the intercept on the shear stress axis and is related to cohesion, and the slope of the line is the apparent viscosity which is related to the consistency or workability of the system. The following general observations can be made:

1. Superplasticizers reduce both the yield value and plastic viscosity [24] of cement pastes. At higher dosages (for example 0.8% for SNF) [5, 6], the yield value approaches zero and the system becomes essentially Newtonian.

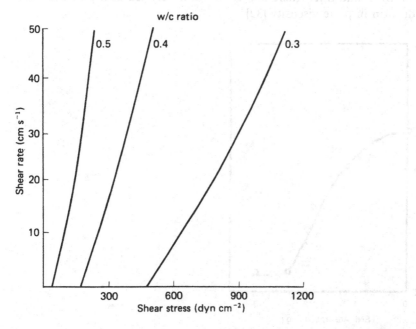

Figure 2.4 Shear-stress–shear-rate relationships for cement pastes at various water--cement ratios

2. The cement composition affects the rheological behavior of the system; cement pastes having low C_3S/C_2S and C_3A/C_4AF ratios have a higher viscosity when the superplasticizer addition is delayed [27].
3. There is a relationship between the amount of superplasticizer adsorbed on to the cement and the apparent viscosity [28, 29]. The second reference indicates that the relationship is linear, while the relationship in the first reference is linear over at least part of the curve (Fig. 2.5).
4. At a fixed addition level of SNF, the Blaine surface area of the cement is directly proportional to the apparent viscosity [30].

2.3.2 Zeta potential

The zeta potential is the difference in potential between that of the total dispersed system and that of the layer at the interface of the dispersed particles (in this case cement) and the dispersing medium (water). Many studies have been made of the effect of superplasticizers on the zeta potential of the cement–water system from which the following conclusions can be drawn:

1. Cement has a positive zeta potential which is diminished and eventually becomes negative on the addition of a superplasticizer [31].
2. With SNF and SMF there is a correlation between zeta potential and reduction in paste viscosity [32].

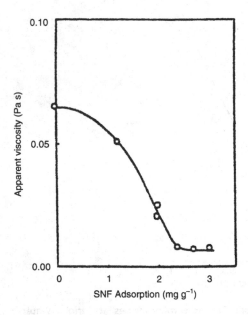

Figure 2.5 Dependence of viscosity on the adsorption of SNF in cement paste.

3. Also with SNF and SMF, there is a correlation between dosage and zeta potential up to a maximum negative value, which corresponds to a minimum paste viscosity [33].
4. For the sulfonate-based polymer types of superplasticizer, the zeta potential increases as the molecular weight increases, up to a maximum, as shown in Fig. 2.6 [34].
5. The zeta potential of the polyacrylate-based materials is significantly lower than that for SNF or SMF as shown in Fig. 2.7 [35]. In fact for a similar lowering of viscosity, the zeta potential for the polyacrylate products can be half that for SNF or SMF [36].

2.3.3 Adsorption

It had been noted early in the study of water-reducing agents of all types, including superplasticizers, that these admixtures do not remain in solution but are substantially and strongly adsorbed by the hydrating cement. The adsorption is studied by adding the admixture to the cement paste and after a period of time, filtering the system. The filtrate is analyzed to give the quantity of admixture left in solution, and the amount adsorbed is calculated by difference. The following observations can be made:

1. Earlier work [37] by one of the authors indicated that superplasticizers of the SNF and SMF type were less strongly adsorbed onto the hydrating cement than normal water-reducing agents and this was used to explain why there was less retardation by the superplasticizers. This

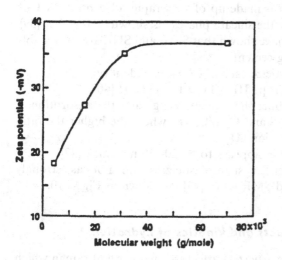

Figure 2.6 Zeta potential vs admixture concentration of different molecular weights.

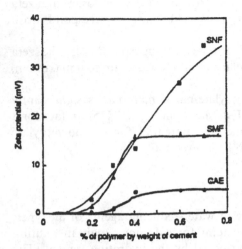

Figure 2.7 Zeta potentials of pastes containing CAE, SMF and SNF as a function of polymer dosage.

theory was flawed, in that later work showed that the part of the SNF that was effective (the polymer) was very strongly adsorbed, and it was the lower-molecular-weight by-products that remained in solution. When the polymer alone is studied [38], the isotherms shown in Fig. 2.8 for three different cements are obtained.

2. SMF and SNF superplasticizers are adsorbed rapidly onto hydrating cement but this net effect is made up of very rapid adsorption by C_3A and slower adsorption by the silicate phases, as shown in Fig. 2.9. [39].

3. Desorption experiments have shown that SMF and SNF are irreversibly adsorbed on to hydrating cement.

4. The amount of superplasticizer adsorbed is dependent on:
    (a) Cement type, where Type III > Type I > Type II [40].
    (b) Fineness, where the finer the cement, the greater the adsorption.
    (c) The ratios of C_3S/C_2S and C_3A/C_4AF, where the higher the ratio the greater the adsorption [41].

5. The adsorbed layer of SNF appears to be 20–50 nm thick [42].

6. The new polyacrylate-based superplasticizers are not as strongly adsorbed as the SNF and SMF types [43], as shown in Fig. 2.10.

2.3.4 *Effects on the products and kinetics of hydration*

There is no doubt that superplasticizers affect the manner and rate in which the individual components in cement react with water and with each other.

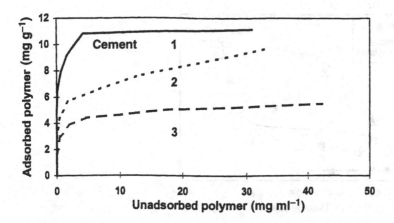

Figure 2.8 Adsorption isotherms of polymer fraction of SNF on different cement brands (polymer adsorption is refered to I g of cement).

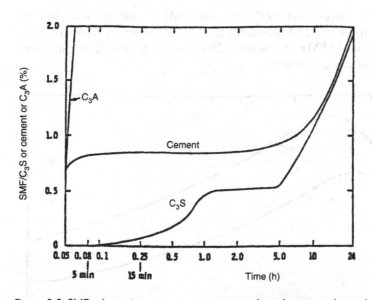

Figure 2.9 SMF adsorption on cement compounds and cement during hydration.

The chemistry is very complicated, even in the absence of superplasticizers, but the following is a summary of what has been established by the many workers in this field

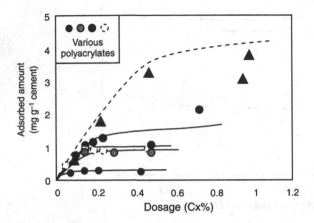

Figure 2.10 Relation between adsorbed amount and dosage.

(a) Interactions with C₃A

1. It is generally agreed that both SNF and SMF retard the hydration of C_3A [44, 45]. Figure 2.11 illustrates the heat evolution of C_3A hydrated in the presence of SMF. Tests up to 28 days with SNF [46] indicate that the C_3A remains less hydrated.

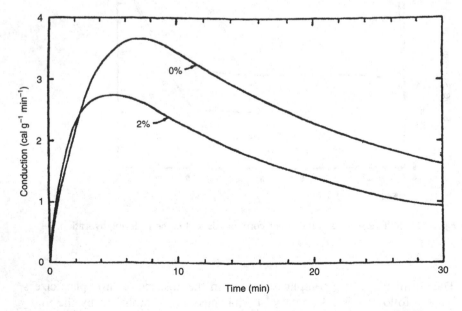

Fig. 2.11 Conduction calorimetric curves of C_3A hydrated in the presence of SMF.

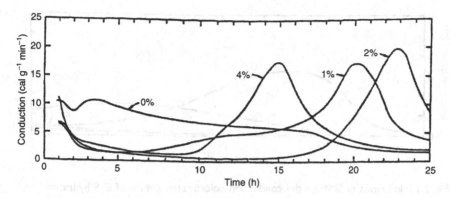

Fig. 2.12 Conduction calorimetric curves for $C_3A + gypsum + H_2O$ containing SMF.

2. SMF [47] and [48] retard the C_3A hydration to the hexagonal phase as well as the subsequent conversion to the cubic form.
3. In studies with SNF [38] of varying molecular weight, there is an indication that the higher-molecular-weight fraction causes less retardation of the hydration process than the lower-molecular-weight materials.
4. In the presence of sulfates, both SNF [49] and SMF [47] retard the C_3A hydration, as shown in Fig. 2.12 for SMF and the conversion of ettringite to the monosulfate hydrate is delayed, as illustrated in Table 2.2.
5. Recent work [50] has shown that in the presence of sulfate ions, SNF forms an organo-mineral compound with C_3A, allows a greater dissolution rate of the $CaSO_4$, and alters the morphology of the hydration products.

(b) Interaction with the C_3S phase

1. Both SNF and SMF retard the hydration of C_3S [51–53] as shown in Fig. 2.13 for SMF [54].

Table 2.2 Amounts of ettringite and monosulfate hydrate formed in the $C_3A-CaSO_4$.
$2H_2O$ system

Hydration time (h)	Ettringite (%)		Monosulfate hydrate	
	Reference	SMF (4%)	Reference	SMF (4%)
2	28.3	—	0	—
6	17.6	24.4	5.5	3.5
24	12.3	—	13.2	—
48	10.0	18.8	15.0	9.9

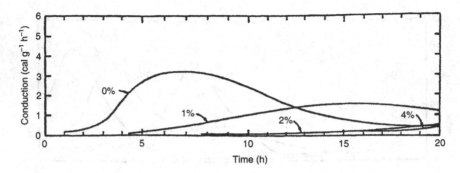

Fig. 2.13 Influence of SMF on the conduction colorimetric curves of C₃S hydration.

2. The early retarded hydration is followed by a period of hydration proceeding faster until at 28 days the degree of hydration is equivalent to or ahead of a control [55].
3. The type and quantity of sulfate and alkalis influences the degree of retardation of the C_3S phase.
4. The molecular weight of SNF influences the degree of retardation of C_3S; the higher the molecular weight, the greater the retardation [38].

(c) Interaction with cement

The interaction of superplasticizers with Portland cement is the most complicated situation of all because of reactions between the various components of the cement and the competition, for example between the superplasticizer and gypsum for reaction with C_3A. However, in general:

1. The hydration is retarded in a similar manner to the individual components as shown in Fig. 2.14 [54].
2. The C_3S phase is not as strongly retarded as for the individual material because the C_3A strongly adsorbs a large proportion of the superplasticizer preferentially.
3. The formation of ettringite is accelerated.
4. The higher the molecular weight of SNF, the greater the retardation of cement hydration.

2.3.5 Interpretation in terms of a mode of action

Superplasticizers operate by adsorption onto the initial hydrates of C_3A, C_2S and C_3S. In the case of C_3A there is evidence that this is more than just a physical effect, and that an organo-mineral compound may be formed.

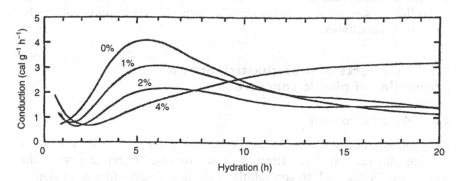

Fig. 2.14 Conduction calorimetric curves for Portland cement hydrated in the presence of SMF.

In the case of SNF and SMF, the ionized sulfonate groups on the adsorbed superplasticizer molecules are strongly negatively charged and the repulsion between the cement particles overcomes the weaker Van der Waal forces of attraction, resulting in a dispersed system. The polyacrylate materials are similarly adsorbed and cause dispersion of the cement particles in part by the same electrostatic mechanism through the ionized carboxyl groups. However, the long flexible side chains attached to these polymers, especially those which are ethoxylated, act as physical barriers, preventing the cement particles from coming within the range of the Van der Waal forces. This steric hinderence mechanism acts as an additional means of causing and maintaining dispersion.

There is some retardation of cement hydration but at 28 days the products of C_3S hydration are essentially the same as in an unsuperplasticized cement system. The C_3A/gypsum reaction products may be changed morphologically to a cubic rather than a hexagonal form.

2.4 Effects of superplasticizers on the properties of concrete

The properties of concrete can be considered in terms of:

- The **initial plastic state** of the fresh concrete when properties such as the workability as measured by slump or flow table test or air content can be determined by the relevant standard method.
- The **later plastic state** when the concrete may be transported, handled and placed; changes in air content and workability may occur and setting time may be a factor in the finishing operation.
- The **hardened state** at a relatively early age (usually 28 days) where mechanical properties are determined as a basis of structural requirements.

- The **later hardened state** during its in place life when the concrete should fulfill its structural or aesthetic role without deterioration under the service conditions.

2.5 The effects of superplasticizers on the properties of plastic concrete

2.5.1 Air entrainment

Different types of superplasticizers alter the air content of fresh concrete in varying degrees and the effect is also dependent on the way the superplasticizer is used. In general, the following observations are noted:

1. When the superplasticizer is used to produce highly workable concrete, more air will be entrained than a control concrete. This is particularly true of SNF and polyacrylate types (where air-detraining agents are sometimes intentionally added into the formulations) and less so with the SMF type.
2. When superplasticizers are used to reduce the water–cement ratio, normally any increase in air content will be minimal, especially in high-cement-content mixes. When it is a requirement to intentionally air-entrain such mixes, the dosage required to obtain a given air content is often considerably increased, presumably because of the reduced aqueous medium in the concrete.
3. The stability of entrained air in superplasticized concrete has been studied in the laboratory [56] and the SMF type appears to lose air more

Table 2.3 Air stability in ready-mixed concrete

Mix design						
Cement	330 kg m^{-3}					
Water	150 kg m^{-3}					
Sand	800 kg m^{-3}					
Stone, No. 67	1080 kg m^{-3}					
ADVSP	260 ml 100 kg^{-1}					
AEA	18 ml 100 kg^{-1}					
Concrete temperature	24°C					

Air entrainment						
Sampling time (h)	Slump (mm)	Plastic air (%)	Hardened air (%)	Spacing factor (mm)	Chord length (mm)	Specific surface (mm^{-1})
0.25	180	7.2	7.4	0.12	0.14	28.3
0.75	150	6.2	6.9	0.15	0.16	24.1
1.25	130	5.8	6.1	0.18	0.17	23.8

rapidly than SNF. Full-scale trials carried out with a polyacrylate type of superplasticizer on ready-mixed concrete [57] indicated that the particular type of material evaluated produced a very stable air void system, as shown in Table 2.3.

2.5.2 Workability

Superplasticizers cause dramatic increases in workability as measured by slump or flow table spread, or alternatively allow very large decreases in water–cement ratios to be made while maintaining workability. Figure 2.15 illustrates the relationship between dosage and water reduction for a typical superplasticizer [58], while Fig 2.16 shows how initial slump, dosage, and flow table spread are related [59]. A major drawback to the practical use of superplasticizers to produce highly workable, almost self-compacting concrete, has been the short duration of the extreme workability, often only 30–40 min. This is illustrated for an SMF type of superplasticizer at various temperatures in Fig. 2.17, where the severe slump loss at higher temperatures is very apparent [60].

There is some indication that the higher-C_3A-content cements lead to more rapid slump loss of superplasticized concrete [61].

The newer polyacrylate-based materials hold promise as a way of holding the high slump over a longer time period, as shown in Fig. 2.18 for one product of this type [62], where it can be seen that the extreme workability is maintained for over two hours. It has been suggested that this is due to this type of superplasticizer operating by the 'steric hinderence' mode of action rather than by electrostatic repulsion [63].

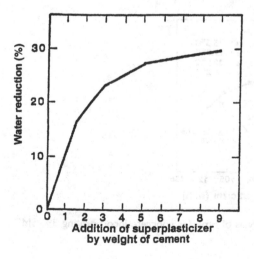

Fig. 2.15 Water reduction obtained by adding a superplasticizer

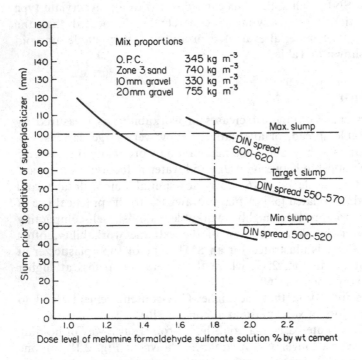

Fig. 2.16 The relationship between initial slump, the flow table spread and the addition level of a superplasticizer.

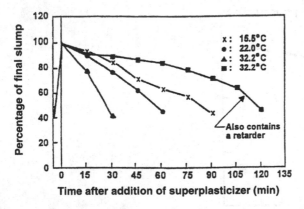

Fig. 2.17 Slump tests after various intervals of time for concrete incorporating 3% SMF by weight of cement (Mailvaganam).

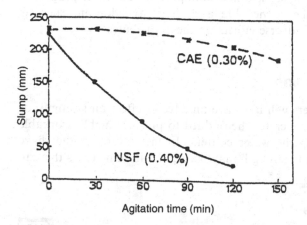

Fig. 2.18 Slump loss at 21 °C of superplasticized concretes with OPC and CAE or SNF ploymer-based admixtures. The figures on the slump-loss curves indicate the percentage of the superplasticizer active ingredient by mass of cement.

2.5.3 Setting time

The effect that superplasticizers have on the setting times of concrete depends on a number of factors including the type of superplasticizer, cement composition, and particularly whether there is a simple addition of the admixture to the concrete or if a reduction in water–cement ratio is made. In general it can be stated that:

1. With a direct addition of superplasticizer to obtain highly workable concrete, initial and final setting times are invariably increased in the order SMF < SNF < polyacrylates. At normal dosages this increase rarely exceeds two hours for materials that are not intentionally formulated to retard.
2. When the water–cement ratio is reduced to give a similar slump to a control mix, the setting time is normally very similar to the control; perhaps a small decrease in the case of SMF and SNF [64] and a slight increase in the case of polyacrylate-based materials [57], normally no greater than one hour either way.

2.6 The effects of superplasticizers on the properties of hardened concrete

The major physical attributes of concrete that make it a construction material of choice are its high compressive strength and stiffness and its ability to protect reinforcing steel. It is also important that these properties are maintained during the service life and that any addition, such as superplasticizers, have no adverse effects on these attributes.

2.6.1 Compressive strength

In general, superplasticizers will not have an adverse effect on strength, or strength development of concrete when added to produce highly workable concrete without reducing the water content. In fact several studies have indicated that an increase in strength may occur, and in some cases this can

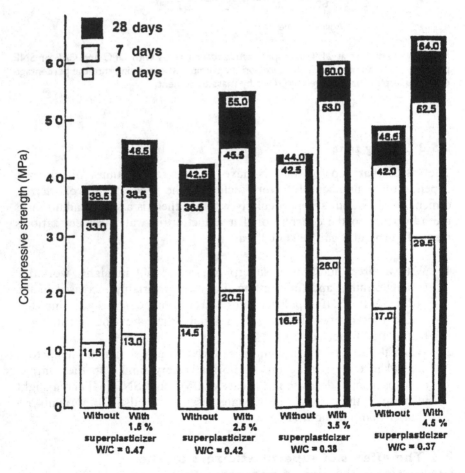

Fig. 2.19 Strength development of high-strength flowing concrete containing melamine-based superplasticizer compared to concrete made with 400 kg of normal Portland cement per m³ in the stiff to low workability ranges (25–100 mm slump).

be substantial [65], as shown in Fig. 2.19 for an SMF-based material. Table 2.4 (column 3) shows data for a NSF superplasticizer where strengths are essentially the same as a control.

When superplasticizers are used to reduce water content of the concrete mix, the increase in strength is normally considered to follow the Abram's

Table 2.4 Freeze-thaw resistance of superplasticized concrete (Rixom)

Mix no. Description	1 Control	2 Addition of admixture normal workability	3 Addition of admixture self-compacting concrete
Mix design:			
10 mm gravel			
(rounded irregular) (kg)	8	8	8
Zone 3 sand (kg)	4	4	4
OPC (kg)	2	2	2
Water (litre)	1.1	0.9	1.0
Admixture (% by weight			
of cement)	0	2.5	2.5
Properties of plastic concrete:			
Slump (mm)	60	60	Collapse
Air content (%)	1.8	1.7	2.2
Compacting factor	0.87	0.90	—
Density (kg m^{-3})	2394	2430	2394
Properties of hardened			
concrete (N mm^{-2}):			
1 day	6.3	12.9	6.0
7 days	31.2	40.2	32.4
28 days	41.2	64.3	42.0
Adsorption (% of dry cube			
weight)			
10 min	1.3	0.8	1.1
30 min	1.9	1.0	2.0
60 min	2.9	1.8	2.8
24 h	5.2	2.6	5.0
ISAT method			
Initial surface adsorption			
test			
BS 1881, Part 5, 1970			
Permeability (ml m^2 s^{-1})			
10 min	0.51	0.15	0.40
30 min	0.42	0.08	0.31
60 min	0.26	0.03	0.20
Freeze-thaw dilation test			
(dilation at 50 cycles, %)	0.10	0.030	0.075
DHO test (number of cycles			
to cause 0.5 mg mm^{-2} scaling	6	32	12

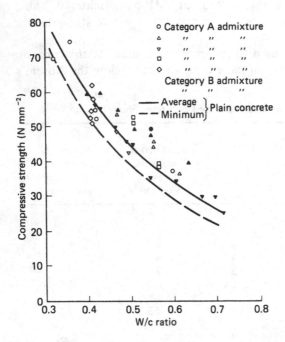

Fig. 2.20 The relationship between water–cement ratio and compressive strength of concretes containing superplasticizers.

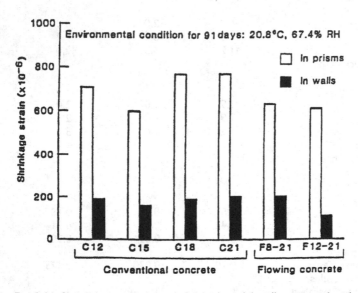

Fig. 2.21 Shrinkage strains in each finished model wall compared with prisms at 91 days.

Table 2.5 Mixture proportions and properties of fresh and hardened concrete (water reduced)

Mix Series	Type of concrete	Mixture proportions and properties of fresh concrete					Properties of hardened concrete				Modulus of elasticity (GPa)
		Water content (kg m^{-3})	W/c Ratio	Entrained air (%)	Slump (mm)	Density (kg m^{-3})	Compressive strength of 150×300 mm cylinders (MPa)			Flexural strength of $90 \times 100 \times 40$ mm prisms (MPa)	
							7 days	28 days	91 days	28 days	
1	Reference (Type I cement)	147	0.49	5.2	75	2348	26.8	32.8	37.8	6.1	32
2	Type I cement and SP-M	120	0.40	5.6	80	2362	37.3	44.0	48.5	7.0	37
3	Type I cement and SP-N	120	0.40	6.0	70	2350	35.5	39.3	47.6	7.0	37
4	Type I cement and SP-L	120	0.40	5.6	80	2360	36.3	42.6	49.9	6.6	36
5	Reference (Type II cement)	147	0.49	4.9	85	2354	25.6	36.6	42.4	6.0	32
6	Type II cement and SP-M	120	0.40	5.6	90	2362	36.3	47.6	55.0	6.9	37
7	Type II cement and SP-N	121	0.40	5.3	75	2377	36.9	47.6	55.8	7.2	37
8	Type II cement and SP-L	121	0.40	4.8	75	2385	35.0	47.6	55.8	7.3	36
9	Reference (Type V cement)	144	0.48	5.4	90	2352	19.1	32.2	38.0	5.0	32
10	Type V cement and SP-M	117	0.38	5.4	75	2364	31.9	40.3	46.2	6.2	36
11	Type V cement and SP-N	118	0.38	5.3	80	2381	33.0	42.0	48.5	5.7	35
12	Type V cement and SP-L	118	0.38	5.2	85	2379	32.8	42.4	50.3	6.2	35

* Coarse aggregate was crushed limestone with a maximum size of 19 mm. Fine aggregate was natural sand. Air-entraining admixture was a sulfonated hydrocarbon.
* SP-M, SP-N, and SP-L refer to melamine, naphthalene and lignosulfonate type superplasticizers (SP) respectively.

Table 2.6 Data on creep measurements for control and water-reduced superplasticized concrete

Admixtures			Mixture proportions				Properties of fresh concrete			Creep measurements on 150 × 300 mm cylinders				
Superplasticizer type	Super-plasticizer	AEA*	W/c ratio by weight	Type I cement (kg m⁻³)	FA	CA	Entrained air (%)	Slump (mm)	Density (kg m⁻³)	f'c at 28 days	f'c at 63 days	Stress applied	Stress–strain ratio	Test creep strain (in./in. × 10p⁻⁶)†
	(ml kg⁻¹ of cement)				(kg m⁻³)						(MPa)			
Control	—	0.31	0.49	298	817	1082	5.3	80	2344	34.3	37.4	15.2	0.44	1101‡
Melamine	23.6	1.18	0.40	304	835	1106	5.4	75	2365	45.2	50.8	19.6	0.43	1085‡
Naphthalene	9.1	3.15	0.40	303	832	1102	6.0	80	2357	47.4	51.1	20.3	0.43	1107‡
Lignosulfonate	25.6	0.34	0.40	305	839	1111	3.4	40	2377	46.0	48.7	19.8	0.43	1157‡

* Air-entraining agent.
† Total creep strain is obtained by structuring shrinkage and elastic strain at loading from the strain readings.
‡ At 334 ± 10 days.

law curve, although consideration of the compilation of data in Fig. 2.20 does indicate a trend to higher values for the superplasticized concrete. The increases in strength by using superplasticizers to reduce the water–cement ratio appear to be consistent over several cement and superplasticizer types and continues beyond the 28-day result, as shown in Table 2.5, which also indicates how flexural strength and modulus are similarly affected [56].

2.6.2 Shrinkage and creep

Most of the studies in this area have indicated that superplasticized concrete has shrinkage and creep characteristics similar to plain concrete. Figure 2.21 illustrates the shrinkage results of plain and superplasticized concrete [66], while Table 2.6 shows similar results for creep [67].

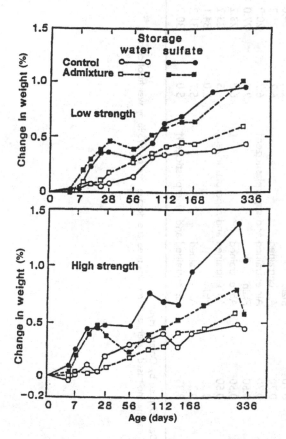

Fig. 2.22 The effect of admixture on the change in weight of water- and sulfate-stored concrete specimens.

Table 2.7 Air-void determinations on hardened concrete

Mix no.	Properties of fresh concrete		Air-Void determinations on hardened concrete *				
	Water–cement ratio (by wt)	Nature of mix	Air content (%)	Paste content (%)	Voids in concrete (%)	Specific surface area (mm)	Void spacing factor (mm)
1	0.70	Plain	2.4	24.6	2.4	10	0.62
2	0.70	Air entrained	6.2	21.2	9.0	14	0.23
3	0.70	Air entrained and superplasticized†	6.7	20.9	9.4	12	0.26
4	0.50	Plain	2.2	27.0	2.0	12	0.59
5	0.50	Air entrained	6.8	23.2	7.7	28	0.13
6	0.50	Air entrained and superplasticized†	6.5	23.2	7.1	21	0.18
7	0.35	Plain	2.0	34.1	1.4	13	0.73
8	0.35	Air entrained	5.7	31.2	6.0	32	0.14
9	0.35	Air entrained and superplasticized†	5.0	30.7	4.8	21	0.23

* Data supplied by Ontario Hydro.
† A napthalene-based superplasticizer was used at a dosage rate of 0.75% by weight of cement.

2.6.3 Freeze-thaw durability

In the early days of the use of superplasticized concrete, some concerns were aired regarding the resistance of air-entrained concrete containing super-plasticizers to freeze–thaw cycling. However, more recent research has indicated the following:

1. In air-entrained superplasticized concrete, the commonly accepted minimum value of the spacing factor of the air void system (0.2 mm) to provide adequate freeze–thaw protection is usually exceeded [68–71].
2. Despite the fact that the minimum spacing factor is exceeded, the concrete freeze–thaw resistance of the concrete does not appear to be adversely affected [71]. Table 2.7 clearly shows how the presence of an SNF superplasticizer increases the spacing factor of the air void system at each of the three water–cement levels evaluated [71].

2.6.4 Sulfate resistance

Research into the susceptibility of superplasticized concrete to sulfate attack has concluded that there is no significant difference between plain concrete and the admixture-containing concrete [71, 72]. Figure 2.22 presents some data for concretes containing an SNF superplasticizer exposed to a 3% magnesium sulfate solution.

References

1 British Patent (1973). 1 386 933.
2 US Patent (1970). 3 537 869.
3 Japanese Patent (1972). 50 329/72.
4 Japanese Patent (1973). 53 192/73.
5 Rixom, M.R. (1974). *Precast Concrete Journal*, **5**, 633–7.
6 Canadian Patent (1975). 961 866.
7 Karsten, R. (1967). *Proceedings of the International Symposium on Admixtures for Mortar and Concrete*, Brussels, 357.
8 Hattori, K. (1967). D.B.P. 1 238 831.
9 British Patent (1969). 1 169 582.
10 Bradley, G. and Howarth, I. (1986) *Cement and Concrete Aggregates*, **8**, 68–75.
11 Cerulli, T. *et al.* (1994). US Patent 5 362 324.
12 Okazaw, S. *et al.* (1993). *Physical Properties of Concrete 2000*, Dhir, R.K. and Jones, M.R. (Eds), Dundee, UK, **2**, 1813–24.
13 Kinshita, M. *et al.* (1994). US Patent 5 362 829.
14 Tanako, Y. and Okazawa, S. (1993). *Chemistry and Dispersing Performance of Concrete 2000*, Dhir, R.K. and Jones, M.R. (Eds), Dundee UK, 351–8.
15 Nmai, C. *et al.* (1990) Presented at ACI Fall Convention Atlanta, GA, Nov 9–14.
16 US Patent (1932), 643 740.

17 Rixom, M.R. (1975). *Proceedings of the Workshop on the Use of Chemical Admixtures in Concrete*, University of New South Wales, 153.
18 US Patent (1956). 2 730 516.
19 Gilbert, E.F. (1968). *Sulphonation and Related Reactions*, Wiley Interscience, New York.
20 British Patent (1969). 1 169 582.
21 Aignesberger, A. (1943). *Cement, Lime and Gravel*, **48**, 188–92.
22 Davis, B. (1975). *Proceedings of the First International Congress on Polymer Concretes*, London, 6.
23 Japanese Patent (1974). 80 133.
24 Young, J.F. (1982). *Concrete Rheology, Materials Research Society Symposium.* 120–51
25 Banfill, P.F.G. (1979). *Cement and Concrete Research.* **9**, 795–6
26 Roy, D.M. and Asaga, K. (1980) *Cement and Concrete Research*, **10**, 387–94.
27 Nawa, T. *et al.* (1989). *Third International Conference of Superplasticizers and Other Admixtures*, Canada, ACI SP-119, 405–24.
28 Asakura, F. *et al.* (1992) *Ninth International Congress on the Chemistry of Cement*, New Delhi, India, **4**, 570–6.
29 Rixom, M.R. and Mailvaganam, N.P. (1986). *Chemical Admixtures for Concrete*, E. & F.N. Spon, London, UK, 22.
30 Nawa, T. and Eguchi, H. (1992). *Ninth International Congress on the Chemistry of Cement*, New Delhi, India, **4**, 597–603.
31 Daimon, M. and Sakai, E. (1995). *Material Science of Concrete IV*, American Ceramics Society, OH, 91–111.
32 Daimon, M. and Roy, D.M. (1979). *Cement and Concrete Research* **9**, 103–10.
33 Al-Kurwi, A. *et al.* (1984). *British Ceramics Society*, **35**, 339–48.
34 Anderson, P.J. *et al.* (1988). *Cement and Concrete Research*, **18**, 980–6.
35 Shonaka, M. *et al.* (1997) *Fifth Canmet/ACI Conference on Superplasticizers and Other Admixtures in Concrete*, Italy, 613.
36 Ohta, T. *et al.* (1997) *Fifth Canmet/ACI Conference on Superplasticizers and Other Admixtures in Concrete*, Italy, 361.
37 Rixom, M.R. and Mailvaganam, N.P. (1986). *Chemical Admixtures for Concrete.* E. & F.N. Spon, London, UK, 21.
38 Ferrari, *et al.* (1997) *Fifth Canmet/ACI Conference on Superplasticizers and Other Admixtures in Concrete*, Italy, 869.
39 Ramachandran, V.S. (1988). *Journal of the ACI*, **80**, 235–41.
40 Burke, A.A. *et al.* (1981). *Second International Conference on Superplasticizers in Concrete*, Ottawa, Canada, 23.
41 Asakura, F. *et al.* (1992). *Ninth International Congress on the Chemistry of Cement*, New Delhi, India, **4**, 570–6.
42 Uchikawa, H. *et al* (1995). *Cement and Concrete Research*, **25**, 353–64.
43 Ohta, A. *et al.* (1997). *Fifth Canmet/ACI Conference on Superplasticizers and Other Admixtures in Concrete*, Italy, 365.
44 Ramachandran, V.S. (1995). *Concrete Admixtures Handbook*, Noyes Publ., NJ, 1153.
45 Slanicka, S. (1980). *Seventh International Congress on the Chemistry of Cement*, Paris, France, 161–6.

46 Henning, O. and Goretzki, L. (1982). *RILEM International Conference on Cement at Early Ages*, Paris, France, **1**, 151–5.
47 Massazza, F. *et al.* (1977). *CEMBUREAU Report #3*.
48 Quon, D.H.H. and Malhotra, V.S. (1981) *Developments in the use of Superplasticizers, ACI SP-68*, 151–5.
49 Massazza, F. and Costa, U. (1980). *Seventh International Congress on the Chemistry of Cement*, Paris, **4**, 529–35.
50 Fernon, V. *et al.* (1997). *Fifth Canmet/ACI Conference on Superplasticizers and Other Admixtures in Concrete*, Italy, 361.
51 Hayek, N. and Diereks, P. (1973). *SKW Symposium*, Trostberg, Germany, 14.
52 Singh, N.B. and Prabha Singh, S. (1993). *Journal of Scientific and Industrial Research*, **52**, 661–75.
53 Odler, I. and Abdul-Maula, S. (1987). *Cement and Concrete Aggregates*, **9**, 38–43.
54 Ramachandran, V.S. (1981) *Third International Congress on Polymers in Concrete*, Japan, 1071–81.
55 Lukas, W. *Fifth International Melment Symposium*, Munich, Germany, 17–21.
56 Johnson, C.D. *et al.* (1979). *Proceeding TRB Symposium on Superplasticizers in Concrete, Transportation Research Record No. 720*, Washington D.C.
57 Jeknavorian, A. *et al.* (1997). *Fifth Canmet/ACI Conference on Superplasticizers and Other Admixtures in Concrete*, Italy, 55–81.
58 Aignesberger, A. and Kern, A. (1981) *ACI SP-68* (ed. Malhotra, V.M.)
59 Roeder, A.R. (1976) Private communication.
60 Malhotra, V.M. and Malanka, D. (1979). *ACI SP-62*, 209–44.
61 Perenchio, W.F. *et al.* (1974). *ACI SP-62*, 137–56.
62 Collepardi, M. (1994). *Proceedings of Advances in Cement and Concrete*, Durham, 257–91.
63 Collepardi, M. (1998). *Cement and Concrete Composites*, in press.
64 Ghosh, R.S. and Malhotra, V.M. (1978) *Canmet Report MRP/MRL 78-189(J)*, Canmet, Energy, Mines and Resources, Ottawa, Canada.

65 Johnson, C.D. *et al.* (1979). *Proceedings TRB Symposium on Superplasticizers in Concrete, Transportation Research Record No. 720*, Washington, D.C.
66 Lane, R.O. and Best, J.F. (1978). *Proceedings International Symposium on Superplasticizers in Concrete*, Canmet, Canada Energy, Mines and Resources, Ottawa, Canada, **1**, 379–402.
67 Burg, R.G. and Ost, B.W. (1994). *PCA R & D Bulletin RD 104T*, Skokie, Illinois.
68 Perenchio, W.F. *et al.* (1978). *Proceedings International Symposium on Superplasticizers in Concrete*, Canmet, Canada Energy, Mines and Resources, Ottawa, Canada 1, 325–46.
69 Mielenz, R.C. and Sprouse, J.H. (1978). *Proceedings International Symposium on Superplasticizers in Concrete*, Canmet, Canada Energy, Mines and Resources, Ottawa, Canada, **1**, 403–24.
70 Roberts, L.R. and Scheinder, P. (1982). *ACI SP-68*, 189–213.
71 Malhotra, V.M. (1982) *Cement Concrete and Aggregates, ASTM*, **4**, 3–23.
72 Brooks, J.J. *et al.* (1979). ACI SP-GL, 293–314.

Chapter 3

Air-entraining agents

3.1 Background and definitions

The air-entraining admixtures are organic materials, usually in solution form, which when added to the gauging water of a concrete mix, entrain a controlled quantity of air in uniformly dispersed microscopic bubbles. This type of air should not be confused with 'entrapped air' which is often present in concrete in the form of irregularly shaped cavities and which can be due to inadequate compaction or flaky aggregates.

There are three major reasons for intentionally entraining air into concrete: durability, cohesion and density.

3.1.1 Durability

During the 1930s it was observed that certain stretches of road in the northeast states of America were more able to withstand the effects of freeze–thaw conditions and the presence of de-icing salts than other roads in the area [1]. An investigation revealed that the more durable roads were less dense and that the cement had been obtained from mills where beef tallow had been used as a grinding aid. It was concluded that the beef tallow had functioned as an air-entraining agent and had enhanced the durability of the concrete. This led to a more controlled investigation, and in 1939 an air-entrained concrete carriageway was intentionally produced by the New York Department of Public Works [2].

The effect of the entrainment of a minor amount (4–6% by volume) of air on the ability of concrete to withstand freeze–thaw cycling is remarkable [3] and is illustrated in Fig. 3.1.

Air entrainment of concrete carriageways, aircraft runways, and parking garages is now accepted as normal practice and several hundred thousands of cubic meters of this type of concrete have been placed. In North America the use of air-entrained concrete is universally accepted and therefore finds greater use than in Europe.

Fig. 3.1 The effect of 300 freeze–thaw cycles (according to ASTM C666 Procedure B) on air-entrained (right) and plain concrete (left).

3.1.2 Cohesion

Concrete which is produced using fine aggregates deficient at the fine end of grading, e.g. sea dredged aggregates, exhibit a tendency to bleed and segregate. The presence of a small amount of entrained air (2–4% by volume) leads to an improvement in cohesion, or mix stability. Alternatively, with mixes which are adequate in this respect, a reduction in sand content can be made when air is entrained without loss of cohesion. The amount that can be removed is approximately equal on a volume basis and leads to a reduction in water–cement ratio to minimize the effect of entrained air on compressive strength.

3.1.3 Density

The two applicational areas above are normally associated with minor quantities (less than 8% by volume) of entrained air. However, using different chemical types of materials, much larger quantities (up to 30% by volume) can be entrained to lower the density, enhance the thermal insulation properties, or to produce lightweight concrete in conjunction with lightweight aggregates. When air entrainment is used in the above broad applications to achieve a given end result, there are a number of side effects which need to be considered: (1) the presence of microscopic air bubbles acts as a 'lubricant' and increases the workability, or allows a consequential reduction in the water–cement ratio; (2) the compressive and tensile strengths decrease with increasing air content when the mix design is unchanged; (3) the yield of the concrete is increased for a given weight of mix ingredients. Those aspects will be quantified later.

3.2 The chemistry of air-entraining agents

The literature describes many different chemical surfactants as suitable for the formulation of air-entraining agents for concrete. However, in practice, the major proportion of commercial products are based on a relatively small number of raw materials and these are set out below:

1. Neutralized wood resins
2. Fatty-acid salts
3. Alkyl-aryl sulfonates
3. Alkyl sulfates
5. Phenol ethoxylates.

3.2.1 Neutralized wood resins

There are several materials that make up this category, including wood-derived tall oil rosins and pine stump extracts sold under the well known Vinsol^{TM} trade mark. They consist of complicated mixtures containing greater or lesser amounts of abietic acid (Fig 3.2) together with pimaric acid, and phenolic compounds such as phlobaphenes.

Finished products are aqueous solutions of 5 to 20% active material in the form of salts. Depending on the composition, they can be produced by simple neutralization with bases, or where esters are present, by saponification at higher temperatures.

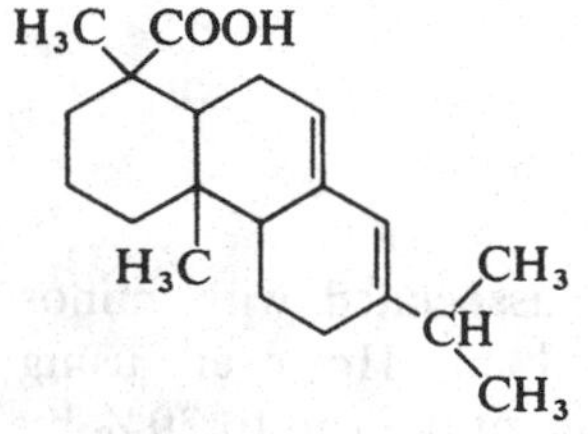

Fig. 3.2 Abietic acid.

3.2.2 Fatty-acid salts

In order to satisfy the requirements of both performance in use and the ability to form stable aqueous solutions of adequate strength, the fatty acids shown in Table 3.1 are used as air-entraining agents for concrete in the form of their alkali-metal salts. These fatty acids are present in a distribution of chain lengths in naturally occurring fats and oils, such as tall oil and coconut oil and are used in the form of such a mixture. Typical formulations contain 5–20% by weight of fatty-acid salt. These products, unlike the neutralized wood resins, are compatible in solution with certain lignosulfonates and hydroxycarboxylic acid salts to form admixtures possessing both air-entraining and water-reducing capabilities.

3.2.3 Alkyl-aryl sulfonates

The alkyl-aryl sulfonates find application in both the production of lightweight concrete and for enhancing the freeze–thaw durability of normal concrete. The usual raw material is orthododecylbenzene sulfonate, which is a basic surfactant used in a variety of industrial and domestic detergents. The formula is shown in Fig. 3.3.

The hydrocarbon base is petroleum derived and does, in fact, contain a distribution of chain lengths with the predominant species being C_{12}. In addition, there can be a greater or lesser degree of chain branching. The sulfonation process utilized can vary from direct reaction with sulfuric acid to SO_2/SO_3 mixtures, but always results in some excess sulfuric acid. On neutralization, a proportion of sodium sulfate is produced which is preferably kept to a minimum for admixture formulations.

3.2.4 Alkyl sulfates

The literature describes the use of several materials of this type and these are shown in Table 3.2. These products are also compatible with many water-reducing agents to produce air-entraining water-reducing agents.

Table 3.1 Fatty acids used as air-entraining agents

Fatty acid	Formula	Reference
Oleic acid	CH_3—$(CH_2)_7$—$CH{=}CH$—$(CH_2)_7COOH$	[4]
Capric acid	$C_9H_{19}COOH$	[5]

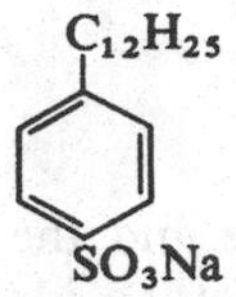

$C_{12}H_{25}$

SO_3Na

Fig. 3.3 Orthododecylbenzene sulfonate.

Table 3.2 Alkyl sulfates used as air-entraining agents

Material	Formula	Reference
Sodium dodecyl sulfate	$C_{12}H_{25}SO_4Na$	[6–10]
Sodium tetradecyl sulfate	$C_{14}H_{29}SO_4Na$	[11]
Sodium cetyl sulfate	$C_{16}H_{33}SO_4Na$	[7, 9, 10]
Sodium oleyl sulfate	$CH_3(CH_2)_7CH{=}CH{-}(CH_2)_8SO_4Na$	

3.2.5 Phenol ethoxylates

Phenol ethoxylates differ from the previous four categories in that they are non-ionic materials. Although not widely used, they are very effective at low addition levels and solutions of 2–4% by weight in water perform satisfactorily at low dosage level. The most common material is nonylphenol ethoxylate, and limited studies [12] have indicated that the higher value of n in Fig. 3.4 is the most effective.

Although the vast majority of commercially available air-entraining agents are simple solutions of materials within one of the above categories, it is possible to produce mixtures and this is occasionally carried out.

3.3 The effects of air-entraining agents on the water–cement system

Numerous studies have been made on the effect of additions of air-entraining agents to cement pastes which enable an insight to be gained into the mechanism by which these materials produce the stable microscopic air

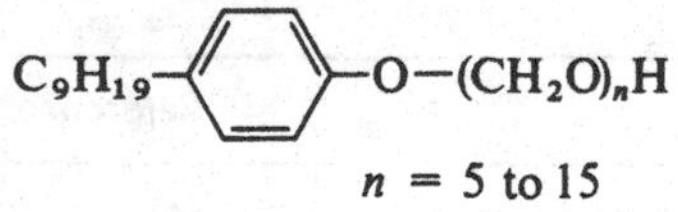

C_9H_{19}—⟨⟩—$O{-}(CH_2O)_nH$

$n = 5$ to 15

Fig. 3.4 Nonylphenol ethoxylates.

void system, and to some extent, their effect on the properties of concrete. However, unlike the water-reducing agents, the comparison of behavior between that in cement pastes and that in total concrete systems is not so useful because of the influence of the aggregate component in determining the air content and the resultant rheological characteristics. The data can be conveniently described under the following headings:

1. Rheology.
2. Air content and characteristics.
3. Distribution between solid and aqueous phase.
4. Effect on cement hydration reactions.
5. Interpretation as a mechanism of action.

3.3.1 Rheology

The effect that air-entraining agents have on the rheology of fresh cement pastes can be considered from the point of view of changes due to the admixture itself, and those due to the presence of entrained air.

An interesting program of work was carried out to isolate the individual effects [13] by preparing the cement paste/admixture mixes in two ways, as given below.

(a) Without air entrainment

The apparatus shown in Fig. 3.5 was used, and was filled completely with cement paste. Admixture additions were made by injecting with a hypodermic needle through the rubber cap. Pastes prepared in this way had air contents of less than 0.6% by volume, were free from any premature stiffening tendencies and were homogeneous.

(b) With varying degrees of air entrainment

The same apparatus was used, but quantities of paste were removed to give an air space in the vessel. On rapid agitation the volume increased, dependent on the air content required. Paste viscosities were measured, using a Stormer viscometer, which is a type of concentric cylinder viscometer. Although it is possible to obtain results in absolute terms, for comparative purposes the times for 100 revolutions of the rotor under a fixed applied torque were recorded.

In the absence of entrained air, the results shown in Fig. 3.6 were obtained, where the similarity of behavior of the three anionic materials, sodium dodecyl sulfate, sodium resinate and a petroleum sulfonate, in increasing the paste viscosity is seen. The non-ionic material, phenol

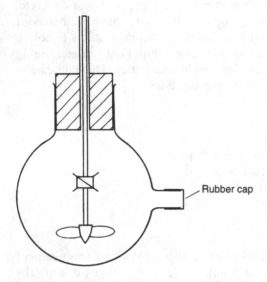

Rubber cap

Fig. 3.5 Mixing apparatus for experiments on air-entrained and non-air-entrained pastes (Bruere).

ethoxylate, has almost no effect on the paste viscosity when no air is entrained.

When sodium resinate was incorporated into pastes made from two cements having significantly differing surface areas, the results given in Table 3.3 were obtained, indicating that the sodium resinate produces a much greater increase in viscosity in paste made from the fine cement than in pastes made from the coarse cement.

When the viscosities of air-entrained pastes were measured by the same means, the results shown in Fig. 3.7 were obtained for sodium abietate at 0.05% by weight of cement. It can be seen that the magnitude of the effect due to the presence of the admixture itself is small in relation to the effect of the air it causes to be entrained.

3.3.2 Air content and characteristics

(a) Dosage and admixture type

The effect of varying the quantity of several air-entraining agents is shown in Fig. 3.8 [10]. The similarity of behavior of sodium dodecyl sulfate and sodium resinate is again illustrated.

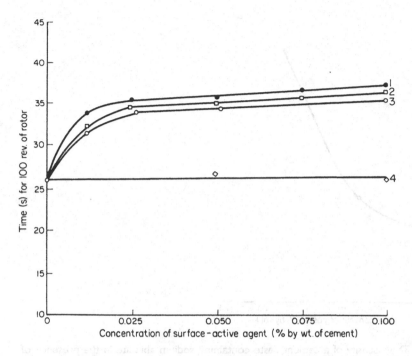

Fig. 3.6 Viscosities of cement pastes containing varying concentrations of surface active agents in the absence of entrained air (Bruere). I = sodium dodecyl sulfate; 2 = sodium abietate; 3 = petroleum sulfonate; 4 = phenol ethoxylate.

Table 3.3 The effect of air entrainment on paste viscosity for two different cements

Cement no.	Surface area (cm² g⁻¹) (Blaine)	Sodium abietate (% by wt of cement)	Time for 100 rev. of rotor(s)	Increase in time for 100 rev. (%)
I	3600	0	26	38
I	3600	0.05	36	
2	4350	0	74	76
2	4350	0.05	130	

Table 3.4 summarizes similar data, but also includes the sodium dodecylbenzene sulfonate type and the resultant effect on the specific surface area and computed spacing factor of the bubbles [14].

The specific surface area of bubbles entrained by each air-entraining agent is largely independent of its concentration and there are indications that sodium dodecyl sulfate, apart from being more effective as an air-entraining

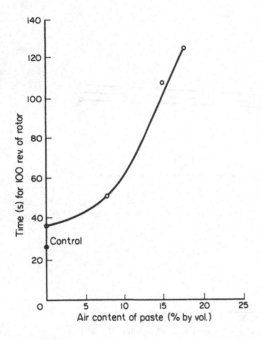

Fig. 3.7 The viscosity of a cement paste containing sodium abietate in the presence of entrained air (Bruere).

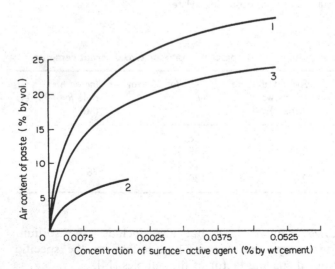

Fig. 3.8 Variations of air-entraining capacities of surface active agents in cement pastes with varying concentrations of agents (Bruere). 1 = sodium dodecyl sulfate; 2 = sodium tetradecyl sulfate; 3 = sodium abietate.

Table 3.4 The effect of various air-entraining agents at different concentrations on the specific surface area and computed spacing factor of air bubbles in cement paste

Surface active agent	Concentration (% by wt of cement)	Air content (% by volume)	Specific surface area of bubble $(mm^2\ mm^{-3})$	Computed spacing factor (mm)
None	—	0.3	31	0.813
Sodium dodecyl	0.005	6.5	68	0.112
sulfate	0.05	14.0	73	0.074
	0.010	17.5	78	0.066
	0.025	21.3	78	0.058
	0.050	27.2	78	0.048
Sodium dodecyl-	0.005	4.0	56	0.188
benzene sulfonate	0.010	5.6	61	0.135
	0.025	11.2	64	0.094
	0.050	13.1	61	0.091
Neutralized wood	0.05	4.6	53	0.170
resins	0.010	10.1	56	0.112
	0.025	17.4	53	0.094
	0.050	22.6	54	0.084

agent than either the sodium dodecylbenzene sulfonate and neutralized wood resins, also results in a high specific surface area of bubbles in closer proximity.

The effect of altering the fatty-acid chain length of sodium salts of linear fatty acids on the air entrained in mortars is illustrated in Fig. 3.9 [5]. The superiority of the 9–11 carbon chain length fatty-acids is clearly shown and, in practice, fatty-acid fractions with a high C_{10} content are chosen; the C_9 and C_{11} fatty acids do not occur naturally in appreciable quantities.

(c) Water–cement ratio

An increase in the water–cement ratio of cement pastes leads to greater air entrainment and a decrease in the specific surface area of bubbles. However, the spacing factor is relatively unchanged, as shown in Table 3.5 [14].

(c) Cement type

The characteristics of the cement used to prepare air-entrained cement pastes have a marked influence on the amount of air entrained and are illustrated in Tables 3.6 and 3.7 for pastes prepared under standard conditions at a water–cement ratio of 0.45 [10]

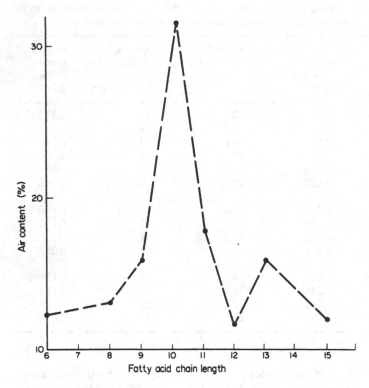

Fig. 3.9 The effect of increasing chain length of the sodium salts of linear fatty acids on air-entraining capacity (Rixom).

The large differences, however, can largely be attributed to differences in viscosities of the different pastes at the same water–cement ratio. When the pastes were prepared to the same consistencies, similar levels of air entrainment were obtained.

Table 3.5 The effect of water–cement ratio of cement pastes on the air content, specific surface area and computed spacing factor

Surface active agent	Water–cement ratio (by wt)	Air content (% by vol.)	Specific surface area of bubble $(mm^2 mm^{-3})$	Computed spacing factor (mm)
0.025% sodium dodecyl sulfate	0.40	16.7	81	0.064
	0.45	21.4	69	0.066
	0.50	25.8	56	0.069

Table 3.6 The relationship between particle size of cement and the level of air entrainment

Particle size of cement (BSS mesh)	Air content of paste (% by vol.)
20–52	44.1
52–100	32.0
140–200	24.8
Passing 200	21.0

Table 3.7 Variations in cement characteristics have a considerable effect on the level of air entrainment at constant water–cement ratio

Cement batch number	Surface area of cement $(cm^2 \ g^{-1})$ (air permeability)	Total alkali content of cement	SO_4 content of cement	Air content of paste (% by vol.)
1	3300	1.09	1.99	14.7
2	3730	0.57	2.56	6.1
3	3480	0.12	1.85	11.0
4	3500	0.25	1.82	5.7

(d) Temperature

Over the range 18–35°C, the effect of temperature is not great and is illustrated in Fig. 3.10 for a paste containing 0.0125% sodium dodecyl sulfate at a water–cement ratio of 0.45 [10]. It can be concluded that the degree of air entrainment obtained in cement pastes of a given consistency is largely a function of the type and dosage of air-entraining agent incorporated, whilst other variables have only a minor effect.

3.3.3 Distribution between solid and aqueous phases

A study of the foaming capacities and stabilities [10] of a variety of air-entraining agents in a solution of cement extracts showed that commonly used anionic air-entraining agents, such as sodium dodecyl sulfate and sodium resinate (1) were visually precipitated from solution, (2) retained their ability to form stable foams after precipitation with only minor amounts of admixture left in solution, and (3) lost the major part of their ability to form stable foams after filtration. It was further shown from studies in cement pastes firstly that the admixture should be adsorbed on the solid particles of the paste with the non-polar ends of the molecule pointed towards the water phase, imparting a hydrophobic character to the cement

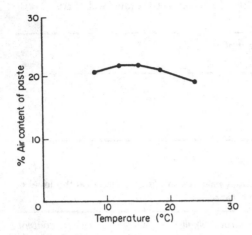

Fig. 3.10 The effect of temperature on the air-entraining capacity of sodium dodecyl sulfate in a cement paste (Bruere).

particle to which the air bubbles can adhere, and secondly that the residual concentration of admixture in the mixing water, although not necessarily high, must be sufficient to generate bubbles during mixing.

In cement paste, the residual concentration of calcium resinate is approximately $0.5 \, g\,l^{-1}$ in the aqueous phase, which compares well with the aqueous solubility of calcium caprate, as shown in Fig. 3.11. It will be noted that the C_9 fatty acid has a solubility of about $1 \, g\,l^{-1}$, whilst the C_{11} is about $0.05 \, g\,l^{-1}$. The solubility of calcium dodecyl sulfate is approximately $0.1 \, g\,l^{-1}$, so it is indicated that the optimum value lies in the region of 0.1–$0.5 \, g\,l^{-1}$ solubility of the calcium salt. It is interesting to note that the abilities of the soluble sodium and insoluble calcium salts to entrain air in cement pastes are very similar [6] (Figs 3.12 and 3.13).

3.3.4 Effects on the hydration chemistry of cement

There is little published data on the effect of air-entraining agents on the chemistry and morphology of cement hydration. However, the limited studies [15] indicate that the normal hydration pattern under isothermal conditions for ordinary Portland cement shown in Fig. 3.14 is modified as follows:

1. For sodium-oleate-based air-entraining agents, the C_3S peak is not affected, but the C_3A peak is accelerated and splits into two up to a 10 times normal dosage level. It is believed that the ettringite and

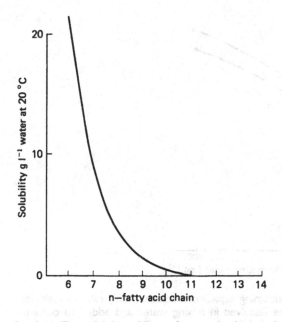

Fig. 3.11 The solubility of linear fatty-acid calcium salts in water.

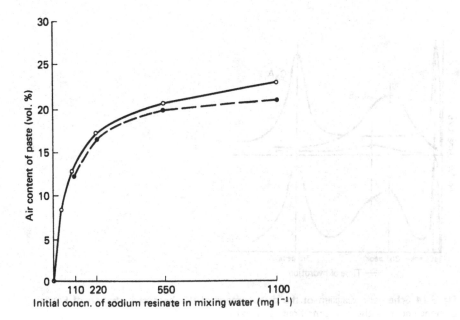

Fig. 3.12 Comparison of the air-entraining capacities of sodium and calcium abietates. O = sodium abietate dissolved in mixing water and added to cement; ● = made with filtrates from mixtures of sodium abietate and cement paste extract (Bruere).

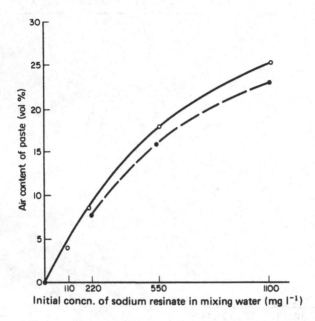

Fig. 3.13 Comparison of the air-entraining capacities of sodium and calcium resinates (wood resins). O = sodium resinate dissolved in mixing water and added to cement; ● = made with filtrates from mixtures of sodium resinate and cement paste extract.

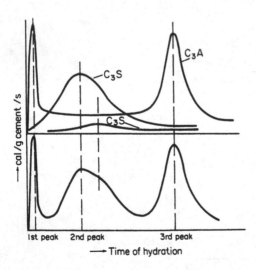

Fig. 3.14 Schematic diagram of the development of the heat of hydration of Portland cement under isothermal conditions (Bruere).

monosulfate reactions are retarded due to an impermeable layer of a calcium-oleate–aluminate-hydrate salt.

2. For the other anionic air-entraining agents, such as neutralized wood resins and sulfates or sulfonates, high dosages lead to a retardation of the C_3S peak, whilst the C_3A peak is accelerated, and sometimes splits into two.

3. Non-ionic materials, such as ethoxylates, do not appear to alter the heat output pattern of ordinary Portland cement.

Any changes in morphology of hydration products have not been published.

3.3.5 Interpretation as a mechanism of action

Air-entraining agents are predominantly anionic surfactants which, on addition to cement pastes, are adsorbed on to the cement particles with their polar groups orientated towards the particles. This 'sheath' is of limited solubility and only a minor, but finite, proportion remains in solution as the calcium salt.

The weak surfactant solution forms bubbles on agitation in the aqueous phase, which are stabilized as microscopic spheres from coalescing into large bubbles by the orientation of insoluble surfactant across the air/liquid interface and by adhering to the hydrophobic surface created on the cement particle by the adsorbed surfactant. This is shown diagrammatically in Fig. 3.15 [16].

The bubbles are less than 0.25 mm diameter and probably do not exist in the fresh paste with diameters less than 10 μm because the high pressure present in such small bubbles would cause the air to be dissolved.

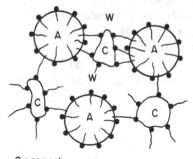

C = cement
A = air
W = water

Fig. 3.15 The interactions between cement, air, water and molecules of air-entraining agent (Kreijger).

Air-entrained pastes possess a higher viscosity than pastes with little or no air content, mainly because of the bridging effect of cement particles by air bubbles increasing the structure of the system.

There is no evidence to suggest that the presence of air-entraining agents of the type normally available commercially alter, in any way, the eventual hydration products of the cement.

3.4 The effects of air-entraining agents on the properties of plastic concrete

In considering the effect that air-entraining admixtures have on the properties of concrete in the plastic state, it is necessary to consider the variables that influence the volume of air entrained, the stability of the air void system during any subsequent handling, placing and compaction processes, and the effect that the entrained air has on the workability, water content, mix stability and mix design procedures relevant to a concrete mix. It is clear that such differences could be considerable in view of the significant alteration that air-entraining agents make to the cement paste fraction of the cement; a normal addition level of air-entraining agent will incorporate an additional, say, 4% air by volume, but this air is made up of bubbles predominantly less than 100 μm diameter, only about 0.003 mm apart, and totalling some 250 000 cm^{-3} of paste.

3.4.1 Volume of air entrained

Most air-entraining agents are formulated to give a total of 3–6% air by volume in most normal concrete mixes at their recommended dosage levels. However, it is often found that the recommended starting dosage results in either too much or too little entrained air, and adjustments have to be made. In some cases, the adjustment in dosage is considerable and can be due to a number of mix, ingredient or environmental factors. Alternatively, the particular application may require less air entrainment (to improve mix stability) or more air entrainment (for density, or improved freeze–thaw resistance reasons) than the normal levels given below. In view of this, it is useful to have an understanding of the way the many variables can influence the volume of air entrained by this category of admixture, and these are discussed below.

(a) Dosage

In any mix, the greater the quantity of air-entraining agent added, the larger will be the volume of entrained air. Figures 3.16 and 3.17 [17, 18] indicate that there is a maximum air content possible by increasing the admixture dosage and Fig. 3.17, in particular, suggests that this maximum is dependent on the characteristics of the cement.

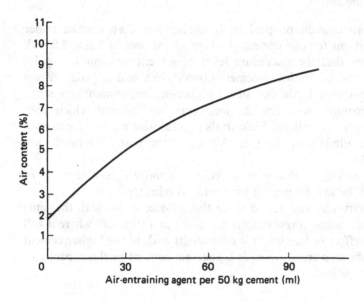

Fig. 3.16 Air content as a function of admixture addition level (Johnson).

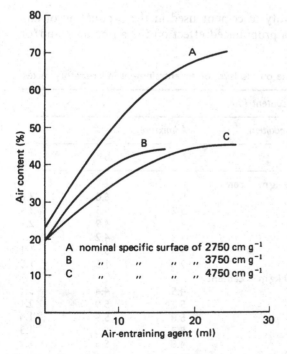

Fig. 3.17 Air content as a function of admixture addition level for cements of different fineness (Mayfield).

(b) Mixing techniques

The change in air content on prolonged mixing has been studied under laboratory conditions for two cements with results shown in Table 3.8 [19]. The data indicate that the maximum level of air entrainment is rapidly achieved in the case of the lower-cement-content mix and is progressively lost as mixing continues. In the case of the higher-cement-content mix, some 5–10 min are required to reach the maximum air content which then diminishes on continued mixing. Field trials in a full-size mixer were carried out on the same admixtures using the 340 kg m^{-3} mix and the results are given in Table 3.9.

The two studies suggest that it is unlikely that more than about 1% of entrained air will be lost by mixing for up to 30 minutes.

The mix capacity was also varied using the full-size mixer with the same concrete mixes and admixtures. Results are given in Table 3.10 where it will be seen that the effect of batch size is only slight and, in the higher cement content mixes, shows a trend towards higher air content as the capacity of the mixes is approached.

(c) Cement characteristics

The characteristics and quantity of cement used in the production of air-entrained concrete can have a pronounced effect on the air content and/or

Table 3.8 The effect of mixing time on the level of air entrainment in laboratory mixes

Mixing time (min.)	Air content (%)			
	No addition	Admixture		
		A	B	C
	256 kg m^{-1} cement			
2	1.8	5.3	6.0	3.8
5	1.1	5.2	5.5	3.4
10	1.0	4.2	4.9	2.1
15	1.0	3.3	4.2	1.9
30	1.0	2.5	3.4	1.1
60	1.0	2.0	1.5	1.0
	340 kg m^{-3} cement			
2	1.3	4.5	4.4	3.1
5	1.0	5.2	5.9	2.1
10	0.9	5.8	5.9	1.4
15	0.8	5.7	5.8	1.3
30	0.8	4.8	5.5	1.2
60	0.8	4.0	4.1	1.0

Table 3.9 The effect of mixing time on the level of air entrainment in a full-size mixer

Mixing time (min.)	No addition	Admixture		
		A	B	C
15	1.0	3.8	—	3.3
30	—	3.2	6.0	3.2
45	—	3.0	5.5	3.0
60	—	2.7	4.9	2.5
75	—	2.7	4.1	2.6
90	—	2.4	3.3	2.4

Table 3.10 Slightly higher air contents are obtained as the volume of mix approaches that of the mix capacity, particularly for the higher cement content mix

Percentage of mixes capacity	Air content (%)			
	No addition	Admixture		
		A	B	C
	256 kg m^{-3}cement			
20	1.5	5.1	5.1	2.7
40	1.6	6.1	5.1	4.2
60	1.5	6.1	6.0	3.8
80	1.5	5.8	6.0	3.5
100	1.5	6.0	6.2	3.9
	340 kg m^{-3}cement			
20	0.7	3.0	3.0	1.9
40	0.8	4.4	3.9	2.2
60	1.0	4.5	4.1	2.4
80	1.0	4.3	4.1	2.9
100	1.1	4.6	4.5	3.0

the dosage of admixture required to obtain the necessary air content. This is illustrated by the evaluation [19] of 12 different cements in an identical mix using a standard dosage of admixture. Results are given in Table 3.11.

Large variations in air content due to a change in cement source are, therefore, quite possible and, although it is not possible to quantify the effect of all cement variables, the following data are relevant.

FINENESS

The fineness of cement is a major factor in determining the quantity of air-entraining admixture required to incorporate a given amount of entrained

Table 3.11 Cement type and source can influence the volume of air obtained in both plain and air-entrained mixes

Cement code	Air entrained (%)			
	No addition	Admixture		
		1	2	3
A	3.1	5.2	5.3	3.6
B	1.2	5.3	4.4	3.2
C	1.2	5.4	5.5	3.0
D	1.1	5.1	5.4	2.0
E	1.4	3.9	3.5	2.8
F	2.3	6.2	5.4	4.5
G	2.4	7.9	6.3	4.3
H	1.0	5.4	4.5	3.6
I	1.3	6.2	5.2	3.0
J	1.8	7.8	8.1	4.6
K	1.0	6.9	5.9	3.6
L	1.7	7.2	5.8	4.3

air. Table 3.12 summarizes the results for three cements differing only in their specific surface area [18]. Differences in cement fineness that are normally experienced could lead to a doubling or halving of admixture requirements, but it is worth noting that in the study from which the above data are derived, an addition level of 10.0 ml per batch would have produced concrete conforming to a 3–6% requirement for all three cements.

CEMENT CONTENT

The amount of entrained air decreases with increasing cement content [20] and typically an increase in cement content of $90 \ kg \, m^{-3}$ will reduce the volume of entrained air in concrete by about 1% of the volume of concrete.

Table 3.12 The amount of air-entraining agent required to obtain 4% air is increased for higher surface area cements

Cement	Specific surface area	Quantity of admixture required (ml $0.0368 \ m^{-3}$ batch) for 4% air
A	2750	6.5
B	3750	10.0
C	4750	14.0

ALKALI CONTENT

An investigation [21] of mortars containing cements differing only in their alkali content indicated that when the alkali (as Na_2O) in the water in contact with the cement reaches about 0.8% by weight of water, the amount of air-entraining agent required to achieve a given air content is minimized. This is shown in Fig. 3.18.

This effect also applies to alkali added as part of the admixture formulation, it is particularly important in cements with low alkali analysis and is illustrated in Table 3.13 [22].

(d) Workability

If the addition of the air-entraining agent is maintained at a constant level, a more workable mix will entrain more air than a less workable one. However, for very workable concrete of slump greater than 180 mm, the air will be more rapidly lost before placing.

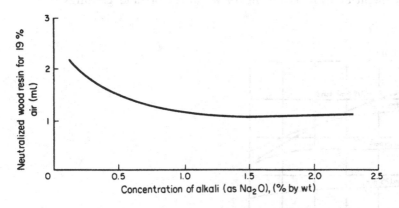

Fig. 3.18 The effect of alkali content of cements on the air-entraining capacities of neutralized wood resins in concrete (Greening).

Table 3.13 The alkalinity of the admixture influences the air content obtained, particularly with cements of low alkali analysis

Air-entraining admixture	pH	Na_2O (%)	Air content (%) of cement
1	9.4	0.35	5.5
2	10.6	0.38	7.4

(e) Temperature

The temperature of the concrete has a significant influence on the amount of air entrained in concrete by a standard addition level of admixture; the higher the temperature, the lower the air content. A typical set of results is shown in Fig. 3.19 [23]. The effect is more marked at higher slump values.

(f) Aggregate type and content

COARSE AGGREGATE

There is no evidence to suggest that coarse aggregate shape or geological origin affect the amount of air entrainment obtained. The only exception is where crushed rock aggregates contain an appreciable quantity of dust which could influence the fine aggregate gradings considered below.

FINE AGGREGATE

The amount of air that is entrained in concrete increases with an increase in the sand content and this is shown for a variety of sand gradings in two

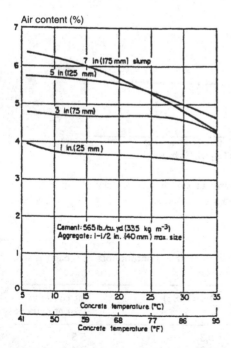

Fig. 3.19 Relationship between temperature and air content of concrete.

concrete mixes in Fig. 3.20 [24]. As a guide, an increase in the sand content of 5% will lead to an increase in air content of 1–1.5%.

A relationship exists between the size of bubbles that can be accommodated in a system containing fine particles of 300–600 μm diameter varies from about 30 to 100 μm. Since a large proportion of the bubbles in air-entrained concrete are less than 100 μm diameter, it is clear that this particle size is of considerable importance in determining the amount of air entrained. Therefore, even at constant sand content, an increase in the properties of particles of this size will lead to an increase in air content. This effect is shown for a number of sand contents in Fig. 3.21 [24].

(g) Fine fillers and pozzolans

When fine particle material of less than 20 μm diameter is included in the mix design, the amount of air-entraining agent must be increased to obtain the required air content [25]. The effect is largely independent of coarse aggregate type. Some results are shown in Fig. 3.22, where it is apparent that the effect is considerable in the case of fly ash and pumice [26]. Although these figures are relevant to a very high replacement of cementitious materials, even at normal replacement levels of 20–30% fly ash, the concrete will require three or four times the quantity of air-entraining agent in comparison to a concrete containing no fly ash to entrain the same volume

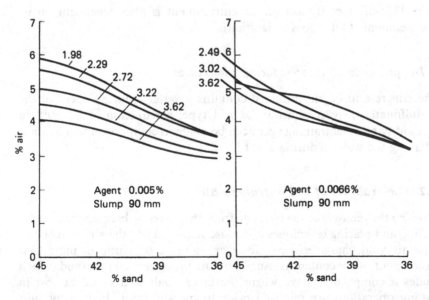

Fig. 3.20 The effect of sand content on the air entrainment of concrete at two addition levels of air-entraining agent, the fineness modulus (FM) of the sand also being varied (Craven)

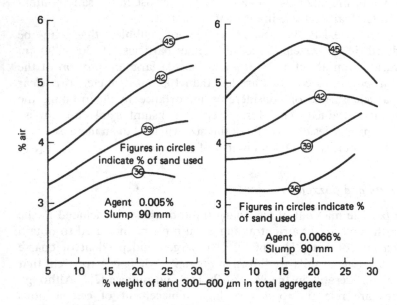

Fig. 3.21 The effect of fine aggregate of 3–600 μm diameter on the air entrainment of concrete (Craven).

of air. The effect of fly ash on air entrainment is also dependent on its carbon content (LOI – loss of ignition).

(h) The presence of water-reducing admixtures

If the concrete to be air entrained contains a water-reducing agent of the lignosulfonate or hydroxycarboxylic acid type, a reduction of 50–60% in the quantity of air-entraining agent can be made in comparison to a mix not containing the water-reducing agent [27, 28].

3.4.2 The stability of the entrained air

Following the removal of the concrete from the mixer, subsequent transport, handling and placing techniques can cause reductions in the air content and in the air void characteristics. However, it must be borne in mind that measurement of air content is usually by the pressure meter method, which includes a compaction stage where the large voids that could be lost in handling operations are removed prior to measurement. In view of this, losses during handling and transport are usually less than 0.5% of the concrete volume, as shown in Fig. 3.23 [29].

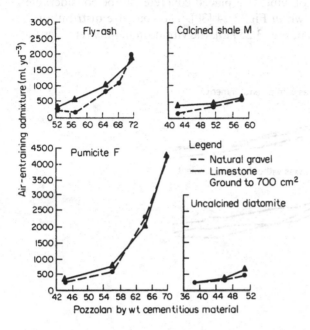

Fig. 3.22 The effect of pozzolan materials on the air-entraining agent dosage required to maintain the air content.

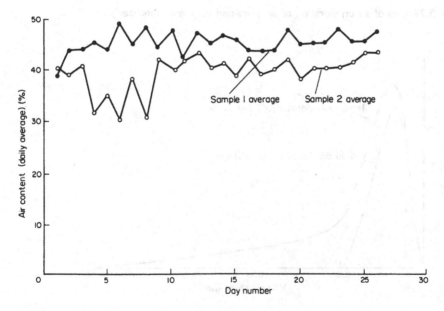

Fig. 3.23 Loss of air on handling of air-entrained concrete (McCurrich)

Losses of entrained air on vibrating placed concrete can be considerable, and typical results are shown in Fig. 3.24 [30]. The void size distribution is also altered and, as shown in Fig. 3.25 [30], the vibration mainly removes the larger diameter voids.

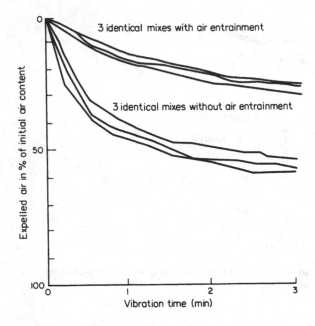

Fig. 3.24 Loss of air on vibration of air-entrained concrete (Johansen).

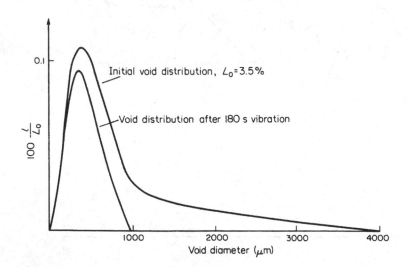

Fig. 3.25 Change in void size distribution on vibration of air-entrained concrete (Johansen).

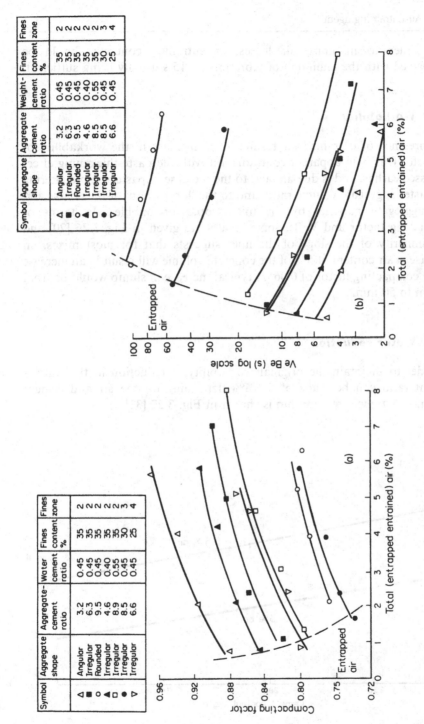

Fig. 3.26 Effect of entrained air on (a) the compacting factor and (b) the VeBe times of various concretes (Cornelius).

In order to minimize air losses, air-entrained concrete should be compacted with the minimum of vibration, 5–15 s usually being sufficient.

3.4.3 Workability

The presence of entrained air results in an increase in the workability of concrete. This is in apparent contradiction with the paste-thickening effect discussed earlier, and is due, in fact, to the increase in paste volume altering the paste–aggregate volume ratio, and to the 'lubrication' of the coarse and fine aggregate particles by the microscopic air bubbles. In terms of compacting factor and VeBe, some results are given in Fig. 3.26 [20] and the similarity of the slope of the lines suggests that for most mixes, an increase in air content of 5% of the concrete volume will result in an increase in the compacting factor of 0.06. A typical increase in slump would be from 12 mm to 50 mm.

3.4.4 Water reduction

In order to maintain the original workability, a reduction in the water–cement ratio can be made of 5–15% depending on the air and cement contents. A typical relationship is shown in Fig. 3.27 [31].

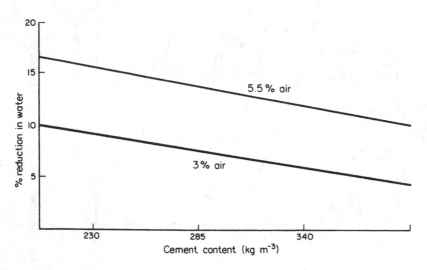

Fig. 3.27 Reduction in water content is dependent on the amount of air entrained and the cement content.

3.4.5 Mix stability

The presence of entrained air clearly makes the concrete more cohesive and allows a reduction in the quantity of fine aggregate, without increasing the tendency to segregate. The quantity of sand that can be removed from a concrete that is satisfactory in terms of cohesion prior to air entrainment is approximately 20 $kg\,m^{-3}$ for each 1% of additional air content.

Bleeding is always reduced by air entrainment, both under static conditions and during vibration, and is particularly useful for producing concrete from fine aggregate deficient in the lower particle size range where bleeding is often a problem. Some comparative data are given in Table 3.14 for plain mixes, mixes containing air-entraining agents, and mixes containing both air-entraining agents and water-reducing admixtures [32].

3.4.6 Mix design requirements

The addition of an air-entraining agent to concrete results in three changes which must be taken into consideration for mix design purposes: there is the increase in yield due to the volume of entrained air which will reduce the cement content per unit volume, the increase in cohesion allowing a reduction in sand content, and the increase in workability allowing a reduction in the water content. The following sequence of mix design procedures will accommodate these changes.

Table 3.14 Bleeding of plain concrete, air-entrained concrete and air-entrained concrete containing a water-reducing agent

Water-reducing admixture	Air-entraining admixture	Slump (mm)	Air content (%)	Sand–aggregate ratio	$ml\,cm^{-2}$ of surface	% total mix water
None	None	87	1.6	0.45	0.48	14.2
None	AEA-1	100	5.3	0.42	0.30	9.7
Class 1	None	112	4.8	0.43	0.19	6.3
Product A	AEA-1	112	5.5	0.42	0.17	6.1
Class 2	None	112	4.7	0.43	0.25	8.6
Product B	AEA-1	112	5.5	0.42	0.18	6.4
Class 2	None	87	3.0	0.43	0.30	10.2
Product D	AEA-1	100	5.4	0.42	0.23	7.8
Class 2	None	100	3.8	0.43	0.28	9.5
Product E	AEA-1	112	5.7	0.42	0.16	5.8

1. Recognized mix design procedures (or existing standard mixes) are used to produce a plain concrete meeting the requirements of workability, strength and cohesive character using available materials.
2. The fine aggregate content of the mix is reduced by 20 kg sand m^{-3} for each 1% of additional air required.
3. The air-entraining admixture, at the manufacturer's recommended addition level and premixed in part of the gauging water, is added to the mix, and sufficient additional gauging water included to achieve the desired workability.
4. Alterations in the admixture level may be necessary to achieve the required air content which will result in consequential changes in the gauging water requirements.
5. Determine plastic density.
6. Correct batch figures for yield to obtain composition per cubic meter.
7. Prepare cubes or cylinders for compressive strength determination.

This sequence is shown in Table 3.15, together with the resultant properties of the plastic and hardened concrete.

The combined effects of sand and water reductions bring the cement content to approximately that of the comparative plain concrete so that 28-day strengths are similar. However, this is not consistent across all cement contents; low-cement-content mixes tend to show an increase in strength, whilst higher cement content will show a slight decrease. This is illustrated by the data in Table 3.16 [31].

Table 3.15 Mix design procedure for air-entrained concrete

	Control mix (kg m^{-3})	Modified air-entrained mix	
		1st adjustment (kg)	(kg m^{-3})
Composition			
20–25 mm gravel (rounded/angular)	1330		1330	1338
Sand (zone 2)	591	(−68)	523	526
Ordinary Portland cement	328		328	330
Water	181	(−15)	166	167
Air-entraining agent (ml)	Nil		295	297
Properties of concrete			
Water–cement ratio	0.55		0.51	
Slump (mm)	38		45	
Air content (%)	1.8		5.5	
Compressive strength: 7 days (N mm^{-2}) 28 days	27.4 32.5		28.6 32.9	
Plastic density (kg m^{-3})	2430		2361	

Table 3.16 Effect of air-entrainment on strength

Total water–cement ratio	Plain concrete		Air-entrained concrete (4%)	
	Cement content* (kg m⁻³)	28-day* strength (N mm⁻³)	Cement content* (kg m⁻³)	28-day strength* (N mm⁻³)
0.44	383	35.0	341	29.4
0.49	341	31.5	298	27.3
0.53	312	28.0	284	25.9
0.58	284	25.2	256	21.7
0.62	270	22.4	241	19.6
0.67	256	19.6	227	17.5
0.71	241	17.5	213	15.4
0.75	227	14.0	199	14.0

*This is American data and although useful to show the comparative strengths for plain and air-entrained concrete, in practice, with higher cement contents and lower workability used in the UK, values should not be used as the basis for mix design.

3.5 The effects of air-entraining agents on the properties of hardened concrete

3.5.1 Structural design parameters

(a) Compressive strength

It is always assumed that the entrainment of air in concrete leads to a considerable reduction in compressive strength. However, it was shown earlier that because of mix design changes as a result of, or made possible by, the presence of the air bubbles, concretes containing 3–6% by volume of air and at constant cement content within the range 200–400 kg m⁻³ will have only slight, if any, losses in strength. Indeed, at the lower cement content there could be an increase in strength (Table 3.16).

It is useful, however, to be able to predict the 28-day strength of special concretes containing, perhaps, much higher air content for insulation or density reasons, or alternatively in mixes where no water or sand reductions are made. In these cases, the volume of additional air entrained should be considered in terms of an equivalent volume of water and added to the water already in the mix (after allowing for aggregate absorption). The new water-and-air–cement ratio can then be used to estimate the 28-day strength from standard curves of the type in Fig. 1.37. Table 3.17 gives the change in water–cement ratio for calculation purposes for each 1% of air entrained.

Table 3.17 Factors to be added to the water–cement ratio to calculate the anticipated strength of air-entrained concrete

Cement content of plain mix (kg m⁻³)	Add to paste water–cement ratio of air-entrained mix to estimate strength for each 1% air entrained
100	0.100
200	0.050
300	0.033
400	0.025
500	0.020
600	0.017

(b) Flexural and tensile strength

There appears to be no published data on the effect of air entrainment of the tensile strength of concrete, but information is available on the flexural strength [33]. This is shown in Fig. 3.28 as a relationship between compressive strength and flexural strength for air-entrained and plain concrete mixes. This graph suggests that the same relationship exists for both types of concrete. However, other work has indicated that the flexural strength of an air-entrained concrete may be higher for a given compressive strength in comparison to a plain concrete [34]. The flexural strength is certainly not adversely affected.

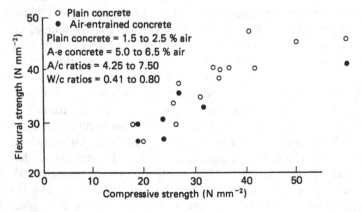

Fig. 3.28 The relationship between the flexural and compressive strengths of plain and air-entrained concretes (after Shacklock).

(c) Stiffness

The relationship between the compressive strength and the modulus of elasticity for a number of similar concretes is shown in Fig. 3.29 [35] and Fig. 3.30 [33]. Again it is clear that the presence of entrained air does not alter the normal relationship.

The available data, therefore, suggest that by making the relevant mix design changes to compensate for the effect of entrained air on strength, the engineering properties of the concrete will be similar to a plain concrete of comparable strength.

3.5.2 Durability aspects

The important factors relevant to the ability of concrete to play its structural, aesthetic and protective roles have been given earlier (Section 1.4). The effect that air-entraining agents have on these factors, with the

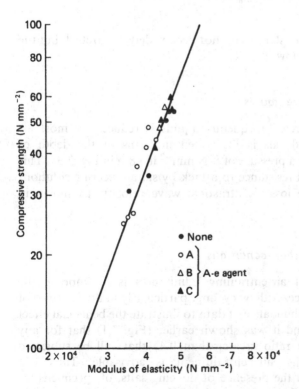

Fig. 3.29 The relationship between the compressive strength and modulus of elasticity of plain and air-entrained concretes (Warris).

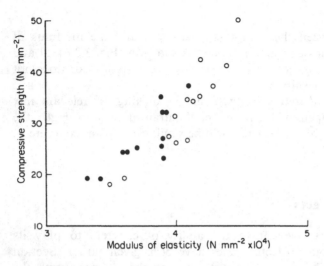

Fig. 3.30 The relationship between the compressive strength and modulus of elasticity of plain and air-entrained concretes (after Shacklock).

exception of freeze–thaw resistance, has not been widely researched, but the available data are given below.

(a) Resistance to aggressive liquids

The permeability of concrete to aqueous liquids is reduced by most air entraining agents [35] and this is illustrated in terms of the depth of penetration of water under a pressure of 8 N mm^{-2} in 48 h in Fig. 3.31. This is reflected in an improved resistance to attack by sulfate-bearing solutions which is indicated by the loss in ultrasonic wave velocity, as shown in Fig. 3.32 [37].

(b) Resistance to freeze–thaw conditions

The major application of air-entraining admixtures is to improve the resistance of concrete to freeze–thaw cycling, particularly in the presence of de-icing salts. There is an abundance of data to illustrate the beneficial effect under these conditions, and it was shown earlier (Fig. 3.1) that for any concrete of water–cement ratio greater than 0.45 that will be subjected to freeze–thaw conditions, air entrainment is advisable. Table 3.18 summarizes the scaling, in the presence of de-icing salts, of specimens air entrained with a short-chain fatty-acid-based air-entraining agent in comparison to a control concrete [37].

The degree of protection afforded is a function of the characteristics of the air void system entrained, and the significance of the variables involved requires an understanding of the mechanism of freeze–thaw resistance operating.

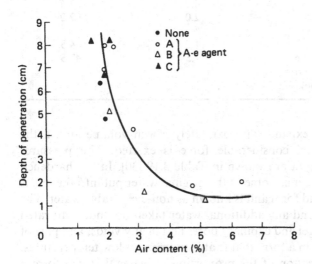

Fig. 3.31 The relationship between permeability and air content (Warris).

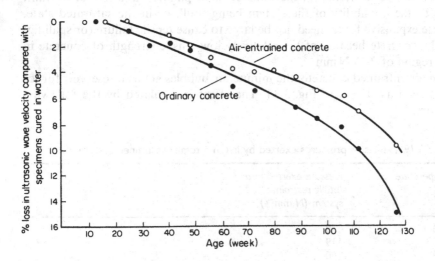

Fig. 3.32 Deterioration of plain and air-entrained concrete in sulfate solutions (Wright).

Table 3.18 Air entrainment improves the freeze–thaw resistance of concrete in the presence of de-icing salts

	Control	Air-entrained
Mix design		
10 mm gravel (kg)	8.0	8.4
Zone 2 sand (kg)	4.0	3.6
OPC (kg)	2.0	2.0
Properties		
Air content	1.0	4.5
Freeze–thaw resistance (number of cycles to cause 0.5 mg mm^{-2} scaling in 3% NaCl solution)	7.5	42.5

When water freezes it expands approximately 9% in volume and, if the expansion is restrained, a considerable force is exerted. The pressures exerted under total restraint are shown in Table 3.19 [30]. In the hardened cement paste of a concrete mix, some of the original water put into the mix is held in a chemically bound form and is known as non-evaporable water. The remainder of the water, and any additional water taken up under saturated conditions resides in the gel and capillary pores and in any voids, and part of this 'evaporable' water is in a form that is able to freeze at low temperatures. Table 3.20 gives an indication of the proportion of water that can form a solid phase at −20 °C [38]. When this water freezes in the capillaries it is apparent that restraint against expansion is not total but, nevertheless, significant forces are set up within the capillary structure, as shown in Fig. 3.33 due to friction at the walls of the capillaries, and to end restraint due to the possibility of the system being 'full', or in the saturated state. These expansive forces need not be large to cause tensile failure (or spalling) in the concrete because of the relatively low tensile strength of concrete in the region of 2–6 N mm^{-2}.

In air-entrained concrete, the minute air bubbles act as a reservoir for ice expansion as shown in Fig. 3.34. The effect is regulated by the following factors:

Table 3.19 Expansive pressures exerted by ice in a totally restrained system

Temperature (°C)	Pressure exerted in a totally restrained system (N mm^{-2})
0	0
−10	119
−20	190

Table 3.20 The proportion of water that can form solid ice at −20°C as a function of paste water–cement ratio

Original water–cement ratio of paste	% evaporable water able to freeze at −20°C
0.45	23
0.50	35
0.60	52

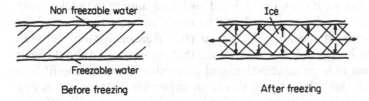

Fig. 3.33 Expansive forces are created within the capillary structure during freezing.

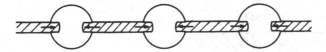

Fig. 3.34 The minute air bubbles act as reservoirs for ice expansion.

1. **The amount of air space available in relation to the expansion of that part of the water likely to freeze.** In theory, the amount of void space necessary is about 4% of the paste volume but, in practice, because of the distribution of void sizes formed and because the total volume of the voids could not be filled by the ice which would already be in the solid form on entering the void, greater air contents are required. The recommendations shown in Table 3.21 are made.
2. **The bubbles need to be empty and not full of water.** This is important to allow for expansion and this is ensured by orientation of the molecules

Table 3.21 Recomended air contents for various maximum aggregate sizes

Maximum size of aggregate (mm)	Air content (% of concrete volume)
40	4 ± 1.5
20	5 ± 1.5
10	7 ± 1.5

of the air-entraining agent hydrophobic 'tails' into the voids as illustrated in Fig. 3.35, preventing the 'wetting' of the high-contact-angle void internal surface presented. When the ice expands into the void, the higher pressures easily overcome this effect. This explains, at least in part, why non-hydrophobic materials, like the phenol ethoxylates, although resulting in air entrainment, do not bestow significant improvements in freeze–thaw resistance to concrete in comparison to materials producing hydrophobic bubbles, such as the neutralized wood resins and fatty-acid soaps (Table 3.22 [39]).

3.   **The length of the water column in the capillary up to a point where expansion is possible** (L in Fig. 3.35). This is to minimize the wall friction to allow the freezing water column to move and is a function of the spacing factor of the air void system (the average distance between adjacent air bubbles). It has been shown that the separation should be at least 0.4 mm to be at all effective, and preferably nearer to one-fifth to one-tenth of this value. Again, this is an important factor in explaining differences in the performance of air-entraining agents of different types, as shown in Table 3.23.

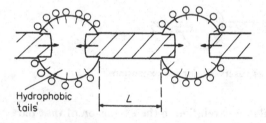

Fig. 3.35 The hydrophobic 'tails' of the air-entraining agent molecules prevent the filling of the air bubbles with water.

Table 3.22 Freeze–thaw resistance of concrete air-entrained by different chemical types of admixture

Air-entraining admixtures	Air content (%)	Relative classification of freeze–thaw resistance *
None	2.0	5
Sodium oleate	5.6	86
Sodium lauryl sulfate	5.8	46
Neutralized wood resins	5.2	57
Phenol ethoxylate	5.2	7

* An average of several techniques measuring effect on modulus and compressive strength after freeze – thaw cycling.

Table 3.23 The effect of spacing factor on freeze–thaw durability

Air-entraining admixture	% air by volume	Average spacing factor (mm)	Durability to freeze–thaw cycling*
None	1.0	1.01	17
Non-ionic air-entraining agent	5.0	0.35	24
Neutralized wood resin	5.1	0.15	70

*An average of several techniques measuring effect on modulus and compressive strength after freeze–thaw cycling.

4. **Rate of freezing.** This is important in allowing time for the freezing water column to exude from the capillary into the void and so the more rapid the fall in temperature, the more severe the effect. This is the major reason for the severity of damage in the presence of de-icing salts, sodium and calcium chloride; the melting of ice at the surface causes latent heat to be extracted from the top layer of the concrete, resulting in very rapid freezing in the capillaries near the surface. The thermal gradient in the concrete leads to the worst tensile stress situations.

It should be pointed out that deterioration under freeze–thaw conditions can also be caused by a mechanism other than the direct freezing of the non-evaporable water. The capillaries contain dissolved salts, such as hydroxides, sulfates and carbonates. As part of the water is frozen, the concentration of salts in the remaining water increases and water will flow by osmotic pressure from the gel pores to the capillary pores, setting up an additional disruptive pressure.

(c) The protection of steel reinforcement

The formation of the passive layer at the concrete/reinforcement interface referred to earlier (Section 1.4) is due to the alkaline nature of the concrete. The alkalinity is due to calcium, sodium and potassium hydroxides which, over a period of time, react with atmospheric carbon dioxide to form carbonates. This reduction in alkalinity in reflected in a diminished protective capacity towards the steel reinforcement.

Some carbonation tests have been reported on plain and air-entrained concretes using two aggregate types, where the depth of carbonation on indoor and atmospheric exposure for 5 years have been measured [40]. The results are given in Table 3.24. In all cases, the air-entrained mixes have shown less carbonation than similar controls, suggesting that air-entrained concrete should provide a better reinforcement protection in the long term

Table 3.24 Depth of carbonation of air-entrained and plain concrete

	Concrete type					
	OPC/sand and gravel 350 kg m^{-3}			OPC/lightweight aggregate 350 kg m^{-3}		
Admixture	None	Neutralized wood resins	Petroleum sulfonate	None	Neutralized wood resins	Petroleum sulfonate
Atmospheric exposure*	1.00	0.69	0.65	3.65	1.55	3.19
Indoor exposure*	1.00	0.79	0.48	2.29	1.75	1.38

*The figures are expressed as relative to the control mix OPC/sand and gravel.

(d) Maintenance of compressive strength

In a previous section it was shown that by suitable changes in mix design, concrete could be produced containing entrained air with similar 28-day compressive strengths to plain concrete. There are no published data however, for the strength development of air-entrained concrete over considerable periods of time. Neither, of course, is there any evidence to the contrary and, indeed, the relatively small number of recorded compressive strengths of up to 1 year indicate no fall-off in strength in comparison to control concretes, as shown in Fig. 3.36 [34, 41, 42].

Experiments on the effect of different curing conditions on the compressive and flexural strengths of plain and air-entrained concrete [35] showed that air-entrained concrete has less tendency to lose moisture under drying conditions, which means that if concrete curing conditions are not ideal, air-entrained concrete should develop strength more normally than plain concrete [43].

(e) Volume deformations

1. **Shrinkage.** Air-entraining agents do not increase the drying shrinkage of concrete, and mixes designed to have similar strength and workability characteristics can be considered to have similar drying shrinkage, whether air entrained or not. Typical data are shown in Table 3.25. This has been confirmed by other workers [35, 42, 43].

   A program of work on cement pastes [45] showed that subsequent rewetting and drying in the presence of neutralized wood resins showed

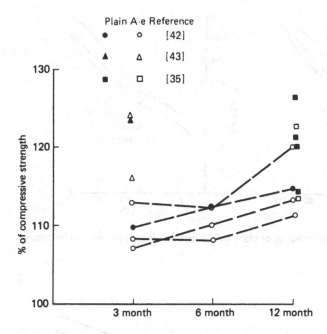

Fig. 3.36 The increase in compressive strength of plain and air-entrained concrete up to a period of 1 year.

Table 3.25 Drying shrinkage of air-entrained concrete is no greater than that of plain concrete

Air-entraining admixture	Concrete proportions				Air (%)	Slump (cm)	Drying shrinkage $\times 10^{-6}$		
	Cement (kg m^{-3})	Water (l m^{-3})	W/c ratio	Fine agg. (%)			7 days	28 days	56 days
None	300	150	0.50	40	1.7	6	220	546	688
A	300	135	0.45	36	4.5	6	225	560	689
B	300	135	0.45	36	4.4	6.5	253	586	698
C	300	135	0.45	36	4.7	6.5	215	525	661

no increase in shrinkage in comparison to a plain paste. The results are given in Fig. 3.37.

2. **Creep.** The creep of air-entrained concrete at a constant stress–strength ratio is said to be not greater than that of comparative plain concretes [46]. However, at higher air contents (greater than 6%) there may be a very marginal increase in creep after some time under load, although at earlier ages this trend does not appear to be observed [47]. Figure 3.38 illustrates this effect.

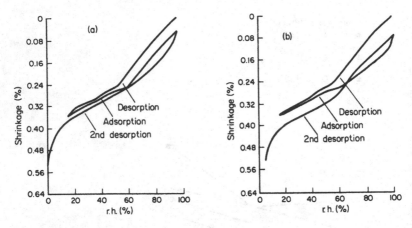

Fig. 3.37 Length changes (shrinkage) of plain and air-entrained cement pastes (Feldman).

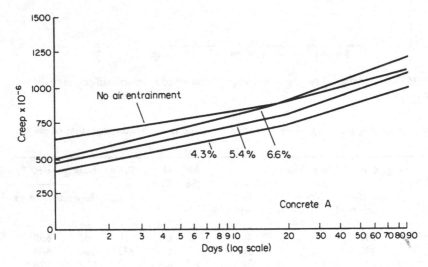

Fig. 3.38 The creep of plain and air-entrained concretes at 21°C and a stress–strength ratio of 45% (Nasser).

Air-entraining admixtures, therefore, produce concrete which is more durable to conditions of freezing and thawing, particularly in the presence of de-icing salts, more resistance to sulphate attack, provides better protection to embedded reinforcement and is more tolerant of poor curing conditions. There appears to be no great difference in the way air-entrained concrete behaves in terms of compressive strength development and volume deformations.

References

1 Jackson, F.H. (1944). *ACI Journal*, **14**, 509.
2 Della-Libera, G. (1967). *Proceedings of the International Symposium on Admixtures for Mortar and Concrete*, Brussels, 142.
3 Neville, A.M. (1963). *Properties of Concrete*, Pitman, London.
4 Hewlett, P.C. (1978). Private communication.
5 Rixom, M.R. (1975). *Proceedings of the Workshop on the use of Chemical Admixtures in Concrete*, University of New South Wales, 149–76.
6 Bruere, G.M. (1971). *Journal of Applied Chemistry*, **21**, 61–4.
7 Taylor, W.H. (1974). *Precast Concrete*, **5**, 83–4, 89–90, 96.
8 Bruere, G.M. (1967). *Proceedings of the International Symposium on Admixtures for Mortar and Concrete*, Brussels, 8–13.
9 Vivian, H.E. (1960). *Proceedings of the Fourth International Symposium on the Chemistry of Cement*, Washington, 917–8.
10 Bruere, G.M. (1955). *ACI Journal*, **51**, 905–19.
11 Bruere, G.M. (1960). *Australian Journal of Applied Science*, **11**, 289–94.
12 Anon. (1965). *Admixtures for Concrete*, CVR Report No. 31, 45–7.
13 Bruere, G.M. (1958). *Australian Journal of Applied Science*, **9**, 349–59.
14 Bruere, G.M. (1967). *Proceedings of the International Symposium on Admixtures for Mortar and Concrete*, Brussels, 14–22.
15 Kreijger, P.C. (1967). *Proceedings of the International Symposium on Admixtures for Mortar and Concrete*, Brussels, 27–32.
16 Kreijger, P.C. (1967). *Proceedings of the International Symposium on Admixtures for Mortar and Concrete*, Brussels, 33–7.
17 Johnson, D.L. (1968). *ACI Journal*, **65**, 402–41.
18 Mayfield, B. (1969). *Civil Engineering and Public Works Review*, **1**, 37–41.
19 Scripture, E.W. (1949). *ACI Journal*, **45**, 653–62.
20 Cornelius, D.F. (1970). *Report LR363*, Road Research Laboratory.
21 & 22 Greening, W.R. (1967). *JPCA Research and Development Laboratories*, **45**, 22–36.
23 Bloem, D.C. (1946). *ACI Journal*, **42**, 629–39.
24 Craven, M.A. (1948). *ACI Journal*, **44**, 205–15.
25 Anon. (1974). ACI Publications SP.46, 99–108.
26 Anon. (1962). *Army Engineer Waterways Experiment Station*, Mississippi, No. AD 756 299.
27 Larson, T.D. *et al.* (1963). *ACI Journal*, **60**, 1739–53.
28 Mielenz, R.C. (1968). *Proceedings of the Fifth International Symposium on the Chemistry of Cement*, Tokyo, 10–18.
29 McCurrich, L.H. (1976). Private communication.
30 Vivian, H.E. (1960). *Proceedings of the Fourth International Symposium on the Chemistry of Cement*, Washington, 915.
31 Anon. (1965). *Grace Technical Bulletin No. 8*.
32 Vollick, C.A. (1959). *ASTM Special Publication No. 266*, 194–5.
33 Shacklock, B.W. (1959). *Civil Engineering and Public Works Review*, **54**, 77–84.
34 Kreijger, P. (1954). *Lecture Notes*, University of Delft, Holland.
35 Warris, B. (1967). *Proceedings of the International Symposium on Admixtures for Mortar and Concrete*, Brussels, 11–15.

36 Wright, P.J.F. (1953). *Proceedings of the Institute of Civil Engineering*, **2**, 337–58.
37 Rixom, M.R. (1975). *Chemistry and Industry*, **7**, 162–5.
38 Powers, T.C. (1945) A working hypothesis for further studies of frost resistance of concrete, *Proceedings of the ACI*, **41**, 245.
39 Krieijger, P.C. (1967). *Proceedings of the International Symposium on Admixtures for Mortar and Concrete*, Brussels, 237–44.
40 Nishi, T. (1967). *Proceedings of the International Symposium on Admixtures for Mortar and Concrete*, Brussels, 112–7.
41 Kobayashi, M. (1967). *Proceedings of the International Symposium on Admixtures for Mortar and Concrete*, Brussels, 80–2.
42 Pais-Cuddou, M. (1967). *Proceedings of the International Symposium on Admixtures for Mortar and Concrete*, Brussels, 191–205.
43 Sutherland, A. (1974). *Air Entrained Concrete*, Publication 45022, Cement and Concrete Association.
44 Keene, P.W. (1960). *Technical Report TRA 331*, Cement and Concrete Association.
45 Feldman, R.F. (1975). *Cement and Concrete Research*, **5**, 25–35.
46 Anon. (1970). *Shrinkage and Creep in Concrete*. ACI Bibliography No. 10, 1966–70.
47 Nasser, K.W. (1973). *Behaviour of Concrete under Temperature Extremes*. ACI Publications SP 39, 139–48.

Chapter 4

Concrete dampproofers

4.1 Background and definitions

Concrete dampproofers are integral admixtures that alter the concrete surface so that it becomes water repellent, or less 'wettable'. This is illustrated in Fig. 4.1, which shows a close up of a water drop on a surface of a concrete that has had a dampproofer incorporated into it. This water repellency conferred on the concrete is only effective in preventing water from entering the surface when the applied pressure is small, e.g. rainfall in windy conditions, or capillary rise. The latter effect is shown in Fig. 4.2. In view of this, these materials are used normally for improving the quality of concrete pavers, tiles, bricks, blocks and cladding panels where the additional benefits of reduced efflorescence, the maintenance of clean surfaces and the more even drying out of adjacent bricks and panels are also obtained.

Fig. 4.1 Dampproofed concrete exhibits a high contact angle to water.

Fig. 4.2 Dampproofed concrete bricks exhibit almost no capillary rise. 2A = Damp-proofed. 2 = No dampproofer.

In water-retaining structures or basement concrete subject to high hydrostatic pressure, materials of this type are generally not beneficial. However, some dampproofing admixtures do contain water-reducing admixtures and will result in a reduction in permeability under an applied hydrostatic head. In addition, the reduced capillary size and quantity will increase the hydrostatic pressure required to enter the concrete surface (see later).

4.2 The chemistry of concrete dampproofers

The chemical materials used to produce dampproofers are able to form a thin hydrophobic layer within the pores and voids and on the surfaces of the concrete in one of three ways: (1) reaction with cement hydration products;

(2) coalescence from globular particle form (emulsion); (3) incorporation in a very finely divided form. Table 4.1 summarizes the chemical types used in each category.

4.2.1 Materials which react with cement hydration products

Dampproofers based on liquid fatty acids, such as oleic, caprylic and capric, are used as major components in fatty-acid mixtures. A typical example is shown in Table 4.2 [3]. The mixtures are added directly to the concrete mix without predilution, or addition to the gauging water.

Stearic acid is widely used in this application and can be added directly to the mix in powder form, premixed with an inert filler, such as talc or silica, which aids dispersion throughout the mix or, for the same reason, as an emulsion in water.

Butyl stearate is also added as an emulsion and because of its slower reaction with the hydration products, it is claimed that better dispersion

Table 4.1 Chemical types of concrete dampproofers

Method of hydrophobic layer deposition	Materials used in formulation	Reference
1. Reaction with cement hydration products	Stearic acid $C_{17}H_{35}COOH$	
	Oleic acid $C_{17}H_{33}COOH$	[1]
	Vegetable and animal fats	[2]
	Butyl stearate	[4, 9]
	Caprylic ($C_7H_{15}COOH$) and	[3]
	Capric ($C_9H_{19}COOH$) acids	
2. Coalescence from emulsion	Wax emulsion	[2, 5]
3. Finely divided form of material	Calcium stearate	[6, 9]
	Aluminum stearate	[8]
	Hydrocarbon resin	[7]
	Bitumen	[8]

Table 4.2 Typical composition of liquid fatty acid dampproofers

Fatty acid	(%)
$C_5H_{11}COOH$	1.1
$C_7H_{15}COOH$	73.3
$C_9H_{19}COOH$	21.5
$C_{11}H_{23}COOH$	4.1

throughout the mix is obtained, so that less material is required, than in the case of stearic acid.

Certain vegetable and animal fats have been used as dampproofers and again emulsions or pastes are preferred and typical formulations are given in Table 4.3 [10]. The fat can be white grease, tallow or soya bean oil and although they all produce hydrophobic concrete, different effects on compressive strength are obtained.

4.2.2 Materials which coalesce on contact with cement hydration products

Very finely divided wax emulsions are effective concrete dampproofing agents and are formulated so that the emulsion breaks down after contact with the alkaline concrete environment and forms a hydrophobic layer. Waxes of melting point 57–60°C are used with an emulsifying agent based on sorbitan monostearate or ethoxylated sorbitan monostearate [2]. The properties of a commercial product are given in Table 4.4 [5].

4.2.3 Finely divided hydrophobic materials

This type of dampproofing admixture is widely used in the concrete products industry, in particular the calcium and aluminium stearates. The calcium stearates can be produced by grinding stearic acid with lime or cement to produce a material containing 10–30% calcium stearate.

Inert materials, such as hydrocarbon resins and coal tar pitches in fine powder form, are claimed [7, 8] to maintain their ability to produce hydrophobic surfaces, even after autoclaving. They would typically be ground to pass the 200 μm sieve.

Table 4.3 Typical composition of fat-based dampproofers

Component	Paste by weight	
	Emulsion	Paste
Finely divided silica	11.2	30.0
Ca(OH)$_2$	2.0	5.0
CaCl$_2$	2.2	10.0
CaCO$_3$	1.0	1.0
Fat	2.0	20.0
Water	25.0	20.0
Mineral spirits	—	10.0

Table 4.4 Typical characteristics of wax emulsion type of dampproofer

Appearance	Milky white emulsion
Specific gravity	0.98
pH	6.5–7.0
Viscosity (cP)	6–8
Size of wax particles (μm)	0.5–1.0
Solids content (%)	30

4.3 The effects of dampproofers on the water–cement system

There is not a great deal of published data on the effect that dampproofing admixtures have on cement pastes in the way that water-reducing and air entraining agents have been studied, but some observations can be made.

4.3.1 Bleeding of cement pastes

Wax emulsions have been shown [11] to cause a considerable reduction in the bleeding rates and capacities of cement pastes, and results are given in Fig. 4.3 in comparison to bentonite (B) and kaolin (K) additions. Although not as effective as bentonite, some of the wax emulsions are clearly very beneficial in this role.

4.3.2 Hydration of cement pastes

Addition of dampproofers based on caprylic, capric or stearic acids, stearates or wax emulsions do not have any effect on the setting characteristics of hydration products of Portland cement. However, the unsaturated fatty acid salts, such as oleates, although not affecting the tricalcium silicate hydration, have a marked effect on the ettringite and monosulfate reaction [12] and this is illustrated in the isothermal calorimetry results in Fig. 4.4. It is possible that a calcium oleoaluminate hydrate complex is formed involving the double bond of the oleic acid.

4.3.3 Effects on the capillary system of hardened paste

Hardened Portland cement contains a distribution of pore and capillary sizes, depending on the initial water–cement ratio and the maturity of the paste.

A typical distribution of pore radii in the hardened cement paste of concrete was shown in Fig. 1.40 which indicated that the majority of pores lie in the region of 0.05 and 1.0 μm diameter and it is through these pores that water passes by applied pressure or capillary rise, as shown in Fig. 4.5(a).

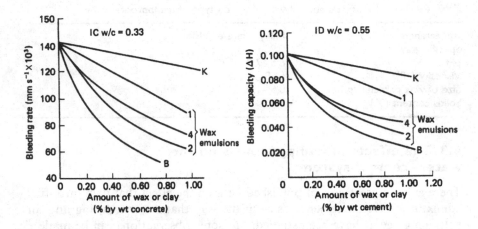

Fig. 4.3 Wax emulsions reduce the bleeding rate of concrete (Bruere).

It is believed that in the presence of dampproofing admixtures, the surfaces of the concrete, and the internal surfaces of the pores become coated with either a layer of molecules in the case of stearic acid and other fatty acids (Fig. 4.5b) or a layer of coalesced or separate particles of material in the case of waxes and bitumens, etc. (Fig. 4.5c). The end result in both cases is the production of hydrophobic surfaces exhibiting high contact angles to water, as shown in Fig. 4.6.

In Fig. 4.5(a), the pressure P required to force water into the surface is given by the expression:

$$P = \frac{-2\gamma \cos\theta}{r}$$

where γ is the surface tension of water = 72 dyn cm^{-1} (0.072 N m^{-1}), θ is approximately 120° for surfaces coated with waxes or fatty acids, r is the radius of capillary (0.5 m for the largest). Therefore

$$P = \frac{-2 \times 72 \cos 120°}{0.5 \times 10^{-4}} \text{ dyn cm}^{-2}$$

$$= \frac{2 \times 72 \times 0.5}{0.5 \times 10^{-4}} \text{ dyn cm}^{-2}$$

$$= \frac{2 \times 72 \times 0.5}{0.5 \times 10^{-4} \times 10^{3}} \text{ cm head of water}$$

$$\approx 1400 \text{ cm head of water.}$$

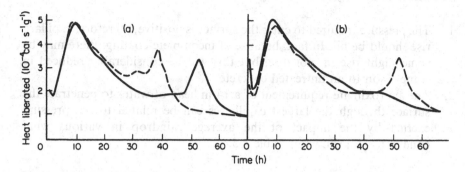

Fig. 4.4 The influence of sodium oleate on cement hydration (Edwards).

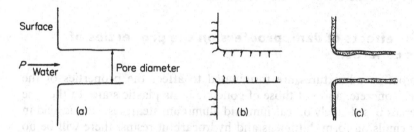

Fig. 4.5 The molecules or particles of the dampproofing admixture line the capillaries with a hydrophobic sheath.

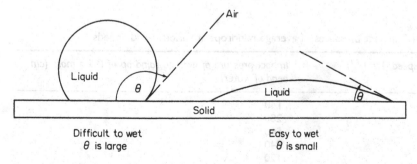

Fig. 4.6 High and low contact angles.

The effect of the high contact angle surface is two-fold:

1. The pressure required to enter the surface is positive, therefore, capillary rise should be nil. In fact, because of incomplete coating; there may be some slight rise in moisture, but this will be considerably reduced in comparison to an untreated concrete.
2. The approximate requirement of a 14 m head of water to penetrate the surface through the largest capillaries can be related to the pressure exerted by the impact of the average raindrop in various wind conditions [13] shown in Table 4.5.

Dampproofed concrete, therefore, should not show significant uptake of surface water in conditions of rain and wind up to about 100 km h^{-1}. In fact, on prolonged exposure, some wetting does occur, because of defects in the hydrophobic coating, and the presence of larger voids in the concrete, up to perhaps 1 mm wide; these are due to incomplete compaction, or to the nature of the concrete in the case of blocks.

4.4 The effects of dampproofers on the properties of plastic concrete

Dampproofing admixtures are formulated to affect the properties of the hardened concrete, and not those of concrete in its plastic state. In the case of materials based solely on calcium and aluminum stearates, stearic acid in solid or emulsion form, bitumens and hydrocarbon resins, there will be no effect on the properties of the plastic concrete with regard to air content, workability, mix design parameters, etc. When water-reducing admixtures or accelerators are included in the formulation, the effect on the concrete will be a function of the particular type of material used (see relevant section). The wax emulsions do appear to have an effect on the properties of the plastic concrete because of the 'lubrication' effect of the very small

Table 4.5 Impact pressures of average raindrops at various wind speeds

Wind speed (km h^{-1})	Impact pressure of average raindrop of 0.5 g mass (cm head of water)
10	140
20	280
40	560
60	840
80	1120
100	1400
120	1680

particles of wax, and the types of emulsifying agent used result in air entrainment in the region of 4–5% by volume.

4.5 The effects of dampproofers on the properties of hardened concrete

4.5.1 Structural design parameters

(a) Compressive and tensile strength

Admixtures of this category do not significantly affect the strength of concrete at 28 days unless a water-reducing admixture has been included in the formulation when the resultant lowering of the water–cement ratio will give a higher strength according to Abram's law. Some results for a wax emulsion type [14] at constant cement content and giving slight air entrainment (approximately 40% by volume) are given in Table 4.6, whilst Table 4.7 gives data for a dry brick mix containing a stearic-acid-based material.

(b) Modulus of elasticity

There are no recorded data to indicate that materials of this type would alter the stiffness of the concrete into which they are incorporated. However, the fact that these materials are associated with the matrix/air interface, and not the cement hydrates themselves, would suggest that the physical properties of the bonding constituents of the hardened cement would remain unchanged.

Table 4.6 Effect of a wax emulsion type of dampproofer on the compressive strength of concrete of varying cement content

Mix type	Cement content $(kg\ m^{-3})$	Compressive strength $(N\ mm^{-3})$ at 28 days
Plain	450	64.9
3% wax emulsion on cement weight	450	62.3
Plain	400	53.6
3% wax emulsion on cement weight	400	54.0
Plain	360	44.0
3% wax emulsion on cement weight	360	42.4
Plain	315	33.5
3% wax emulsion on cement weight	315	33.5
Plain	270	21.1
3% wax emulsion on cement weight	270	22.4
Plain	225	Too harsh to prepare cubes
3% wax emulsion on cement weight	225	15.3

Table 4.7 Effect of a stearic-acid-based dampproofer on the compressive strength of concrete

Mix	% dampproofer on cement	Average compressive strength (N mm^{-2})	
		7 days	28 days
1	0	14.2	17.6
2	1	14.3	18.6
3	2	15.2	18.2

Mix proportions: cement : sand : crushed limestone = 1 : 4.0 : 6.5, water–cement = 0.45 : 1.

4.5.2 Durability aspects

It was explained earlier that materials of this type are added to concrete products to reduce the ingress of rain and ground water for aesthetic and damp-proofing reasons, rather than to prolong the structural capabilities of the construction. The improvements in aesthetic qualities are not short lived, and results [15] for a stearate-based composition over a 10-year period are given in Table 4.8.

The results given in Table 4.8 illustrate the following points:

1. It is necessary to add sufficient dampproofer so that the absorption of the surface is reduced to a negligible level to obtain the best results.
2. The presence of the dampproofing admixture at higher levels completely inhibits disfiguring algal growth.
3. The presence of even small amounts of dampproofer improve the freeze–thaw durability of the concrete [16]. This is further indicated by

Table 4.8 The relationship between the initial surface absorption test and durability of concrete containing various proportions of a stearic acid based dampproofer

% dampproofer wt/wt cement	Initial surface absorption at 28 days (BS 1881)			Description after 10 years as coping on a roof site
	10 min	30 min	1 h	
0	0.48	0.28	0.20	Broken up due to frost, blackened by algae
0.10	0.42	0.25	0.19	Darkened by algal growth
0.25	0.45	0.20	0.17	Slightly darkened by algal growth
0.50	0.43	0.19	0.13	Slightly dirty
1.00	0.33	0.14	0.12	Fairly clean
2.00	0.07	0.04	0.02	Pristine condition

freeze–thaw testing of concrete specimens containing the wax emulsion type of dampproofer shown in Fig. 4.7. In this case, however, some air entrainment and a reduction in the water–cement ratio was obtained which would contribute to the effect [5].

Volume changes such as shrinkage and creep are not significantly affected on single drying, as shown in Table 4.9 [5], but because subsequent moisture uptake is considerably reduced, the shrinkage under drying and wetting cycling (site conditions) will show a reduction.

In view of the way in which these materials function, no improvement in resistance to attack by aggressive gases, e.g. industrial atmospheres, is obtained, although the reduced absorption of aqueous media will improve resistance to attack by aggressive, but neutral media, such as sulfates and other salts. Indeed, the presence of materials such as stearates at high level can completely inhibit the corrosion of reinforcing steel against a fairly high level of chlorides [17] in the concrete. These hydrophobic materials are now

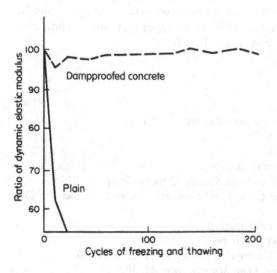

Mix	kg m^{-3} cement content	W/c ratio	Air (%)
Plain	285	0.70	0.9
Dampproofed	285	0.61	4.2

Fig. 4.7 The freeze–thaw resistance of plain concrete and concrete dampproofed with a wax emsulsion type of admixture.

Table 4.9 Drying shrinkage of concrete containing varying proportions of a wax-emulsion-based dampproofer

Specimen no.	Wax emulsion (%)	Drying shrinkage of concrete					
		3 day	7 day	14 day	28 day	56 day	91 day
1	0	1.05	2.13	2.87	4.68	6.98	8.45
2	1	0.96	2.10	2.65	4.66	6.82	8.14
3	2	0.98	2.06	2.47	4.79	6.42	8.00
4	3	0.75	2.09	2.88	4.61	6.61	7.92

being used in the formulation of corrosion inhibitors to get the dual benefit of reducing the ingress of chloride containing water into the concrete and anodic or cathodic protection of the embedded steel [18–21].

The dampproofing admixtures will, therefore, improve the aesthetic qualities of concrete in terms of maintenance of a clean appearance over a prolonged period of time without adverse effects on other properties, and in the areas of freeze–thaw resistance, shrinkage under wet–dry cycling and reinforcement protection, may contribute beneficially.

References

 1 Nurie, R.W. (1953). *Cement and Lime Manufacture*, **26**, 47–51.
 2 Australian Patent (1964). 271 527.
 3 British Patent (1976). 1434 924.
 4 Hewlett, P.C. Private communication.
 5 Anon. (1970). *Onada Technical Bulletin*, Onada Cement Company, Tokyo.
 6 Lea, F.M. (1956). *Chemistry of Cement and Concrete*, Chemical Publication Co. Inc., N.Y. 602–4,
 7 Tar Residuals Ltd. Private communication.
 8 Dennis, R.H. (1970). *Chemistry and Industry*, 377–80.
 9 Anon. (1965). *Admixtures for Concrete*, CUR Report No.31, 39–40.
10 Dory, D.M. (1969). *Cement and Lime Manufacture*, **42**, 107–14.
11 Bruere, G.M. (1974). *Cement and Concrete Research*, **4**, 557–66.
12 Nasser, K.W. (1973). *Behaviour of Concrete under Temperature Extremes*, ACI Publication SP 39, 139–48.
13 Wakenham, H. *et al.* (1945). *American Dyestuffs Report*, **18**, 178–82.
14 Brown, L.C. (1970). Paper presented at the 50th Anniversary Convention of NCMA, Texas.
15 Levitt, M. (1971). *British Journal of Non-Destructive Testing*, **12**, 106–12.
16 Shacklock, B.W. (1971). *Supplement to the Consulting Engineer*, **27**, 9–13.
17 Gouda, U.K. (1970). *British Corrosion Journal*, **5**, 204–8.

18 Bobrowski, G. *et al.* (1992) *Concrete International*, **4**, 65–8.
19 Krauss, P.D. and Nmai, C.K. (1994). *Proceedings of Third International Conference on Durability of Concrete*, Nice, France, 246–62.
20 Bobrowski, G. and Youn, D.J. (1993) *Proceedings of International Conference on Concrete 2000*, 1249–61. Scotland, UK.
21 Berke, N. *et al.* (1996) US Patent 5 527 388.

Chapter 5

Accelerators

5.1 Background and definitions

Concrete accelerators increase the rate of hardening of cement and concrete mixes. The major material used to obtain this effect, calcium chloride, has been used since 1885 [1] and finds application mainly in cold weather, when it allows the early strength gain to approach that of concrete cured under normal curing temperatures and, probably more importantly, reduces the setting time so that finishing operations can proceed without undue delay. In this way, shutter and mold stripping, lifting and handling of precast items and finishing of flatwork can proceed normally. In addition, the concrete is less liable to damage by early age freezing. In this latter respect it should be pointed out that calcium chloride is not an antifreeze in the sense of significantly lowering the freezing point of water in the mix, so that although the time required for protection is reduced, the standard procedures for protection should be followed for cold-weather concreting.

Although the use of accelerators is far greater in winter, they do find application under more normal conditions to speed up the setting and hardening process for earlier finishing or mold turn round.

There has been controversy over the use of calcium chloride in concrete containing embedded metal in view of the possibility of corrosion, particularly where the concrete is of a porous nature. Many countries have made provision in the relevant codes of practice to prevent or limit its use where steel reinforcement is present. This has renewed interest in 'chloride-free' accelerators as replacements for calcium chloride in reinforced concrete. However, calcium chloride remains a most effective material for use in unreinforced concrete for economic production under winter conditions and its effects on concrete, whether beneficial or undesirable, are well researched and quantified. In some areas the newer non-chloride materials, although shown to reduce the likelihood of reinforcement corrosion, have not been widely studied and their other effects on concrete are less known.

The fact remains that calcium chloride has been widely used as an accelerator for plain unreinforced concrete and this area of application, which in accounts for over 60% of calcium chloride usage will continue in the future.

5.2 The chemistry of accelerators

This category of admixture is based mainly on the major raw materials, calcium chloride, calcium nitrate, calcium formate [2] and calcium thiocyanate, with minor amounts of other materials occasionally being included in the formulations, such as calcium thiosulfate [3] and triethanolamine (TEA). TEA is not normally used alone but because it is sometimes used in other categories of admixture to compensate for retarding influences it will be included in this section.

5.2.1 Calcium chloride

Calcium chloride ($CaCl_2$) is produced as a by-product in the Solvay process for sodium carbonate manufacture. The overall process involved is:

$$CaCO_3 + 2NaCl \rightarrow Na_2CO_3 + CaCl_2$$
limestone   brine solution

It is obtained either as a liquor or as flake material of approximately 20% moisture content, and for use as an accelerating admixture is normally supplied as a 33–35% solution.

5.2.2 Calcium formate

Calcium formate ($Ca(HCOO)_2$) is produced as a by-product in the manufacture of a polyhydric alcohol, pentaerythritol:

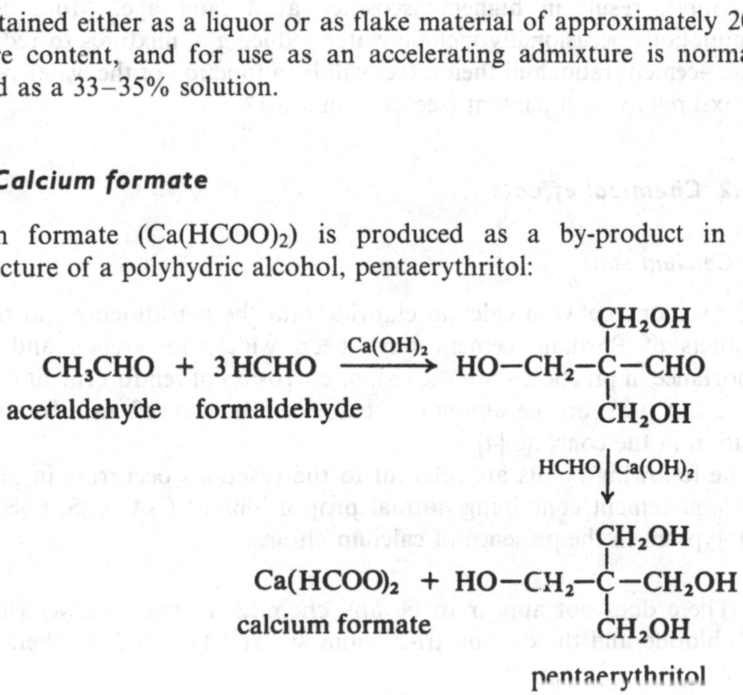

$$CH_3CHO + 3HCHO \xrightarrow{Ca(OH)_2} HO-CH_2-\underset{\underset{CH_2OH}{|}}{\overset{\overset{CH_2OH}{|}}{C}}-CHO$$

acetaldehyde   formaldehyde

$$\downarrow HCHO \,|\, Ca(OH)_2$$

$$Ca(HCOO)_2 + HO-CH_2-\underset{\underset{CH_2OH}{|}}{\overset{\overset{CH_2OH}{|}}{C}}-CH_2OH$$

calcium formate

pentaerythritol

It is obtained as a fine powder and is supplied normally in this form as an accelerating admixture because of its limited solubility in water (about 15% at normal room temperature).

5.2.3 Triethanolamine

Triethanolamine ($N(C_2H_4OH)_3$) is an oily, water-soluble liquid with a 'fishy' odor and is produced by the reaction between ammonia and ethylene oxide:

$$NH_3 + 3\,CH_2\!\!-\!\!CH_2 \longrightarrow N(CH_2CH_2OH)_3$$

It is normally used as a component in other admixture formulations and rarely, if ever, as a sole ingredient.

5.3 The effects of accelerators on the water–cement system

5.3.1 Rheological effects

Most admixtures of this type do not significantly alter the rheology of cement pastes at early ages. The quicker stiffening of accelerated pastes will, of course, result in higher viscosities at a later age. More complex formulations occasionally include water-reducing admixtures to reduce the water–cement ratio, and their effect will be a function of the water-reducing admixture type and content (see Section 1.3.1).

5.3.2 Chemical effects

(a) Calcium salts

The reactions between calcium chloride and the constituents and reaction products of Portland cement have been widely researched and are of importance in practice, since the risk of corrosion of reinforcement depends, at least in part, on the amount of chloride which is left in a free state in solution in the concrete [4].

The following points are relevant to the reactions occurring in pastes of Portland cement containing normal proportions of C_3A, C_3S, C_2S, C_4AF and gypsum in the presence of calcium chloride.

1. There does not appear to be any chemical reaction between calcium chloride and the di- and tri-calcium silicates [5] although their rate of reaction is increased.

2.   Calcium chloride does not react significantly with cement pastes for a period of 2–6 h [1, 5] after mixing, although rapid setting can occur in this period.

3.   The free calcium chloride concentration, after the initial period described in (2), progressively drops to almost nil.

4.   The formation of new reaction products between C_3A, gypsum, and calcium chloride and the disappearance of initial products, has been studied [5, 6]; results are shown schematically in Fig. 5.1 and can be summarized as follows:

(a)   At the first contact between calcium chloride solution and cement particles, both gypsum and chloride react to form very small amounts of calcium trisulfoaluminate (ettringite) and calcium chloroaluminate respectively:

$$C_3A + 3CaSO_4 + 32H_2O \rightarrow C_3A.3CaSO_4.32H_2O$$

$$C_3A + CaCl_2 + 10H_2O \rightarrow C_3A.CaCl_2.10H_2O$$

(b)   The gypsum continues to react to form ettringite whilst the calcium chloride does not react, but can continue to promote the silicate hydration:

$$C_3A + 3CaSO_4 + 32H_2O \rightarrow C_3A.3CaSO_4.32H_2O$$

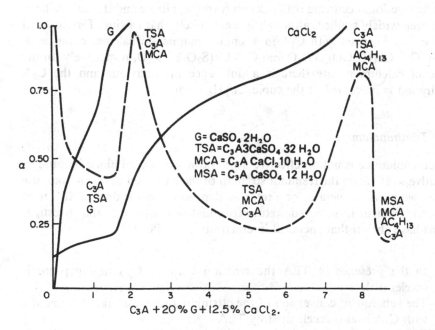

Fig. 5.1 The composition of a paste consisting of C_3A and gypsum in the presence of calcium chloride at various times (Tenoutasse).

(c) When all the gypsum has been consumed, the calcium chloride begins to react again with the C_3A until the chloride in solution is reduced to about nil:

$$CaCl_2 + C_3A + 10H_2O \rightarrow C_3A.CaCl_2.10H_2O$$

There is some uncertainty as to the exact nature of the chloroaluminate formed and although it is likely that the mono-chloroaluminate is predominantly formed, it is possible that some trichloroaluminate is also present as a reaction product.

(d) When the chloride ion has been removed, the remaining C_3A is hydrated to give $C_3A.Ca(OH)_2.12 H_2O$, which then converts the trisulfoaluminate to the monosulfoaluminate:

$$C_3A.3CaSO_4.32H_2O + 2C_3A.Ca(OH)_2.12H_2O \rightarrow$$

$$3C_3A.CaSO_4.12H_2O + 20H_2O + 2Ca(OH)_2$$

The final composition, therefore, consists of: $C_3A.Ca(OH)_2.12$ $H_2O, C_3A.CaCl_2.10H_2O$, and $C_3A.CaSO_4.12H_2O$ in solid solution.

(e) There is also some indication that a reaction can occur between lime and calcium chloride to form $3CaO.CaCl_2.16H_2O$ [7] particularly at low temperatures.

The reactions occurring with calcium formate, nitrate and thiosulfate have not been widely studied, although it seems likely that calcium formate and thiosulfate [6] react with C_3A in a similar manner to calcium chloride to form $C_3A.Ca(HCOO)_2.xH_2O$ and $C_3A.Ca(S_2O_3)_2.yH_2O$, respectively. In the case of calcium nitrate there is a difference in behaviour and the C_3A hydration is promoted to the cubic C_3AH_6 form.

(b) Triethanolamine

Triethanolamine is not an effective accelerator when used alone because of its adverse effect on the resultant strength of the hardened paste and because it can act as an accelerator or a retarder depending on the dosage. It is used as an ingredient in some admixture formulations, however, and investigations have shown that chemical interactions occur [8, 9].

1. In the presence of TEA the reaction between C_3A and gypsum is accelerated.
2. The subsequent conversion of the ettringite to monosulfate by reaction with C_3A is also accelerated by TEA.
3. The formation of the hexagonal aluminate hydrate and conversion to the cubic form is accelerated by TEA.

4. There may be a chelating effect whereby TEA reacts with the ferrite phase of Portland cement [10], as illustrated in Fig. 5.2.
5. There is some evidence for the formation of a surface complex between C_2S and C_3S initial hydrates and TEA.

5.3.3 Effects on cement hydration

(a) Kinetics of reaction

Conduction calorimetric curves of Portland cement hydrated isothermally containing various quantities of triethanolamine are shown in Fig. 5.3 [8]. On initial contact with water each sample evolves heat (not shown in figure) that can be attributed to heat of wetting, hydration of free lime and reaction of C_3A with gypsum to form ettringite. The amount of heat developed increases with the quantity of TEA indicating that the C_3A + gypsum reaction is accelerated.

The second peak, occurring after 9–10 h in a plain cement paste, is mainly due to C_3S hydration and in the presence of TEA is extended, suggesting that the C_3S hydration is retarded, particularly at high amounts of TEA.

Similar isothermal calorimetric curves for calcium chloride, formate [7], bromide and nitrate [11, 12] are given in Figs 5.4, 5.5 and 5.6, which indicate that:

1. Calcium chloride is a more effective accelerator than the other materials.
2. All the materials accelerate reaction of C_3S phase hydration, which is the main strength contributing component of Portland cement.
3. The C_3A phase reactions are either diminished or retarded beyond the area of the curves.

Other data [13] has indicated at constant weight addition, the various calcium salts have the following effectiveness in accelerating the hydration of C_3S:

Bromide > chloride > thiocyanate > iodide > nitrate > perchlorate

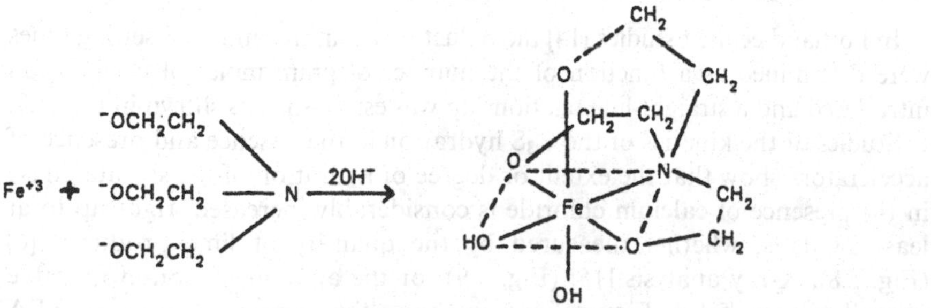

Fig. 5.2 Chelation (solubilization) of the ferric ion in Portland cement by triethanolamine.

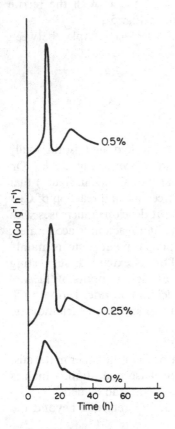

Fig. 5.3 Conduction calorimetric curves of cement hydrated in the presence of triethanolamine (Ramachandran).

This same study also showed the influence of the cation in effecting C_3S hydration and the following order was established:

$$Ca^{++} > Sr^{++} > Ba^{++} > Li^+ > Na^+ > K^+$$

In Portland cement studies [14] the reductions in initial and final setting times were determined as a function of the number of gram moles of calcium ion introduced and a straight-line relationship was established as shown in Fig. 5.7.

Studies of the kinetics of the C_3S hydration in the absence and presence of accelerators show that the extent or degree of hydration of the silicate phase in the presence of calcium chloride is considerably increased, right up to at least 28 days, whether measured by the quantity of lime produced [6] (Fig. 5.8), X-ray analysis [15] (Fig. 5.9), or the amount of non-evaporable water [16] (Fig. 5.10). Figure 5.8 also shows that a small amount of TEA retards the hydration of the C_3S phase for a considerable time, and the trend

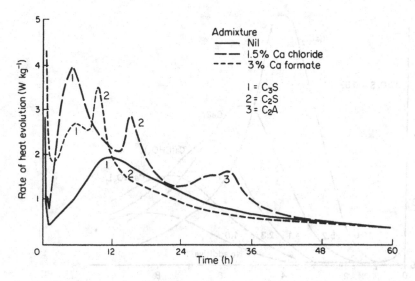

Fig. 5.4 Conduction calorimetric curves of plain and calcium chloride and formate containing pastes (Edmeades).

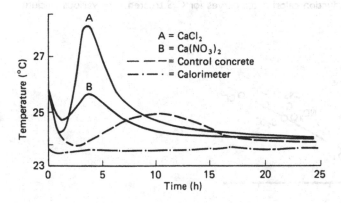

Fig. 5.5 Conduction calorimetric curves of plain and calcium and nitrate containing pastes (Edwards).

of the curves would suggest that the presence of TEA will result in permanently partly hydrated C_3S.

(b) Composition and morphology of resultant hydrates

It has already been shown that the presence of accelerating admixtures produces hydration products of a different type to those from a plain cement paste because of chemical involvement, predominantly with the C_3A phase

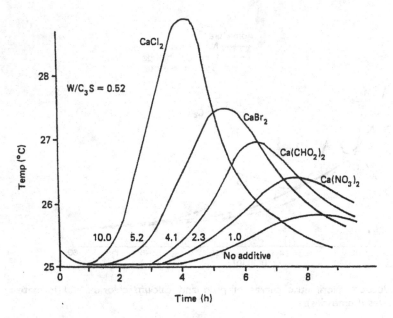

Fig. 5.6 Isothermal conduction calorimeter curves for C₃S treated with various calcium salts.

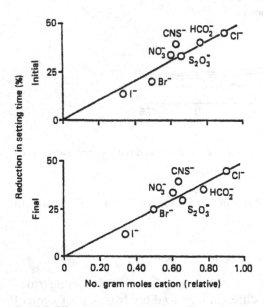

Fig. 5.7 Effect of various calcium salts on the time of setting of Portland cement pastes.

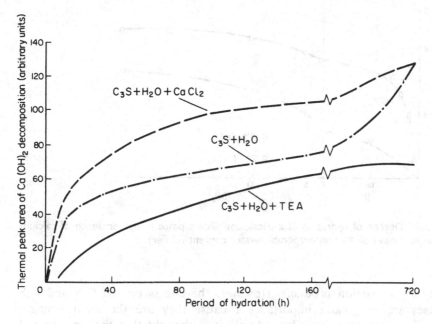

Fig. 5.8 The rate of reaction of tricalcium silicate in the presence of calcium chloride and triethanolamine, measured by line production (Ramachandran).

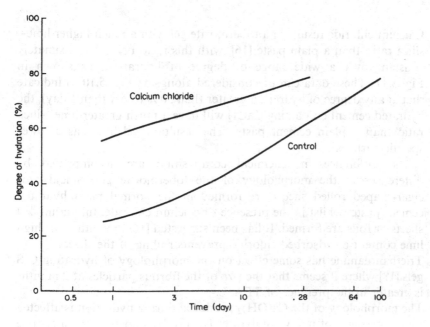

Fig. 5.9 The degree of hydration of cement pastes in the presence of calcium chloride in comparison to a plain paste; measured by X-ray analysis (Young).

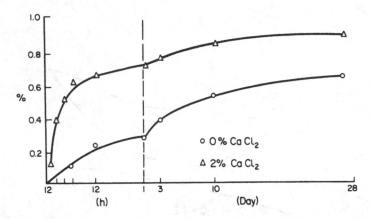

Fig. 5.10 Degree of hydration of a tricalcium silicate paste in the presence of calcium chloride, measured by non-evaporable water content (Odler).

and/or its reaction products. However, the effects on the C_2S and C_3S phases are of greater importance because they are the main strength contributing components. In addition, it is thought that the amount and type of $Ca(OH)_2$ produced, may have some effect on structural properties [17] of the hardened paste. The following observations have been made:

1. Calcium chloride results in a tobermorite gel with a much higher lime–silica ratio than a plain paste [16], with this ratio being approximately constant over a wide range of degree of hydration, as shown in Fig. 5.11. These data can be considered alongside Fig. 5.10 to indicate that at any degree of hydration greater than 30% (older than 1 day), the hydrated cement containing $CaCl_2$ will have a much greater lime–silica ratio than a plain cement paste. The resultant gel also has a lower specific surface.

     The differences in chemical composition are accompanied by differences in the morphology of the tobermorite gel. Spicular or cigar-shaped rolled sheets are formed in the normal plain hydrated cement paste, whilst in the presence of calcium chloride, thin crumpled sheets or foils are formed. It has been suggested [16] that either the high lime content or adsorbed chloride prevents rolling of the sheets.

2. Triethanolamine has some effect on the morphology of hydrating C_3S gels [18] where it seems that the size of the fibrous particles at 2 months is greater in the presence of TEA.

3. The morphology of the $Ca(OH)_2$ produced during hydration is affected by the presence of many admixtures [15,17]. It is possible to categorize

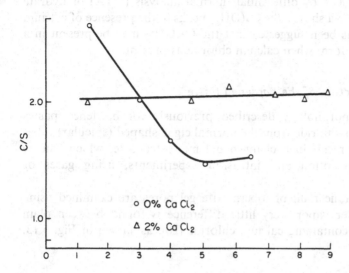

Fig. 5.11 The lime – silica ratio of hydration products of tricalcium silicate hydrating in the presence of 0 and 2% calcium chloride as a function of the degree of hydration (Odler).

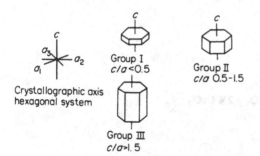

Fig. 5.12 Classification of calcium hydroxide morphology based on a c to a crystallographic axis ratio (Berger).

the type of morphology as in Fig. 5.12. Calcium formate does not appear to affect the morphology of the $Ca(OH)_2$ produced in comparison to a plain C_3S paste which falls into group II of the classification. Calcium chloride, however, changes the form to group III.

The number of $Ca(OH)_2$ crystals formed in a given volume of paste have also been studied and, although many admixtures can have a considerable effect, neither calcium chloride nor calcium formate have a significant effect.

The peaks produced by differential thermal analysis (DTA) of hydrate C_3S samples indicate a shift in the $Ca(OH)_2$ peaks in the presence of calcium chloride [19]. It has been suggested that the $Ca(OH)_2$ may be present in a differently bonded form when calcium chloride is present.

(c) The microstructure of the hardened paste

The changes in morphology described previously for hardened pastes containing calcium chloride from the normal cigar-shaped (spicular) tubes to crumpled foils result in a changed gel microstructure, which can be studied using adsorption and intrusion experiments, using gases or mercury.

When the hydraulic radii of tobermorite gel pores are examined using adsorption of water vapor, very little difference is found between plain pastes and pastes containing calcium chloride [20], as shown in Fig. 5.13.

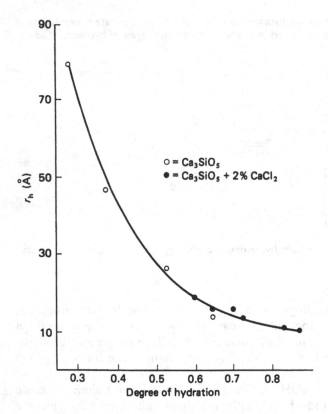

Fig. 5.13 Hydraulic radius of hydrated Ca_3SiO_5 calculated from the adsorption side of water vapour isotherms (Skalny).

However, nitrogen adsorption reveals considerable differences, as shown in Fig. 5.14, in terms of surface area distribution of a 28-day hydrated C_3S specimen [20], or of specific surface area as a function of hydration time [21] in Fig. 5.15.

This difference between H_2O and N_2 adsorption data has been attributed to either the accessibility of water to interlayer spaces in the tobermorite gel or to the presence of 'ink bottle' pores with narrow necks and wide bodies. The considerable increase (4–5 times) in pore surface and pore volume available to nitrogen in pastes containing calcium chloride suggests that the crumpled pore type of morphology is more open than the spicular type [20].

The type of data produced by H_2O and N_2 adsorption is relevant to gel pores having radii up to about 50 Å, but larger pores and capillaries exist in the hardened cement paste and probably are more significant in determining the porosity or permeability of the hardened paste in concrete to gases and liquids.

Mercury intrusion data for larger pores of 6.5×10^{-3} to 10 µm are shown in Figs 5.16 and 5.17 at various hydration times and at equal hydration, respectively [22]. It can be seen that the total intrusion at the limit of the instrument (6.5×10^{-3} µm) is less for chloride containing pastes at all ages, but when considered at equal degrees of hydration a higher proportion of coarse pores is formed.

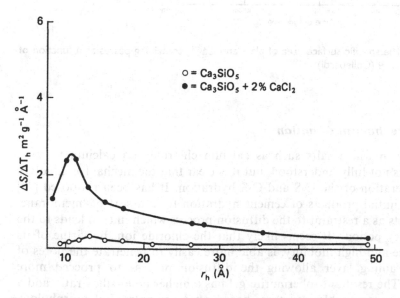

Fig. 5.14 Surface area distribution of a 28-day hydrated Ca_3SiO_5 sample measured by nitrogen adsorption (Skalny).

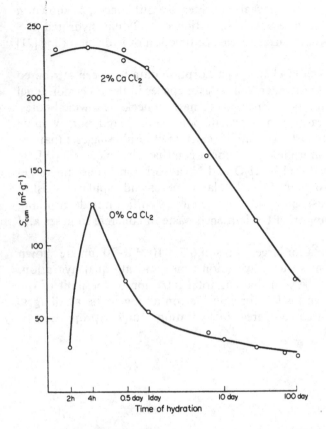

Fig. 5.15 The specific surface area of plain and CaCl$_2$ containing pastes as a function of hydration time (Collepardi).

5.3.4 Mechanism of action

The way in which salts such as calcium chloride and calcium formate operate is not fully understood, but it is clear that the mechanism involves an acceleration of the C$_2$S and C$_3$S hydration. It has been proposed [23] that the initial products of cement hydration form a sort of 'membrane' which acts as a restraint to the diffusion process which in turn leads to the 'dormancy period'. It seems likely that the chloride ion, by virtue of its small size and high mobility, is able more easily to penetrate the pores of the restraining layer allowing the diffusion process to proceed more rapidly. The resultant tobermorite gel has a higher lime–silica ratio and a more open, accessible structure, based on a crumpled foil morphology rather than the usual spicular. The considerable reaction with, and

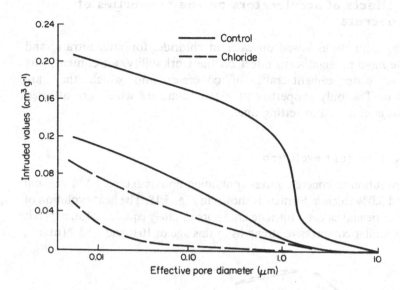

Fig. 5.16 Mercury intrusion porosimetry curves of C₃S pastes showing differences in capillary porosity distribution.

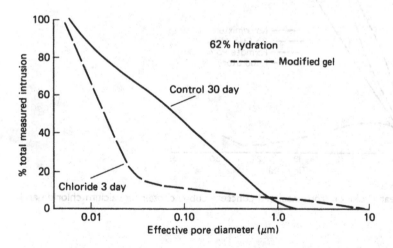

Fig. 5.17 Porosity distribution curves at equal hydration with intruded volume plotted as a percentage of total intrusion (Young).

modification to, the C_3A hydration is not relevant to the acceleration process.

Triethanolamine accelerates the reactions associated with the C_3A hydration, but retards, possibly permanently, the C_3S hydration, which is reflected in reduced ultimate strength.

5.4 The effects of accelerators on the properties of plastic concrete

Accelerating admixtures based on calcium chloride, formate, nitrate, and thiocyanate have no significant effect on the workability, air content, mix stability, or water–cement ratio of concretes into which they are incorporated. The only properties of plastic concrete which are affected are the heat evolution and setting time.

5.4.1 Effect on heat evolution

The heat evolution of concrete mixes containing no admixture, 1.5% calcium chloride and 3.0% calcium formate is shown in Fig. 5.18. The heat evolution of calcium chloride and calcium formate are approximately equal at 24 h, which is reflected in similar compressive strengths at this age of 10.0 and 12.5 N mm^{-1},

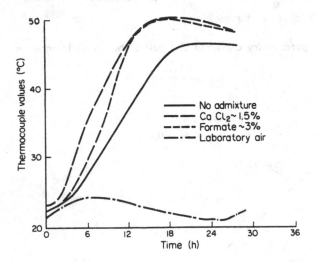

Fig. 5.18 Heat evolution from insulated concrete cubes containing calcium chloride and formate.

Table 5.1 The effect of calcium chloride and formate on mortar stiffening times

Accelerating admixture type	Stiffening time (h) for penetration resistance of 0.5 N mm^{-2}	3.5 N mm^{-2}
None (control)	$3\frac{1}{4}$	5
3.2% calcium chloride	1	$1\frac{3}{4}$
2.0% calcium formate	$2\frac{1}{4}$	$3\frac{1}{4}$

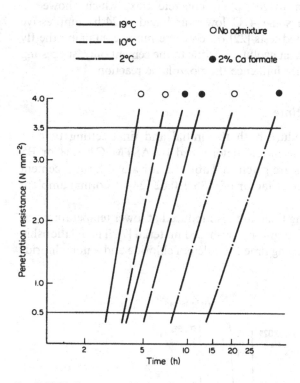

Fig. 5.19 Stiffening times of concrete with and without calcium formate accelerator (to BS 5075) (Edmeades).

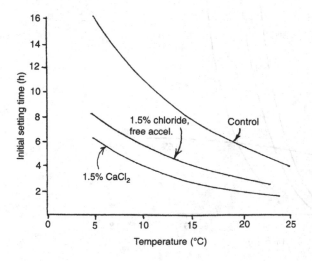

Fig. 5.20 Effect of temperature on setting time with accelerators.

respectively, in comparison to the plain concrete mix, which showed a maximum temperature rise some 4°C lower and had a 24 h compressive strength of 5.5 mm^{-2}. Limited tests [22] on concrete mixes containing the fly ash, indicate that increased heat evolution is due to the cement only, suggesting that calcium chloride may not influence the pozzolanic reaction.

5.4.2 Effect on setting time

Accelerating admixtures reduce both the initial and final setting time of mortar sieved from concrete mixes, determined by ASTM C403:68 or BS 5075 (1975). Typical results are given in Table 5.1 for a 300 kg m^{-3} cement content mix with a compacting factor of 0.85 ± 0.02 [24] at normal ambient temperature.

The initial and final setting times are also reduced at lower temperatures, as shown in Fig. 5.19 for a calcium-formate-based material [7]. The relationship between temperature and setting time for calcium chloride and a non-chloride

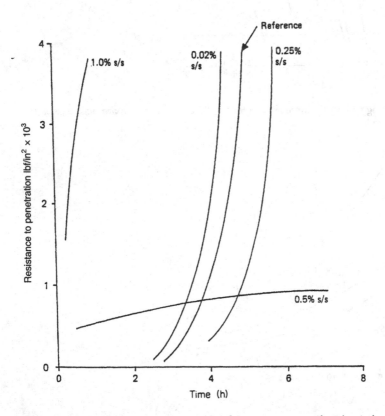

Fig. 5.21 Time of setting characteristics of concretes treated with triethanolamine.

accelerator (calcium nitrate) is shown in Fig. 5.20, where it can be seen that calcium chloride can compensate for a 15°C drop in temperature and the non-chloride accelerator can compensate for a 10°C drop [25].

The composition of the Portland cement in the concrete can also influence the effectiveness of both calcium chloride and calcium nitrate and a regression analysis [26] for data for 10 cements produced the equation

$$\text{Setting time (minutes)} = 1349 - [28\%C_4AF] - [1.8 \text{ cm}^2 \text{ g}^{-1} \text{ (Blaine)}]$$
$$- [13\%C_3A] - [84\%Ca(NO_3)_2] - [140\%CaCl_2]$$

Triethanolamine is very dosage sensitive, as shown in Fig. 5.21, where TEA at 0.02% has an accelerating effect, at 0.25% is a mild retarder, at 0.5% is a strong retarder, and at 1.0% causes flash set [27].

5.5 The effects of accelerators on the properties of hardened concrete

5.5.1 Structural design parameters

(a) Compressive strength

The compressive strength of concrete containing an admixture of the calcium chloride type will have a higher 28-day compressive strength than a plain concrete mix having the same mix composition. This is true of concretes cured over a range of temperatures [22], as shown in Fig. 5.22. The difference is greater at lower temperatures and Table 5.2 summarizes the increases at 28 days that can be expected; these are largely independent of cement content. While other accelerator types have not been as widely researched, the data available for calcium nitrite [28] and calcium formate [29] indicate that 28-day strengths of concrete incorporating these materials are generally higher, as shown in Tables 5.3 and 5.4.

(c) Flexural and tensile strength and modulus of elasticity

The data available on the effect of accelerators on the flexural and tensile strengths and the modulus of elasticity are limited [22, 30, 31] and are relevant only to calcium chloride and triethanolamine. The data generally point to either no effect or a slight reduction in all three properties. Also, in view of the increase in compressive strength, it is reasonable to assume that for a given compressive strength, the presence of calcium chloride or triethanolamine will reduce the flexural and tensile strengths and the modulus of elasticity.

5.5.2 Durability aspects

(a) Permeability

The permeability and porosity of concrete containing calcium chloride in relation to a plain concrete depends on two conflicting variables:

1. The degree of hydration of the concrete, which in the case of the calcium-chloride-containing concrete will initially be considerably increased, and the larger volume of hydration products will lead to a reduced permeability.
2. The adverse effect that calcium chloride has on the capillary porosity distribution. At later ages (after perhaps 1 year) the degree of hydration in both calcium-chloride-containing concrete and a plain concrete will be similar and, under these circumstances, the concrete will be more porous and allow easier access to aggressive gases and liquids. Pore volume distributions of cement pastes at an advanced state of hydration of more than 90% [32] are shown in Fig. 5.23, where the greater porosity is indicated and the data can be used to calculate an increase in average hydraulic radius from 28 Å (2.8 nm) in the chloride-free sample, to 108 Å (10.8 nm) in the chloride-containing sample.

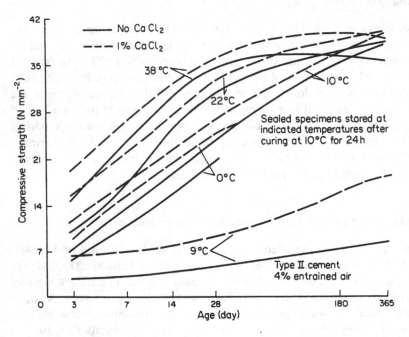

Fig. 5.22 1% CaCl₂ increases the compressive strength of concrete stored at various temperatures (Shideler).

Table 5.2 The increase in compressive strength of concrete containing calcium chloride is greater at lower temperatures

Curing temperature (°C)	Increase in compressive strength at 28 days over plain concrete (%)
−10	90
0	25
10	16
20	12
40	7

Table 5.3 The effect of calcium nitrite on strength development

Admixture (%)	Compressive strength (N mm^2)		
	1 day	7 days	28 days
0	9.0	23.5	34.7
2	11.1	31.3	39.5
3	13.5	34.2	40.7
4	15.8	36.8	44.0
5	16.3	36.7	44.8

Table 5.4 Influence of calcium formate on the compressive strength of concrete

Calcium formate	Compressive strength (% of reference)			
	1 day	7 days	28 days	1 year
1.0	136	123	115	110
2.0	152	131	107	103

(b) Resistance to aggressive environment

The resistance of concrete containing calcium chloride to attack by aqueous sulfate is reduced [22, 31, 33]. A comprehensive study of concrete over a 5-year period [24] using various cements and cement content stored in high-sulfate-containing water, gave the results shown in Figs 5.24 to 5.29, for which the following conclusions can be reached:

1. Calcium chloride at 2% addition level contributed to a lowering of sulfate resistance in almost all cases.

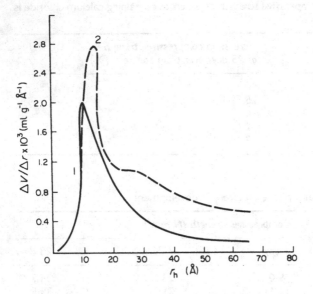

Fig. 5.23 Pore volume distribution of cement pastes from nitrogen adsorption. Curve 1 = cement with no admixture; curve 2 = cement paste and 2% CaCl$_2$ (Gouda).

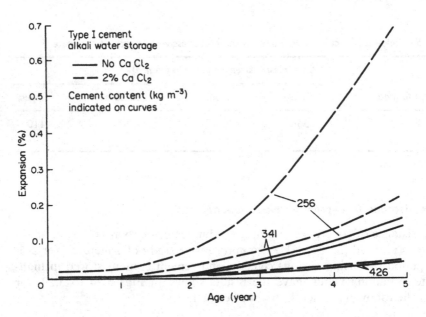

Fig. 5.24 CaCl$_2$ decreases the resistance of concrete to sulfate attack (Shideler).

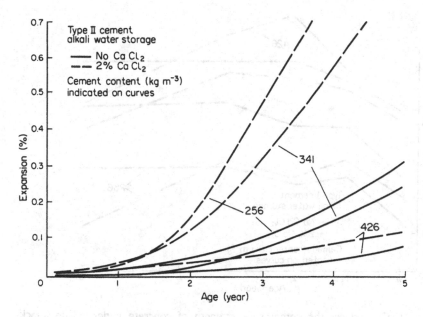

Fig. 5.25 CaCl$_2$ decreases the resistance to attack, particularly in lean concrete (Shideler).

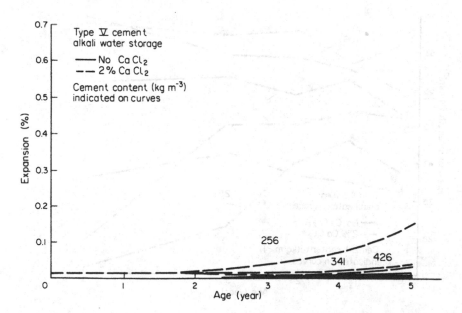

Fig. 5.26 CaCl$_2$ is least detrimental to sulfate resistance with Type V cement (low C$_3$A) (Shideler).

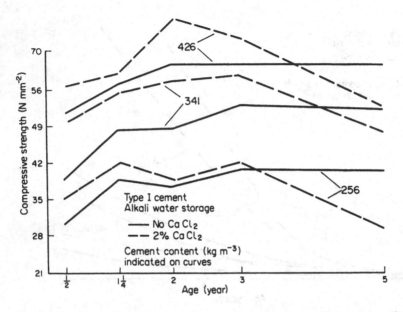

Fig. 5.27 CaCl₂ reduces the compressive strength of concrete under sulfate attack (Shideler).

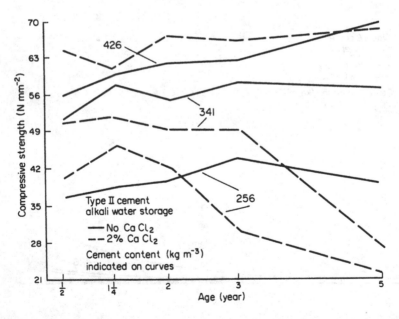

Fig. 5.28 CaCl₂ reduces the compressive strength of lean concrete under sulfate attack (Shideler).

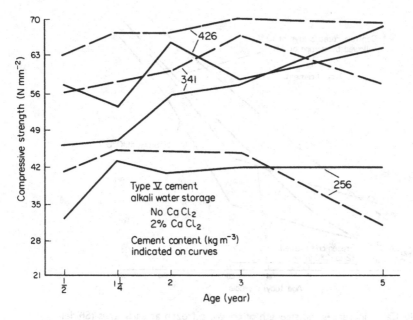

Fig. 5.29 CaCl₂ has little effect on sulfate resistance with Type V cement (low C₃A) (Shideler).

2. The higher C_3A type I and II cements were more adversely affected than the low C_3A type V cement.
3. The low-cement-content mixes were drastically affected with regard to both expansion and compressive strength reduction.
4. Only mixes containing 426 kg m^{-3} low-C_3A cement showed unaffected resistance to sulfate attack over a 5-year period.

(c) Resistance to freeze–thaw conditions

Concrete containing calcium chloride develops strength more rapidly and, therefore, has a greater resistance to damage by freezing at an early age, as shown in Fig. 5.30. There is some indication, however, that at later ages, the more mature concrete is less resistant to freeze–thaw cycling [22, 33].

(d) Protection of steel reinforcement

The effect that accelerators have on the role of concrete in providing protection against the corrosion of steel reinforcement has been the subject of several investigations and considerable controversy. Many studies have shown that the factors below are relevant to the discussion:

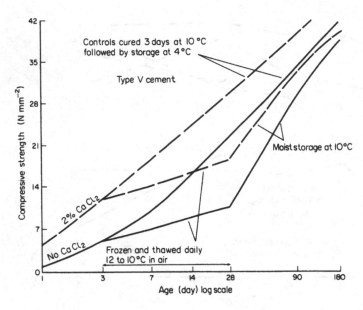

Fig. 5.30 CaCl₂ increases the strength of concrete frozen at early ages (Shideler).

1. The porosity of the concrete at an advanced state of maturity is increased in the presence of calcium chloride and, therefore, will allow a greater opportunity for air and moisture to come into contact with the steel reinforcement, encouraging corrosive effects. In practice, with reinforcement cover meeting the relevant codes of practice, this effect is regarded as of minimal significance.

2. The presence of calcium chloride at concentrations greater than about 1.5% by weight of cement can lead to breakdown of the passive layer of Fe_2O_3 normally present at the steel/concrete interface, rendering the reinforcement more susceptible to corrosion. This does not appear to occur with either calcium-formate or nitrate-based materials and can be shown by recording the potential of a steel/concrete electrode relative to that of a saturated calomel electrode at a constant current density as a function of time. A typical circuit [34] is shown in Fig. 5.31, and results for calcium chloride and a calcium-formate-based material [35] are shown in Fig. 5.32 where the breakdown of the passive layer is seen for 3% calcium chloride. There is some evidence that the presence of chlorides not only renders the steel more liable to corrosion attack, but also alters the crystallographic nature of the initial corrosion products from the normal orthorhombic form of Fe_2O_3 to a tetragonal form [36].

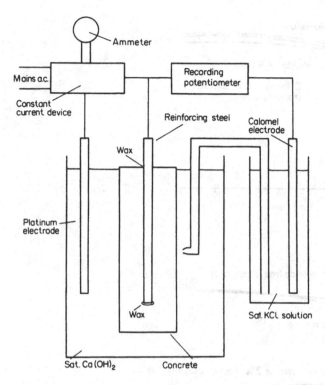

Fig. 5.31 Typical circuit for recording the potential of a steel/concrete electrode relative to that of a saturated calomel electrode (Gouda).

3. Visual inspection of steel surfaces which have been embedded in concrete and then broken open indicates that the chloride ion has a chemical action on the steel which, combined with the local electro-chemical cell action imposed by the distribution of narrow and wide pores over the steel surface and local concentrations of air over the surface, causes a 'pitting' type of corrosion of the steel surface. Table 5.5 gives results for the corrosion of narrow steel wires embedded in concrete for periods of 2 years in terms of the maximum depth of pits, the mean depth of the three deepest pits and the average number of pits for each group of embedded wires [37]. It can be seen that in concretes containing ordinary Portland cement, only slight pitting occurs when 2% by weight of calcium chloride is included, but severe pitting occurs with a 5% addition. In mixes containing sulfate-resistant cement, pitting is considerable even at 2% addition level.
4. Any corrosive attack of normal diameter reinforcing steel is not usually enough to cause significant reductions in the tensile strength of the steel

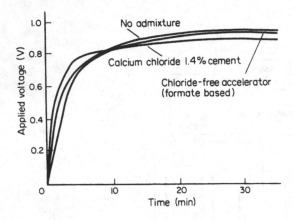

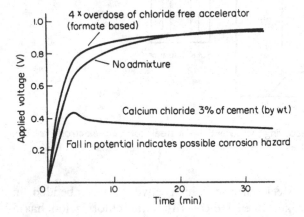

Fig. 5.32 Potential–time curves for plain concretes and concretes containing calcium-chloride and calcium-formate-based accelerators (McCurrich).

in the early stages. The problem lies in the spalling of concrete away from the reinforcement because of the expansive nature of the products of corrosion. The resultant exposure of the reinforcement allows further corrosion which could reduce the tensile strength of the steel sufficiently to cause structural failure.

5. The presence of an excessive number of voids in the concrete [38] or porosity due to poor compaction or a deficiency of fine aggregate [39] leads to an increase in the amount of reinforcement corrosion.

6. The bond between concrete and reinforcing steel appears to be greater in concrete containing calcium chloride and increases with calcium

Table 5.5 Effect of CaCl₂ on corrosion of high tensile steel wires in normally cured prestressed concrete specimens stored outdoors

Cement	Flake CaCl$_2$ (%)	Pitting values* on wires cleaned in 'Clarke's solution'		
		6 months	12 months	24 months
Ordinary Portland	0	<2.5: —: —	2.5: —: —	2.5: 2.5: —
	2	5.1: 5.1: —	17.8: 17.8: 2.5	17.8: 10.2: 2.5
	5	50.8:38.1:22.9	99.1: 73.7:15.2	127.0:121.9:25.4
Sulfate-resisting Portland	2	86.4:58.4: 5.1	76.2: 71.1:10.2	101.6: 81.3:17.8
	5	53.3:48.3:35.6	116.8:101.6:17.8(1)	127.0:109.2:22.9(3)

* Pitting values are expressed as maximum depth in mm $\times 10^{-2}$: mean depth in mm $\times 10^{-2}$ of the three deepest pits: average number of pits per wire for the group of eight wires in each concrete specimen. Figures in brackets refers to numbers of breaks in wires due to corrosion. Wires were prestressed and 2.03 mm in diameter.

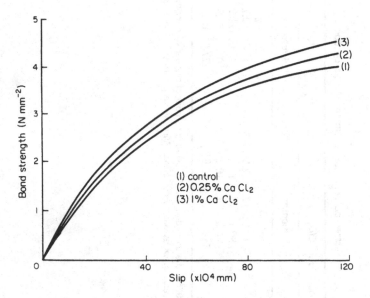

Fig. 5.33 Bond strength against slip for plain concrete and concrete containing calcium chloride (Kondo).

chloride concentration, as shown in Fig. 5.33 [40], which may be a function of the higher compressive strength and shrinkage of the accelerator containing concrete.

7. Earlier work on thiocyanate-based accelerators [41] had indicated that this material increased the potential for corrosion of reinforcing steel. Several subsequent studies [42–46] have shown that there is a threshold level of between 0.5 and 0.7% thiocyanate ion by weight of cement below which corrosion will not occur.

The general conclusion can be reached that in well-designed properly compacted concrete, the addition of up to 1.5% by weight of cement of calcium chloride and 0.5% of thiocyanate ion can be used without any significant detrimental effect on embedded ferrous metals.

Excessive dosages of calcium chloride, particularly in concrete of high porosity, can lead to accelerated corrosion with subsequent spalling and possible structural failure. Recent changes in codes of practice preventing the use of calcium chloride in concrete containing embedded metal are based on the difficulties of controlling addition levels under site conditions and the possibility of porous or poorly compacted concrete being produced.

There is very limited information available on the effect of calcium formate and nitrate but certainly the passive layer at the concrete/steel

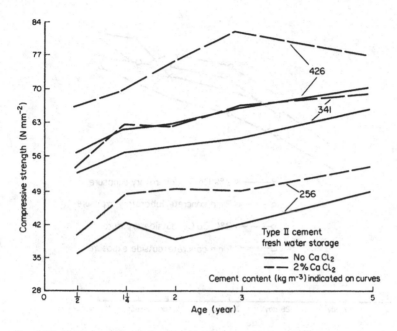

Fig. 5.34 Calcium chloride does not adversely affect the strength of concrete up to the 5 years studied (Shideler).

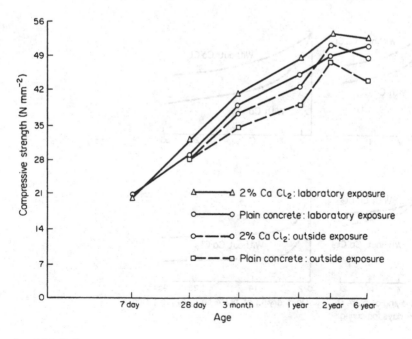

Fig. 5.35 Calcium chloride does not affect the long-term strength development of dense concrete cured under laboratory conditions, or with outside exposure (Blenkinsop).

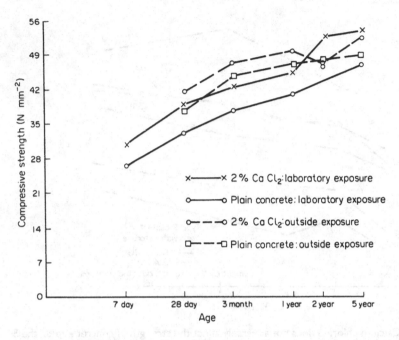

Fig. 5.36 Calcium chloride does not affect the long-term strength development of porous concrete cured under laboratory conditions, or with outside exposure (Blenkinsop).

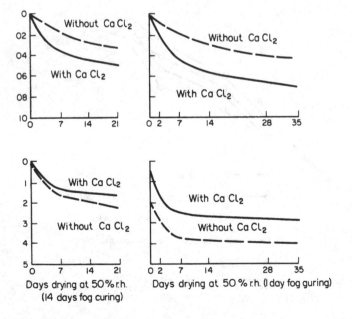

Fig. 5.37 CaCl₂ increases drying shrinkage of concrete, although the moisture loss is decreased (Shideler).

Table 5.6 Drying shrinkage of concretes containing calcium chloride, triethanolamine and calcium formate

Admixture	Concentration (% by weight of cement)	Water–cement ratio	Cement content (kg m⁻³)	Slump (mm)	Drying shrinkage (%) at (day)					
					7	14	28	56	84	168
None*	—	0.60	305	65	0.013	0.020	0.031	0.043	0.049	0.056
Calcium chloride*	2.0	0.60	305	60	0.023	0.033	0.045	0.056	0.062	0.068
Triethanolamine*	0.033	0.60	305	65	0.016	0.029	0.040	0.053	0.056	0.064
None†	—	0.60	305	65	0.016	0.018	0.027	0.039	0.045	0.052
Calcium formate†	3.00	0.60	305	65	0.015	0.025	0.036	0.048	0.054	0.056

* Average of mixes containing five different cements.
† One mix only.

interface does not appear to be attacked. Calcium nitrite functions as a corrosion inhibitor.

(e) Compressive strength development

Although the main purpose of using accelerators is to obtain high early strength and reduced setting times, the increased strength has been found [22, 39] to continue for a period of several years when addition levels of up to 2% of calcium chloride by weight of cement are used, and Figs 5.34 to 5.36 illustrate this effect for specimens stored for 5 years underwater (Fig. 5.34), in laboratory atmosphere (Fig. 5.35), and under outdoor exposure (Fig. 5.36). However, higher percentages of calcium chloride (greater than 4.0%) will lead to reduced strengths in comparison to a plain cement at periods greater than about 1 year.

(f) Volume deformation

SHRINKAGE

The drying shrinkage of concrete containing calcium chloride is increased in comparison to plain concrete, even though the amount of moisture lost is less [22]. This is illustrated in Fig. 5.37 and it is thought that the reduced moisture loss will be due to the more advanced state of hydration in the specimens containing calcium chloride. The increased shrinkage must, therefore, be a characteristic of the type of cement hydration products formed.

Under saturated conditions, such as total water immersion, the amount of expansion of the concrete is reduced when calcium chloride is present.

There are only limited data available on the effect of other accelerating admixtures, although one comparative study [46] suggests that calcium formate

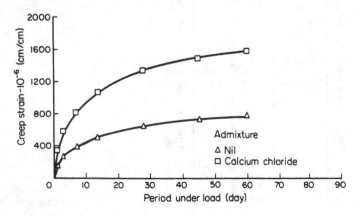

Fig. 5.38 The creep of plain concrete and concrete containing calcium chloride (Hope).

and triethanolamine also increase the drying shrinkage of concrete into which they are incorporated and some of this information is summarized in Table 5.6 in comparison to plain concretes and concretes containing calcium chloride.

CREEP

The creep of concrete under drying conditions is increased by the presence of calcium chloride in the mix [30], as shown in Fig. 5.38 for 1.5% calcium chloride by weight of cement.

References

1 Rosenburg, A.M. (1974). *ACI Journal*, **71**, 1261–8.
2 US Patent (1974). 3 801 338.
3 Murakami, K. (1968). *Proceedings of the 5th International Symposium on the Chemistry of Cement*, Tokyo, Supplement paper II-2.
4 Wolhuter, C.W. (1974). *Proceedings of the 1st Australian Conference on Engineering Materials*, University of New South Wales, 29–43.
5 Tenoutasse, N. (1968). *Proceedings of the 5th International Symposium on the Chemistry of Cement*, Tokyo, 372–8.
6 Ramachandran, V.S. (1972). *Thermochemica Acta*, **4**, 343–66.
7 Edmeades, R.M. (1976). Private communication.
8 Ramachandran, V.S. (1976). *Cement and Concrete Research*, **6**, 623–32.
9 Ramachandran, V.S. (1973). *Cement and Concrete Research*, **3**, 41–54.
10 Chaberek, S. and Martell, A.E. (1959) *Organic Sequestering Agents*, John Wiley and Sons, NY, 325.
11 Edwards, G.C. (1966). *Journal of Applied Chemistry*, **16**, 166–8.
12 Dodson, V. (1990). *Chemical Admixtures*, Van Nostrand Reinhold, NY, 82.
13 Kantro, D.L. (1975). *Journal of Testing and Evaluation*, **3**(4), 312–21.
14 Dodson, V. (1990). *Chemical Admixtures*, Van Nostrand Reinhold, NY, 84.
15 Young, J.F. (1973). *Cement and Concrete Research*, **3**, 689–700.
16 Odler, I. (1971). *Journal of the American Ceramics Society*, **54**, 362–3.
17 Berger, R.L. (1972). *Cement and Concrete Research*, **2**, 43–55.
18 Ciach, T.D. (1971). *Cement and Concrete Research*, **1**, 159–76.
19 Ben-Dor, L. (1975). *Journal of the American Ceramics Society*, **58**, 87–9.
20 Skalny, J. (1971). *Journal of Colloid and Interface Science*, **35**, 434–40.
21 Collepardi, M. (1972). *Cement and Concrete Research*, **2**, 57–65.
22 Shideler, J.J. (1952). *ACI Journal*, **48**, 537–59.
23 Double, D.D. *et al.* (1978). *Proceedings of the Royal Society* A, **360**, 435.
24 Fletcher, K.E. (1971). *Concrete*, **5**, 175–9.
25 Rixom, M.R. (1998) *International Workshop on Supplementary Cementing Materials, Superplasticizers, and Other Chemical Admixtures*, CANMET/ACI, Toronto, Canada.
26 Rixom, M.R. *ACI Fall Meeting*, Atlanta, GA.
27 Dodson, V. (1990) *Chemical Admixtures*, Van Nostrand Reinhold, NY, 93–4.
28 Rosenberg, A.M. *et al.* (1977) *ASTM STP-629*, 89–99.

29 Dodson, V. (1990) *Chemical Admixtures*, Van Nostrand Reinhold, NY, 88.
30 Hope, B.B. (1971). *ACI Journal*, **68**, 361–5.
31 Monofore, G.E. (1960). *ACI Journal*, **56**, 491–515.
32 Baker, C.A. (1966). *Humes Technical Bulletin*, No. 4.
33 Gouda, U.K. (1973). *Journal of Colloid and Interface Science*, **43**, 293–302.
34 Ramachandran, V.S. (1975). *Precast Concrete*, **6**, 149–51.
35 Gouda, U.K. (1966). *British Corrosion Journal*, **1**, 138–42.
36 McCurrich, L.H. (1975). *Proceedings of the Conference on Ready-Mixed Concrete*, Dundee.
37 Murat, M. *et al.* (1974). *Cement and Concrete Research*, **4**, 945–52.
38 Roberts, M.H. (1962). *Magazine of Concrete Research*, **14**, 143–54.
39 Blenkinsop, J.C. (1963). *Magazine of Concrete Research*, **15**, 33–8.
40 Kondo, Y. *et al.* (1959). *ACI Journal*, **55**, 299–312.
41 Manns, W. and Eichler, W.R. (1982). *Betonwerke und Fertigteil-Technik* (Weisbaden), **48**(3), 154–62.
42 Zoob, A.B. and Perenchio, W.F. (1987). *Wiss, Janney, Elstner Associates Report* No. 871434.
43 Anon. (1987). *Webster Engineering Associates Inc. Report* No. 85134. Garfield Heights.
44 Stark, D.C. (1984). *CTL Project Report* No. CR2402–4310, Construction Technology Lab., Skokie.
45 Nmai, C.K. (1987). *Technical Report Master Builders Inc.*, Cleveland.
46 Bruere, G.M. *et al.* (1971). *CSIRO (Australia) Division of Applied Mineralogy*, Technical Paper No. 1.

Chapter 6

Special purpose admixtures

6.1 Introduction

Chemical admixtures are frequently used in concrete for the purpose of altering one or more of its properties in the fresh or hardened state. In addition to mainstream applications of water reduction, retardation, acceleration and air entrainment, chemical admixtures are used for a variety of special purposes. Included in this category are both conventional admixtures, used in unusual ways, and special admixtures designed for specific applications. By their mode of action these chemicals, when used singly or in combination with other admixtures, broaden the scope of applications of concrete. New and specialized types of concrete such as cohesive underwater concrete and roller-compacted concrete requiring varying placement methods, concreting in the depth of winter without conventional heating and protection methods, corrosion inhibition in the presence of chlorides, reducing alkali–aggregate reaction, shrinkage control, the recycling of concrete waste, and the production of highly durable high-performance concrete are now made possible through the use of admixtures.

The widespread deterioration of our infrastructure in the recent decades and the high cost of replacement has now made repair of concrete structures the dominant construction activity. There is an increasing demand for products and techniques that can repair surface damage. Admixtures that provide properties such as volume stability, water tightness, increased chemical resistance, antiwashout characteristics for underwater concrete and quick-setting accelerators used in shotcrete are therefore finding wide use. They are also finding increasing use in a number of mainstream applications in combination with conventional admixtures to ensure the quality of the placed concrete. Consequently, these admixtures are receiving significant attention both in research and product development. The search for better performance and environmental concerns has led researchers and formulators to the development of new products and new methods to evaluate their efficacy. The success of the their efforts will undoubtedly intensify the use of these admixtures in a variety of applications, contributing to enhanced safety and economy in new construction and repair.

6.2 Alkali–aggregate expansion-reducing admixtures

Due to the serious depletion of sound aggregate sources, and increased incidence of damage to large concrete structures resulting from alkali-aggregate reactions there has been a strong drive for the development of a chemical to inhibit the reaction. In the last decade therefore, researchers increasingly turned to the use of chemicals to prevent the reactions. Even though Japan has used these chemicals in the manufacture of precast elements and in a limited number of structures, the use of these materials (chemicals) is still in the experimental stage and the available data have not been developed into a meaningful field practice.

6.2.1 Alkali–aggregate reaction

The most common types of alkali–aggregate reactions (AAR) result from the attack of hydroxyl ion on at least two distinct groups of chemically reactive aggregates: those that are largely composed of silica (or siliceous mineral and glasses) and those that are largely composed of carbonate, particularly dolomitic carbonates [1, 2]. Alkali hydroxides, usually derived from cement, interact with reactive components in the aggregate particles, hydrolyzing the reactive silica to form alkali–silica gel which then absorbs water, and increases in volume. Hydroxyl concentration of pore solutions in concrete made with a high-alkali cement may be 10 times as high than that made with a low-alkali cement and 15 times that of a saturated $Ca(OH)_2$. Other sources contributing to the total alkali contents in concrete include sea and ground waters, some mineral constituents in the aggregate, de-icing salts and admixtures. Figure 6.1 shows the expansion of mortar containing cement with different amounts of Na_2O equivalent [2].

Methods of preventing alkali–aggregate expansion reactions in concrete consist of avoidance of reactive aggregates, use of cement with alkalis less than 0.6% Na_2O equivalent, the use of pozzolanic materials, sealing of the hardened concrete and coating of aggregate particles with an impermeable material. Air entrainment has also been suggested as a means to reduce expansion. The beneficial effect of lowering the water–cement ratio in concrete through the use of superplasticizers is debated. Although lower water–cement ratios lower porosity, permeability and mobility of alkali ions, alkali concentration in the pores will increase and this will aggravate the AAR problem. The effects of the various chemicals that have shown promise in retarding or inhibiting the AAR reaction are discussed below.

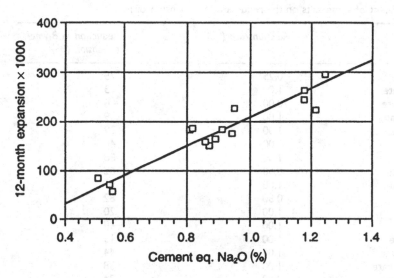

Fig. 6.1 Expansion of mortar containing different amounts of Na₂O. (Reprinted with permission, The Concrete Society.)

6.2.2 Types of admixtures

The different types of admixtures, known to reduce alkali–aggregate reactions, can be divided into two groups: those that are effective in reducing the expansion due to the alkali–silica reaction, and those that lower expansions resulting from the alkali–carbonate reaction. For the alkali–silica reaction, reductions in the expansion of mortar specimens have been obtained with soluble salts of lithium, barium and sodium, proteinaceous air-entraining agents, aluminum powder, $CuSO_4$, sodium silicofluoride, alkyl alkoxy silane, lithium carbonate, lithium fluoride, styrene–butadiene rubber latex and lithium hydroxide water-reducing and set-retarding agents[3–6]. Barium salts used as inhibitors are barium nitrate, acetate and hydroxide. Sodium and potassium nitrate have also been used. Table 6.1 shows data on the effect of some salts on expansion due to AAR. In the alkali–carbonate reaction, lithium carbonate and ferric chloride meet the requirements for suitable materials. The small size and high charge of the cation in the latter salt satisfies the parameters required for decreasing the reaction.

(a) Lithium salts

Lithium hydroxide is effective in reducing the expansion caused by alkali–silica reactions. Mixes containing 1.0% LiOH have shown decreased

Table 6.1 Effect of some salts on the reduction in expansion of mortars*

Material	Addition rate (%)	Reduction in 8-week expansion (%)
Al powder	0.25	75
Ba carbonate	1.00	3
Ca carbonate	10.00	−6
Cr phosphate	1.00	9
Cu chloride	1.00	29
Cu sulfate	1.00	46
Li chloride	1.00	88
Li carbonate	0.50	62
Li carbonate	1.00	91
Li fluoride	0.50	82
Li nitrate	1.00	20
Li sufate	1.00	48
Na chloride	1.00	15
Na carbonate	1.00	44
NH₄ carbonate	1.00	38
Zn carbonate	0.50	34

*McCoy, W.J. and Caldwell, A.G. (1951). New approach to inhibiting alkali aggregate expansion. *Journal of the American Concrete Institution*, **47**, 693–706.

expansion up to 50 days [7]. A systematic investigation on the effect of lithium fluoride and lithium carbonate on the expansion of mortars (with 1% Na₂O equivalent) gave the results shown in Table 6.2 [8]. Although less soluble than LiOH, both LiF and Li₂CO₃ are capable of reducing the alkali– silica expansion, if dosages exceeding 0.5% LiF and 1.0% Li₂CO₃ are used. Both these salts appear to be converted to LiOH in pore solution. In another study involving LiOH, LiNO₂ and Li₂CO₃ it was found that all these compounds were able to decrease the expansion due to AAR [9]. Of these

Table 6.2 Expansion of mortars containing different amounts of LiF or Li₂CO₃

Sample	Dosage (%)	Expansion (%) after 6–36 months				
		6	12	18	24	36
Reference	0.00	0.54	0.62	0.62	0.63	0.63
Ref + Lif	0.25	0.43	0.59	0.64	0.68	0.71
	0.50	0.04	0.06	0.06	0.06	0.06
	1.00	0.02	0.02	0.02	0.02	0.02
Ref + Li₂CO₃	0.25	0.46	0.61	0.62	0.62	0.63
	0.50	0.30	0.50	0.54	0.55	0.58
	1.00	0.03	0.04	0.04	0.04	0.05

Adapted with permission from Table 4.1, Eliminating or Minimizing alkali-Silica, Stock *et al.*, 1993.

compounds, Li_2CO_3 was found to be more effective than the others. The relative inhibiting effect was considered to depend on the Li/Na ratio. However, the effectiveness decreased when LiF and $LiOH.H_2O$ were added in amounts greater than 0.7%, although the compressive strengths of mortars were of same value at different dosages [10].

The work carried out to date indicates that not all lithium salts are effective in diminishing the alkali–aggregate expansion reaction [10]. Optimum dosage requirements and long-term effects have to be evaluated. Before the relative effects of various lithium compounds such as hydroxide, carbonate, nitrate, fluoride, perchlorate, chloride, etc., can be substantiated much more work employing reliable predictive tests has to be carried out.

(b) Conventional admixtures

Information from previous work suggests that air-entrainment offers a measure of protection against alkali–aggregate expansion. An air-entrainment of 3.6% can cause a 60% reduction in expansion [10]. In Fig. 6.2 the

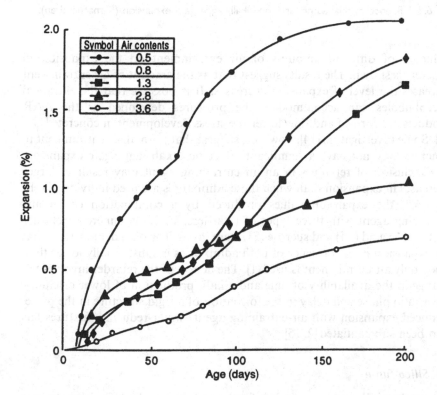

Symbol	Air contents
●	0.5
◆	0.8
■	1.3
▲	2.8
○	3.6

Fig. 6.2 Influence of air entrainment on expansion due to alkali–aggregate reaction (Ramachandran).

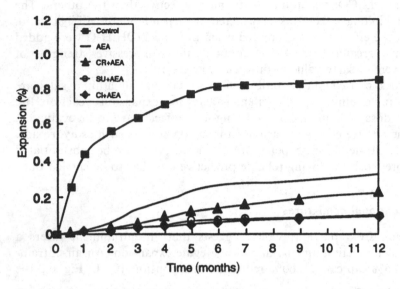

Fig. 6.3 Influence of dual admixtures on alkali–aggregate expansion (Ramachandran).

influence of different amounts of air entrainment on the expansion in concrete is shown. The results suggest that as the amount of air entrainment increases the level of expansion decreases. It is possible that the entrained microbubbles can accommodate the pressures developed as the AAR products are formed and thus lower the stress development in concrete [11–13]. Some investigators [14], however, suggest that normal air entrainment in concrete may not have a significant effect on alkali–aggregate expansion. The inclusion of retarders in an air-entraining agent may result in larger decreases in expansion than when these admixtures are used individually. In Fig. 6.3, the expansion values produced by a combination of an air-entraining agent with three types of retarders, viz., a commercial retarder (CR), citric acid (CI) and sucrose (SU), are given. The data demonstrate that the expansion in the presence of both admixtures is substantially lower than when only air entrainment is used [1]. The action of the retarder involves the change in the availability of lime and alkali, production of lower calcium–silica ratio phase and delay in the formation of a rigid structure in the paste. Reduced expansion with air-entraining agent–water-reducing mixtures has also been substantiated [2, 15].

(c) Silica fume

Although small amounts of silica fume are found to be effective at earlier periods [16, 17], they are somewhat ineffective in the long term [18, 19]. There has also been concern that at larger dosages of silica fume, silica fume itself may become a source that would react with the alkalis in cement [20, 21]. The effectiveness of silica fume depends on a number of parameters such as composition of silica fume (SiO_2 and alkali contents), the percentage of silica fume used, the type of alkali-aggregate expansion reaction (alkali–silica or alkali–carbonate) and the type, fineness and alkali contents of cement. The effectiveness of a combination of silica fume and air entrainment was studied [22]. Both silica fume and air entrainment individually reduced AAR expansion, but the combination reduced the expansion by the maximum amount. Table 6.3 [22] compares the expansion values in mortar bars containing silica fume and an air-entraining agent.

(d) Silanes, siloxanes and silicofluorides

Alkyl alkoxy silanes have been found to be very effective in reducing alkali–aggregate expansion [11] (Fig. 6.4). Of the silanes used in the study, hexyl trimethyl siloxane and decyl trimethoxyl silane were found to be more effective in decreasing the expansion than the others. In the same study, Ohama et al. [11] investigated the effect of sodium silicofluoride, alkyl alkoxy silane, lithium carbonate, lithium fluoride, styrene–butadiene rubber latex and lithium hydroxide on compressive strength and the expansion of mortar containing cement with 2% equivalent Na_2O. The reduction of the level of expansion shown in Fig. 6.4 with the siloxanes was attributed to the formation of an insoluble compound with strong siloxane linkages which impart a degree of water repellence. Siloxanes appear to be more effective in preventing expansion than lithium compounds and other properties of the concrete are not significantly affected. The results of the relative compressive strengths indicate that the mortar with sodium silicofluoride shows the highest strength, being about 50% higher than that of the reference. The silanes, siloxanes and silicofluorides thus appear to be promising chemicals to combat AAR.

Table 6.3 Comparison of expansion of mortars containing silica fume and air-entraining agent (Ramachandran)

Silica fume (%)	Expansion (%)	
	No air entrainment	With air entrainment
0	1.05	0.48
6	0.85	0.45
12	0.62	0.28

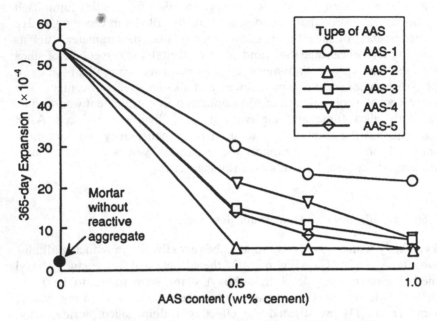

Fig. 6.4 The effect of silanes on alkali–aggregate expansion (Ohama [45]).

(e) Phosphate sealers

A method of treating the reactive aggregate before making concrete, or treating the hardened concrete with calcium phosphate, has been proposed as a method of countering the alkali–silica expansion. The expansion of a reference concrete containing a non-reactive aggregate was compared with that containing a reactive aggregate treated with monocalcium phosphate monohydrate. The aggregate treated with phosphate, even for a minute, had almost no expansion at 28 days, compared to an expansion of about 1% in the reference concrete [21]. This method was less effective for concrete containing reactive carbonate aggregates.

6.2.3 Mode of action

Both lithium and barium salts reduce expansion based on alkali–silicate and alkali–carbonate swelling through the osmotic pressure development mechanisms. Two theories have been postulated [5, 11]. According to the first mechanism, soluble metal silicates, formed by the attack of alkali hydroxides on silica, are said to convert to insoluble lithium silicates in the presence of lithium salts. Consequently, less water is imbibed and this results in reduced expansion. In the second mechanism, the salts are said to provide small high-valency cations which equalize the osmotic swelling differential that exists across a semipermeable membrane in the immediate vicinity of the aggregate [11].

The mechanism of the inhibitive action of LiOH proposed by Stark *et al.* [7] is attributed to the formation of lithium silicate that dissolves at the surface of the aggregate without causing swelling [7]. In the presence of KOH and NaOH the gel product incorporates Li ions; and the amount of Li in this gel increases with its concentration. The threshold level of Na : Li is 1 : 0.67 to 1 : 1 molar ratio at which expansion due to alkali–silica reaction is reduced to safe levels. Some workers [22] have found that when LiOH is added to mortar much more lithium is taken up by the cement hydration products than Na or K. This would indicate that small amounts of lithium are not very effective. It can therefore be concluded that a critical amount of lithium is needed to overcome the combined concentrations of KOH and NaOH to eliminate the expansive effect and that the product formed with Li is non-expansive.

The mechanism of action by which silane and siloxanes reduce expansion has been attributed to water repellence and air entrainment. Phosphate addition or coatings may interfere with the dissolution of silica gel and the formation of gel. It is also possible that phosphate reduces the osmotic potential and the swelling pressure in the gel. The manner in which air entrainment reduced expansion was attributed to the accommodation of alkali–silica gel in the air void system. For example, it was found that air-entrained concrete with 4% air voids could reduce AAR expansion by 40% [23].

Based on the mechanisms by which the salts reduce expansion, the requirements of a suitable admixture would be as follows [7, 11]:

1. It should form relatively soluble hydroxide.
2. It should react to produce an insoluble silicate.
3. The ions must not interfere with or modify the hydration reaction so that the concrete properties are not adversely affected.
4. The ions should not take part in the alkali–aggregate reaction to form other expansive materials.
5. It should provide small, high-valency cations.

6.2.4 Effects on the plastic and hardened properties of mortar and concrete

- Setting characteristics: Li_2CO_3 accelerates both initial and final set times at dosages exceeding 1% by weight of cement. Flash set can occur in some circumstances. $FeCl_3$ also accelerates set, although to a lesser degree than Li_2CO_3. High dosages (> 7%) produce flash set. Proteinaceous materials >1% may retard set and cause heavy air entrainment. Aluminum powder does not affect set, but plastic shrinkage can occur if the gassing reaction is completed well before the initial set of the cement.

- **Compressive strength:** compressive strength is decreased at all levels of Li_2CO_3 addition; the effect increases with dosage and at addition rates exceeding 3%, drastic strength reduction results. The extent of strength reduction that occurs with $FeCl_3$ depends on the dosage. At addition levels less than 7% by weight of cement, no strength reduction results. Dosages greater than 7% produce significant decrease in strength values. Proteinaceous air-entraining agents produce effects (compressive strength) similar to other air-entraining agents. The use of siloxanes and silicofluorides, while providing adequate inhibition of AAR, produces no strength reduction.

The effects on other properties of both plastic and hardened mortar or concrete have not been investigated. Furthermore, since the studies have been done on a laboratory scale, no information pertaining to effects of placing and curing under normal or varied environmental conditions is available.

6.3 Antifreeze admixtures

When fresh concrete freezes, the strength of such concrete is lowered by 20–40%, its resistance to freeze–thaw cycling as given by the durability factor is lower by 40–60% and the bond between reinforcement and concrete is lowered by 70% compared with normally cured concrete [24]. Thus, when concreting is done under cold weather conditions it is important to ensure that the concrete will not freeze while it is in the plastic state. Two options are available for cold weather concreting: (1) maintenance of near normal ambient and concrete temperatures through the heating of concrete ingredients and the provision of heated enclosures and (2) the use of chemical admixtures. This section discusses the use of specific chemicals that permit concreting at below freezing temperatures.

Conventional non-chloride accelerating admixtures are used in cold weather concreting to offset the retarding effects of slow hydration on the rate of strength development. Such admixtures however do not permit concreting at or below freezing temperatures. When concreting is carried out under more drastic Arctic-like weather conditions, special admixtures, called antifreeze admixtures, which affect the physical condition of the mix water, are used. Antifreeze admixtures are capable of depressing the freezing point of water in concrete considerably (Table 7.23) and their use at temperatures as low as −30°C enables an extension of the period of construction activity [24].

Although the use of antifreeze admixtures has been an acceptable practice in Russia for nearly three decades, their use in other countries has been more recent. Freeze-protection, non-chloride, non-corrosive, water-reducing admixtures which can prevent ice formation at sub-zero temperatures, accelerate set time and strength development, are currently used in North America [25]. One of the drawbacks cited for the sodium-thiocyanate-based admixture is the potential for AAR, particularly when used at very high dosages. Antifreeze admixtures are also being sold in Japan [26]. An antifreeze admixture that is both non-chloride and non-alkaline was developed in 1991 in Japan. This admixture contains polyglycol ester derivatives and calcium nitrite-nitrate [26]. In Finland, premixed dry-mix mortars containing antifreeze admixtures are marketed for use at $-15°C$ and are used for joint construction, pointing and repair purposes [27]. The RILEM committee on admixtures now recognizes antifreeze admixtures for cold-weather concreting [28].

6.3.1 Chemical composition and mode of action

Two groups of antifreeze admixtures are generally used to obtain both antifreezing characteristics and accelerated setting and hardening. The first group includes chemicals, such as sodium nitrite, sodium chloride, weak electrolytes (e.g. aqueous solution of ammonia) and non-electrolytic organic compounds such as high-molecular-weight alcohols and carbamide, which lower the freezing point of water in concrete and act as a weak accelerators of setting and hardening. The second group are binary or ternary admixture systems that contain potash and additives based on calcium chloride (CC) and a mixture of CC with sodium chloride (SC), sodium nitrite (SN), calcium nitrite–nitrate (CNN), calcium nitrite–nitrate–urea (CNN + U) and other chemicals that provide effective antifreezing action and accelerate significantly the setting and hardening. Generally, larger dosages are used compared to conventional chemical admixtures [24, 29, 30]. For example, 8% sodium nitrite is used to keep water in the liquid state in concrete at a temperature of $-15°C$.

Antifreeze admixtures function by lowering the freezing point of the liquid phase of concrete and accelerating the hydration of cement at freezing temperatures. Depending on the dosage used, non-chloride admixtures enable concrete and mortar to be placed at sub-freezing temperatures ($-6.7°C/20°F$) and reduce protective measures required for cold-weather work. In addition to the improved quality of concrete [31–36] early strength development permits earlier stripping and reuse of forms and hence completion of work. The significant difference in strength development between a plain concrete and a concrete containing an antifreeze admixture, placed and cured at cold temperatures is compared in Table 6.4.

Table 6.4 Concrete with freeze-protection admixture (Ratinov and Rosenburg).

Property	Plain concrete	Freeze-protection admixture
Set time (−4°C)	—	—
Compressive strength (MPa)		
−4°C (3 days)	3.4	9.24
−10°C (7 days)	8.3	39.3
−10°C (28 days)	18.1	49.9

Compressive strength values (as a percentage of design strength) for concrete cured at −5°C and containing sodium nitrite (SN), calcium nitrite–nitrate–chloride (CNNC), and calcium chloride–calcium nitrate (CC + SC) are shown in Table 6.5. Higher strengths are achieved by binary admixtures at lower dosages than when the individual components are used singly. It has also been shown that with the antifreeze admixtures the strength of concrete with ordinary cement may exceed that obtained with the rapid-hardening cement containing no admixture. Table 6.5 compares the results obtained with the two types of cement at −10°C. Best results [37] are obtained using a mixture of potassium carbonate and a retarder.

In addition to the two basic groups of antifreeze admixtures, others containing superplastizisers such as sulfonated naphthalene formaldehyde (SNF) and sulfonated melamine formaldehyde (SMF) have been developed. The main advantage afforded by such admixtures is that they produce significant water-reduction. The water reduction afforded (about 20–25%) reduces the freezable free water content which usually serves as a 'heat sink' for the heat liberated by the initial hydration reactions. Hence, lower amounts of antifreeze admixtures can be used.

Table 6.5 Strengths of concrete containing some binary antifreezing admixtures at 10°C (Ratinov and Rosenburg)

Admixture type	Percentage of 28-day strength at 28°C	
	Ordinary cement	Rapid hardening
Sodium nitrate + sodium sulfate	85	70
Sodium nitrate + calcium chloride	63	63
Potassium carbonate + retarder	74	48
Calcium nitrate + sodium sulfate	75	43

Selection of an antifreeze admixture will depend on the type of structure, the operating conditions and whether the admixture will be used with other protective methods of winter concreting. The effects of antifreeze admixtures are often specific to certain cement brands and fine aggregates [24]. It is important therefore to determine through laboratory mix trials both the operating range and the particular dosage of the admixture required for the intended application. Cold-weather concreting with antifreeze admixtures does not exclude the use of other admixtures such air-entraining agents, water reducers, retarders and superplasticizers. However, the dosage of the specific admixture to be used in combination with the antifreeze admixture should be established experimentally, because higher than normal amounts may be required.

The application of antifreeze admixtures is technologically simple, convenient and beneficial for winter concreting. Improved cohesiveness, plasticity, minimization of cold joints and sand streaking are some of the advantages. It is estimated that the use of these admixtures permit significantly better cost savings than steam curing or concreting in enclosures or heating of the constituents [24]. The use of these admixtures in combination with some cold-weather protection methods affords advantages in terms of lower amount of admixture, early commissioning of the structure and saving of energy.

When concrete is to be placed in cold weather, it is preferable that accelerators or antifreezers be used in combination with air-entraining agents and water-reducing admixtures. These combinations not only reduce the amount of freezable water in the mix but also generally reduce the quantity of antifreezers and accelerators needed to obtain desired effects compared to the amounts that have to be used when these are used separately. In addition these combinations may be useful in increasing the resistance of concrete to frost action and to corrosive agents.

Despite the 40-year history of use in Russia and the fact that past tests indicate that they are beneficial for use under cold-weather conditions, little consideration has been given to these materials in Western Europe and North America. More basic research is now underway in research organizations in Canada, Europe and the USA. One notable project is that being conducted as part of the Corps of Engineers Construction Productivity Advancement Research Program, a cost-sharing partnership between the Corps and industry [27].

A resume of the use of antifreeze admixtures has been presented to make the reader aware of these admixtures. The writer is, however, cognizant that more specific information will be required for those interested in their application. Readers who require greater detail are directed to reference [24], which is an exhaustive treatment of the subject.

6.4 Antiwashout admixtures

Dewatering of hydraulic structures, such as dams, for repair is difficult and expensive. Recent advances in cohesion-inducing admixtures have allowed placement of concrete underwater without the use of conventional tremies. These admixtures are generally recognized by their highly pseudoplastic (viscosity decreasing as shear increases) flow behavior. This allows the formulation of concrete mortars and slurries that resist sedimentation and sagging at rest but can be easily mixed, pumped and sprayed. The concrete produced with these admixtures is cohesive enough to allow limited exposure to the water, yet has good mobility to move underwater with little washout of cement. Such cohesion-inducing admixtures are referred to as antiwashout or viscosity-enhancing admixtures (AWAs or VEAs respectively).

6.4.1 Categories

Materials that have been used as viscosity-inducing admixtures may be classified according to their physical action in concrete as follows [38–41].

Class A. Water-soluble synthetic and natural organic polymers, which increase the viscosity of the mixing water. They include cellulose ethers, pregelatinized starches, polyethylene oxides, alginates, carrageenans, polyacrylamide, carboxyvinyl polymers and polyvinyl alcohol.

Class B. Organic water-soluble flocculants, which are adsorbed on the cement particles and increase viscosity by promoting interparticle attraction. These materials are styrene copolymers with carboxyl groups, synthetic polyelectrolytes and natural gums.

Class C. Emulsions of various organic materials, which increase interparticle attraction and also supply additional superfine particles in the cement paste. These are materials consisting of acrylic emulsions and aqueous clay dispersions.

Class D. Inorganic materials of high surface area or unusual surface properties which increase the water-retaining capacity of the mix. These include very fine clays (bentonites), pyrogenic silicas, condensed silica fume, milled asbestos and other fibrous materials.

Class E. Inorganic materials, which supply additional fine particles to the mortar pastes and thereby increase the thixotropy, such as fly ash, hydrated lime, kaolin, diatomaceous earth and other raw or calcined pozzolanic materials and various rock dusts. [40, 42].

The increased viscosity of water in the mix results in a greater thixotropy of the concrete and an improved resistance to segregation. Dosage of the admixture for classes A, B and C range from 1 to 1.5% by weight of the water in the mix and they are frequently used in combination with a superplastiziser [43]. The magnitude of the thickening effect produced is dependent on the admixture dosage and the molecular weight of the main component. It is usually discharged into the mixer at the same time as the other materials. Materials in classes D and E influence the void structure by acting as pore fillers, although the increased fines content often also increases lubrication of the mix.

The types of water-soluble polymers used for the thickening cement slurries, mortar and concrete are shown in Table 6.6. Although many polymers shown in Table 6.6 can be used to increase the viscosity of the water in the mix, they are not all pseudoplastic polymers compatible with cement systems. Only a few can be consistently combined with water-reducing admixtures (WRAs) and superplasticizers to produce concretes with cohesive yet highly flowable mixtures [40, 41, 43].

The primary structure of Class A materials is that of a backbone of six carbons chains with side chains. They are high-molecular-weight materials that build viscosities in solution via hydrogen bonding. Certain materials which lack significant hydrophobic substituents (such as the high-molecular-weight polysaccharides) are more tolerant of salt and cations, stable to changes in pH and temperature, and do not generate foam. This has been attributed to the intramolecular hydrogen bonding between the main and side chains and the lack of activity at the air/water interface respectively [41, 44].

Some of the properties which differentiate the efficiency of the various polymers include [40, 41]:

1. Ability to increase viscosity at very low dosages.
2. Reduction in sedimentation while in the plastic state.

Table 6.6 Types of water soluble polymers (Ohama et al.)

Natural	Semi-synthetic	Synthetic
Starches	Hydroxypropyl methylcellulose (HPMC)	Polyvinyl alcohol (PVA)
Guar gum	Hydroxyethyl cellulose (HEC)	Polyethylene oxide (PEO)
Locust bean gum	Carboxymethyl cellulose (CMC)	Polyacrylamide
Alginates	Hydroxypropyl cellulose (HPC)	Polyacrylate
Carrageenans	Starch derivatives	Polyvinylpyrrolidone
Agar	Propylene glycol alginate (PGA)	—
Gum arabic	—	—
Gum tragacanth	—	—
Xanthan gum	—	—
Welan	—	—
Rhamsan gum	—	—
Gellan gum	—	—

3. Control of bleeding.
4. Tolerance of high level of salts.
5. Compatibility with other admixtures (such as WRAs and super-plasticizers) so that rheology of the mix is not significantly altered, set is not drastically extended, or heavy air-entrainment and foaming is not produced.
6. Ability to build sufficient viscosity at low shear rates.

6.4.2 Formulating non-dispersible underwater concrete

For the production of non-dispersible colloidal underwater concrete, minimum water–cement ratios range from 0.36 to 0.40, and cement and fine aggregate contents are usually higher than corresponding mixes placed on land. Often it is difficult to adjust the mixture proportions to achieve desired design parameters for all properties of concrete. Consequently, the properties of colloidal underwater concrete are controlled by the addition of three chemical admixtures. For example, silica fume may be used in conjunction with an AWA and a superplasticizer or conventional water reducers to reduce segregation. The key to a non-dispersible concrete with self-leveling characteristics is the successful optimization of the AWA with the super-plasticizer used to increase the slump. Best results have been obtained with lignosulfonates and melamine formaldehyde sulphonate types of super-plasticizers. The naphthalene formaldehyde sulfonate-based superplasticizers appear to produce erratic results in flowability and cohesion [45, 46].

Auxiliary agents [43] which react with the polymer in the AWA to increase the apparent molecular weight of the polymer, thereby improving its

Table 6.7 Test results for concrete containing polymer and glutaraldehyde auxiliary agent (Khayat)

Polymer (%)	Glutaraldehyde (%)	Transmittance*	Slump (mm)	Strength (28 days)
0.6	—	92	130	34.5
	0.5	95	125	34.8
	1.0	97	125	35.0
0.8	—	95	180	33.5
	0.5	98	170	34.0
	1.0	100	160	34.2
1.5	—	99	225	33.2
	0.2	100	220	33.8
2.0	—	100	220	32.0
	0.1	100	215	32.5
3.0	—	100	145	29.0
	0.1	100	140	29.2

*Measured as a turbidity by photoelectric photometry

cohesion-inducing properties, can be used with the AWA to increase resistance to cement washout. The agent also permits a reduction of the polymer by 0.8% by weight, without any adverse effect on strength (Table 6.7).

6.4.3 Effects produced on plastic and hardened concrete

The following effects are produced when an AWA is used in concrete [40, 43, 44]:

- Cohesiveness, mobility, and self-leveling characteristics are increased.
- Concretes containing a cellulose ether AWA will have an extended set time. Those containing an acrylic AWA do not have extended set time except when combined with superplasticizers (Fig. 6.5).
- There is improved resistance to segregation when the concrete is allowed to free-fall through water. (Fig. 6.6).
- The pH value and turbidity of the water is decreased with increased dosage of the admixture.
- Because of the high water retentivity, these admixtures induce in concrete, there is hardly any bleeding (Fig. 6.7).
- Some class A, B and C admixtures have inherent surfactant properties that lower the surface tension of the aqueous phase of the mix. Consequently, dosages above optimum levels will entrain unwanted air.
- A slight reduction in compressive strength, particularly at early ages, results in medium-cement-content mixes, the extent of reduction depending on admixture dosage, percentage of entrained air, slump or consistency and degree of set retardation.

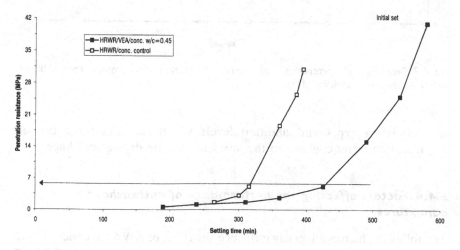

Fig. 6.5 Setting behavior of washout-resistant concrete (Rakitzky [41])

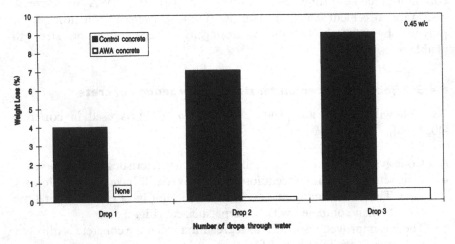

Fig. 6.6 Washout resistance of 7 in (17.8 cm) slump concretes (Rakitsky [41]).

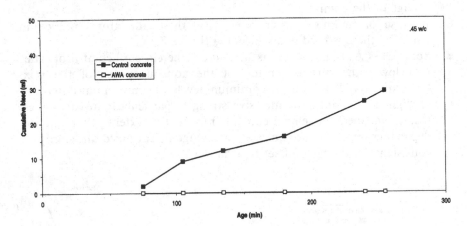

Fig. 6.7 Bleeding characteristics of washout-resistant concrete, total bleed $= 0.0145$ ml cm^{-2} (Rakitsky [41]).

- Admixture types and addition levels which necessitate the use of increased water contents in the mix will increase drying shrinkage.

6.4.4 Factors affecting the performance of antiwashout admixtures

The following factors adversely influence the effect of AWAs in concrete and mortar mixes. Since pumping delays can be time consuming and costly, it is important that the pump operator pay due regard to these items and the engineer designs the concrete mix taking into account these effects [40, 43, 45]:

1. **Cement content:** high cement contents respond poorly to the use of AWAs. Often highly thixotropic mixes are produced even at minimum recommended dosages. Suitable dosages should be evaluated for the specific mix prior to field use.
2. **Temperature:** high temperatures may cause rapid slump loss and it may be necessary to use a retarder in conjunction with the AWA.
3. **Fine aggregate characteristics:** mixes containing a high percentage of fine sand with FMs less than 2.6 give heavy-bodied mixes which impair pumpability due to the high cohesion and friction produced in the pipeline.
4. **Compatibility with other admixtures:** due to the inherent surfactant properties exhibited by most of the materials in classes A, B and C, their use with other admixtures such as water reducers should be evaluated in terms of the side effects on the plastic and hardened concrete, viz. severe set retardation and excessive air entrainment. The molecules of materials with fewer surfactant properties compete with the superplasticizer for adsorption sites on the cement particle. This will result in a higher dosage of the superplasticizer.

6.4.5 Mixing and storage

The range of admixture dosages used in the five classes is as follows: Class A: 0.2–0.5% solid by weight of cement; Class B: 0.01–0.10% solid by weight of cement; Class C: 0.10–1.50% solid by weight of cement; Classes D and E: 1–25% solid by weight of cement [40, 41]. Due to the small amounts of the admixture required, the addition of the correct quantity must be ensured. Suitable automatic dispensing equipment should be used for liquid materials. For the powders, a general volume dispenser such as a volume cup of the correct size or a measured weight of the material by weight of the cement may be used. Some manufacturers market the material packaged in water-soluble packages which contain the prescribed addition rate (in weight) of the admixtures. The packages are added directly at the mixer (Fig. 6.8).

During the batching of the concrete, the AWA is added just after the aggregates are discharged and a brief mixing period follows prior to the addition of cement. Due to their solid form, and because of their tendency to float on the mixing water, longer, more vigorous mixing may be required to ensure thorough distribution of the admixtures. Initial mixing produces a stiff consistency; further mixing results in the production of a wetter consistency as the polymers gradually dissolve. It is advisable to restrict addition of extra water at the initial mixing stages. Materials such as

Fig. 6.8 Ready-to-use water-soluble packages of antiwashout admixture (courtesy Fritz Admixtures, Texas, USA).

polyethylene oxides, cellulose ethers and some synthetic polyelectrolytes are hygroscopic and tend to produce slow-dissolving 'clumps' when they come in contact with moisture. To ensure uniform distribution in the concrete they should preferably be dissolved in water prior to addition into the mix. With polyethylene oxides, gradual dispersion of the powder in a large volume of water containing a little isopropanol, stirring vigorously, prevents the formation of clumps. Some powder forms readily imbibe moisture, while emulsions are susceptible to coagulation or precipitation of the solid phase when subjected to freezing temperatures. These materials should therefore be stored under dry and normal temperature (20–22°C).

6.4.6 Applications

Non-dispersible concrete can be poured into a water-filled form without a tremie pipe to produce dense structural repairs. This type of material has particular advantages over conventional concrete both in terms of the quality of the repair produced and the reduction in placement cost associated with plant and diver manpower requirements. Field tests carried out on a commercially available AWA [46] show that bonding capability, pumpability and flowability around reinforcement are improved over conventional tremie-placed concrete. The quality and strengths of cores show AWA concrete to be suitable material for *in situ* structural concrete construction at considerable water depths [46].

Typical uses include the production of non-dispersible underwater concrete and reduction of the accumulation of bleed water in mass concrete placed in deep forms. Consequently, AWAs are useful in mass concrete work because they prevent the formation of laitance on the surface of the concrete and thereby reduce the excessive cleaning between successive lifts. The admixtures also reduce the voids formed under horizontal reinforcing bars. Therefore, bond to steel increases and potential corrosion problems are reduced. The admixtures are also used in conjunction with WRAs in oil-well cementing grouts to reduce pipeline friction and rapid water loss and grouting of pre- and post-tensioned concrete ducts [47]. New valves and control devices under development in Europe and Japan used in conjunction with AWA will likely advance the field on underwater concrete.

6.5 Corrosion-inhibiting admixtures

In countries where there is heavy snowfall and where severe winter conditions prevail, large quantities of salt are used in snow and ice removal from pavements. The copious use of salt has caused the deterioration of thousands of bridges, other highway structures and parking garage decks. Chloride ion ingress occurs through cracks as salt-laden water enters the concrete and soon reaches the steel reinforcement; active corrosion then produces voluminous rust which exerts tensile stresses, causing further cracking of the concrete. Concurrent expansive cracking and scaling due to corrosion and freeze–thaw action leads to spalling and disintegration of the concrete.

Several preventive methods such as high-quality concrete cover, epoxy-coated rebar, coating of the surface with waterproofing elastomeric membranes and sealers, overlaying with high-density, low-permeability concrete or latex-modified toppings have been used to control corrosion [48–50]. Many of these preventive measures control corrosion by reducing permeability and carbonation and increasing the resistivity of the concrete. Despite the use of these first-line defence measures, concrete cracks, and coatings and sealers wear off, permitting entry of chloride-laden water. Corrosion-inhibiting admixtures are usually used as a second line of defence to prevent corrosion that may occur due to cracking [50]. Thus, the protection afforded to the reinforcement by the highly alkaline concrete may be augmented by the use of corrosion-inhibiting admixtures [49–51].

6.5.1 Material parameters

Materials used as corrosion inhibitors should not be toxic or pose environmental hazards and should [52]:

1. Be effective at the pH and temperature of the environment in which they are used.
2. Have strong electron acceptor or donor properties.
3. Have good solubility characteristics but not leach out from the concrete.
4. Quickly saturate the steel surface.
5. Induce polarization of the respective electrons at relatively low current values.
6. Be compatible with other admixtures used in concrete.
7. Not adversely affect the durability of concrete.
8. Not be strongly adsorbed by the cement paste, so that the major amount of the added admixture is available to inhibit corrosion.
9. Not alter the rate of hydration significantly.

6.5.2 Types of corrosion inhibitors

A corrosion-inhibiting admixture is a chemical compound which, when added in small concentrations to concrete or mortar, effectively checks or retards corrosion. These admixtures can be grouped into three broad classes, anodic, cathodic and mixed, depending on whether they interfere with the corrosion reaction preferentially at the anodic or cathodic sites or whether both are involved [48]. Six types of mechanisms, viz. anodic (oxidizing passivators), anodic (non-oxidizing passivators), cathodic, precipitation (anodic and cathodic effects), oxygen scavenging and film forming (adsorption on the steel surface) have been identified by the National Association of Corrosion Engineers' committee on corrosion inhibitors [49].

Anodic inhibitors such as nitrites, chromates and molybdates are strong oxidizing passivators. They strengthen the protective oxide layer over the steel which otherwise would break down in the presence of chloride ions. The mechanism involves a redox reaction in which the chloride and nitrite ions engage in competing reactions; the inhibitor is reduced and steel becomes oxidized to iron oxide as follows:

$$2Fe^{++} + 2OH^- + 2NO_2^- \rightarrow 2NO + Fe_2O_3 + H_2O$$

Cathodic corrosion inhibitors reduce the corrosion rate indirectly by retarding the cathodic process which is related to anodic dissolution. In this process, access to the reducible species such as protons, to electroactive site on the steel, is restricted. Reaction products of cathodic inhibitors may not be bonded to the metal surface as strongly as those used as anodic inhibitors. The effectiveness of the cathodic inhibitor is related to its molecular structure. Increased overall electron density and spatial distribution of the branch groups determine the extent of chemisorption on the metal and hence its effectiveness. Commonly used cathodic inhibitor materials are bases, such as $NaOH$, Na_2CO_3, or NH_4OH, which increase the pH of the medium and thereby also decrease the solubility of the ferrous ion. Other materials consisting of aniline and its chloroalkyl- and nitro-substituted forms, as well as the aminoethanol group, are effective at 1–2% dosage levels by cement weight in the presence of 1–2% $CaCl_2$ [50]. Most work has been carried out on aniline and its chloro-, alkyl-nitro-substituted forms and mercaptobenzothiazole [51]. Inorganic salts such as $NaOH$, Na_2CO_3 and NH_4OH are generally used at a dosage of 2–4%.

Mixed inhibitors may simultaneously affect both anodic and cathodic processes. A mixed inhibitor is usually more desirable because its effect is all-encompassing, covering corrosion resulting from chloride attack as well as that due to microcells on the metal surface. Mixed inhibitors contain molecules in which electron density distribution causes the inhibitor to be attracted to both anodic and cathodic sites. They are aromatic or olefinic molecules with both proton-forming and electron acceptor functional group such as NH_2 or SH, as in aminobenzene thiol. Mixed inhibitors are similarly used at 1–2% addition rates [50].

Several types of corrosion inhibitors have been investigated in the last 20 years [53–55]; these include calcium and sodium nitrites, sodium benzoate, sodium/potassium chromate, sodium salts of silicates and phosphates, stannous chloride, hydrazine hydrate, sodium fluorophosphate, permanganate, aniline and related compounds, alkalis, azides, ferrocyanide, EDTA and many chelating compounds. However, in terms of field practice and research data, nitrite-based compounds occupy a dominant position.

Admixture formulations containing two or more compounds (multicomponent) in which each component plays a specific role or compliments the corrosion resistance capacities of the other are also used. For example, a mixture of calcium nitrite and calcium formate is used to both accelerate strength of the concrete and inhibit corrosion of the steel during steam curing.

(a) Nitrite-based inhibitors

The most widely used anodic inhibitors are calcium and sodium nitrite, sodium benzoate and sodium chromate. With the exception of calcium nitrite, no other chemical is available in North America as a proprietary product. Nitrites have been used in the USA for more than 14 years and for nearly 40 years in Europe. Calcium nitrite is marketed as a non-chloride accelerator, as well as a corrosion inhibitor. For 25–30% solids in solution, dosage rates range from 2 to 4% by weight of cement depending on the application [50]. Calcium nitrite has been used in bridges, parking and roof decks, marine and other prestressed concrete structures that are exposed to chloride attack.

Sodium nitrite ($NaNO_2$), a fine free-flowing powder, has been used effectively in the absence of chlorides in both normal and steam-cured concretes at dosages of 1–2% by cement weight. Several investigators have confirmed the effectiveness of sodium nitrite in inhibiting or retarding corrosion [52, 55–57]. One of the problems associated with the use of $NaNO_2$ is its detrimental effect on concrete strength. There is also concern that the addition of sodium nitrite to concrete containing alkali-prone aggregates may enhance the alkali–aggregate expansion. If inadequate quantities are used or if the ratio of the inhibitor to the chloride level is small, corrosion becomes intensely localized and the attack is significant, causing sever pitting. Also, the ready leaching renders the protection a stopgap measure. Thus, their use is restricted due to these limitations.

A rigorous 4-year study (involving 1200 samples with 1.3-inch (3.3 cm) concrete cover prepared from 15 mix designs different dosages of nitrite and partly submerged in 3% sodium chloride) demonstrated that calcium nitrite delays the onset of corrosion and when it begins the rate remains lower than that of the unprotected members. The corrosion resistance is better in a lower water–cement ratio concrete. Corrosion was measured by polarization resistance technique, electrochemical impedance and visual inspection by periodic removal of the specimen [58]. The results of corrosion with reference to time are shown for concretes with and without calcium nitrite in Fig. 6.9 [58]. Advantages claimed for calcium nitrite over sodium nitrites include reduced leaching, efflorescence and potential for reducing alkali–aggregate reaction [50].

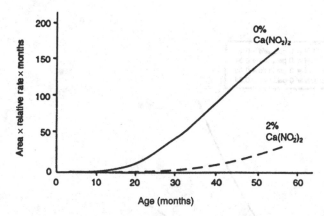

Fig. 6.9 Weighted total corrosion of concrete with 2% calcium nitrite (NACE International, reprinted with permission).

In order to counteract the corrosion of reinforcing steel when it is exposed to salt, it is important to know what dosages of the inhibitor must be added. The dosage required is related to the amount of chloride in concrete and is determined by the chloride–nitrite ratios that will control corrosion. Based on data from various studies, Berke and Rosenberg[59] have identified the chloride–nitrite ratios at which corrosion occurs, using a 30% calcium nitrite solution. The corrosion-inhibiting effect produced by calcium nitrite in concrete in the presence of chlorides is shown in Fig. 6.10. It can be seen that the control of the effects of corrosion is dependent both on the level of chloride ion present at the vicinity of the reinforcement as well as the dosage of the admixture. Studies carried on mini-bridge decks, subjected to daily salt application have shown that calcium nitrite provides more than an order of magnitude reduction in the corrosion rate and that calcium nitrite afforded protection against corrosion if the chloride–nitrite weight ratio (Cl^-/NO^-_2) did not exceed 1.5 [60–62]. Based on these results some guidelines were developed for the dosage requirements of calcium nitrite. These values are given in Table 6.8.

Calcium nitrite which meets the requirements of ASTM C494 is both an accelerator and a corrosion inhibitor. It accelerates the setting times and also rate of development of strength compared to concretes containing no admixture (Table 6.9) [62]. Setting time acceleration and rate of strength development increases as the dosage is increased, but it has little effect on air content and slump. Calcium nitrite is compatible with both silica fume and fly ash concrete. In high-performance concrete containing a silica fume, it plays a significant role in controlling chloride diffusion [63]

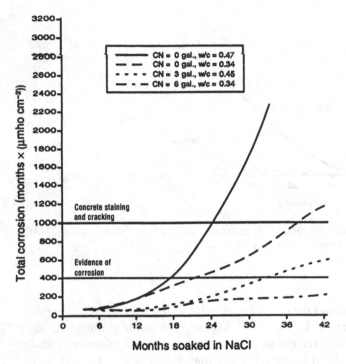

Fig. 6.10 Corrosion vs time in 3% NaCl as a function of water–cement (w/c) ratio and calcium nitrite content (NACE International, reprinted with permission).

Table 6.8 Dosage requirements for calcium nitrite at different chloride levels (Berke and Weil)

Calcium nitrite, gal/yd³	(l m⁻³)	Chloride, lb/yd³	(kg mg⁻³)
2	(9.9)	6.0	(3.57)
3	(14.85)	9.9	(5.89)
4	(19.8)	13.0	(7.74)
5	(24.75)	15.0	(8.93)
6	(29.7)	16.0	(9.52)

Two types of calcium nitrite-based corrosion inhibitors are currently marketed, viz. a set- and strength-accelerating type and a normal-setting type. The former increases the early strength development in concrete. This effect increases with the dosage. Both admixtures are compatible with all types of Portland cements and other admixtures, although both may moderately reduce the amount of entrained air. Freezing for both

Table 6.9 Mechanical properties of concrete containing calcium nitrite (Berke and Weil)

Mix no.	Accelerator (%) (cement basis)	Acceleration (%) (initial set)	Compressive strengths (MPa)		
			1 day	3 days	7 days
1	0	—	10.4	21.2	29.0
2	1.69	24	14.1	22.5	29.9
3	3.39	30	14.9	23.8	32.9
4	5.06	38	15.3	24.3	36.4

admixtures occurs at about $-15\,^{\circ}\mathrm{C}$ but the corrosion-inhibiting properties are restored by thawing and thorough agitation. Mix-water adjustment (using an adjustment factor is 0.635 kg of water per liter) is necessary when using these admixtures.

(b) Nitrite/other chemical combinations

Although molybdate compounds have been advocated for corrosion inhibition purposes they have not been used as inhibitors in concrete practice. Experiments to ascertain the synergistic effect of a calcium-nitrite–sodium-molybdate combination (4.5 parts to 1 part) on corrosion of steel in concrete [64] showed that at the inhibitor–chloride ratio of 1:11 the combined admixture protected steel from corrosion and that it was more effective than when calcium nitrite was used alone.

Sodium benzoate has been used singly in concrete structures exposed to severe corrosion attack and also in combination with sodium nitrite in cement slurries to paint on steel reinforcement before embedment in concrete [65, 66]. Work done by Lewis *et al.* showed that concrete to which 2% benzoate was added produced setting times that were similar to control concretes, but the compressive strengths decreased by about 40% [66, 67]. Lewis *et al.* [66] however have concluded that sodium benzoate has a more persistent inhibitory effect than calcium nitrite.

(c) Phosphate-based inhibitors

Corrosion-inhibitive properties of the compound Na_2PO_3F have been tested by Andrade *et al.*, either by incorporating it in a mortar or as a penetrant [68]. This compound, which is currently available as a proprietary product, is reported to act as an anodic inhibitor, possibly with some cathodic action. The minimum required ratio of phosphate to chloride was suggested as 1:1. The mechanism of action of this admixture is to stabilize the passive layer of iron oxide on the steel and also increase the density of concrete, thus decreasing the permeability of chloride ions; in leaner mixes some retardation occurs. It is compatible with other admixtures and particularly effective in concrete containing pozzolanic materials.

(d) Water-based organic materials

These include water-based materials such as amine, esters and alkylsilanes. Water-based amines and esters have no significant effect on slump and rate of hardening of concrete. It is claimed they delay corrosion of steel both in cracked and uncracked members. Effectiveness of the admixture has been attributed to a dual mechanism of corrosion inhibition, viz. prevention of chloride and moisture ingress and formation of a protective film on the surface of the steel. Ester molecules derived from the admixture are said to line the pores of the concrete thus increasing the resistance to moisture and chloride penetration, while the film-forming characteristics of the amine promote adsorption of the molecules on the steel surface, providing a protective coat which keeps moisture and chlorides out.

Two particular advantages are claimed by the manufacturer: (1) it is effective in all qualities of concrete and (2) it removes the 'guessing' of the chloride loading in the determining the required dosage. The recommended dosage for optimum corrosion protection is one gallon per cubic yard (5 liter/m^{-3}). It is reported that this admixture would need an increase in the amount of air-entraining agent to obtain a specific amount of air. Like most emulsions the admixture needs to protected from freezing and from heat and should therefore be stored above 1.7°C and not exceeding 52°C [69].

It is important to ensure that when using corrosion inhibitors with other conventional admixtures, they are added separately, at different times of the mix cycle. Corrosion-inhibiting admixtures which also accelerate the set of concrete may require the combination of a retarding admixture when ambient and mix temperatures exceed 35°C. In like manner, set-retarding corrosion inhibitors may require the addition of an accelerator to offset the retardation of early strength development (e.g. use of sodium nitrite in conjunction with sodium benzoate).

6.5.3 Research on other corrosion inhibitors

Other compounds that have been tested for their inhibiting action include carboxylic acids, hydrazine hydrate, chelating compounds, aniline and related compounds and petroleum-based compounds. The influence of various chelating agents such as TEA, EDTA, DPTA, HEDTA and Chel-138 on their ability to control corrosion has been investigated [70]. All the chemicals are reported to reduce the compressive strength of concrete. The strengths were particularly low in the presence of TEA and EDTA, compared to the reference. In the presence of 0.1 N NaCl solution at pH 10 and 12, both HEDTA and Chel-138 decreased the critical potential for pitting. This decrease was proportional to the concentration of these additives. Hydrazine hydrate, which behaves as an anodic inhibitor, functions by blocking the anodic sites and enhancing the anodic polarization above a critical concentration depending on the hydrazine–aggressive-ion ratio [71]. Higher concentrations of hydrazine cause instantaneous passivation of steel. Water-soluble carboxylic acids such as malonic acid and a dicarboxylic acid have been shown to function as good corrosion inhibitors even in the presence of 2.5% chloride [71]. However, they cannot be recommended for use in concrete because of the severe setting retardation that occurs even at very low dosages.

Corrosion is the dominant degradation mechanism in the deterioration of concrete structures. It has therefore received a great deal of attention both by researchers and practitioners, resulting in the publication of a significant body of information. However, despite the large amount of published information there is a paucity of data relating [72] to the mechanism of passivation and depassivation of steel in the presence of chloride, the amount of chloride necessary for depassivation, the relative roles of admixtured chloride and that which enters the concrete through external application on corrosion, the amount of corrosion product that is needed to cause destructive corrosion, the range of half cell potential that represents corrosion activity, and the best method of analysis of corrosive chloride. Notwithstanding this lack of information, the proven effect of inhibitors in extending the service life of repaired structures cannot be discounted.

6.6 Calcium-sulfoaluminate-based expanding admixtures

Calcium sulfoaluminate-based admixtures (CSAs) have provided a means of offsetting volume changes due to hardened shrinkage. Shrinkage compensation is obtained at lower addition rates, while chemical prestressing of the reinforcement is achieved at higher dosages. Other allied admixtures of this type and mixtures of $CaSO_4$ and CaO have also been used. Although the calcium-sulfoaluminate-based Type K cement is the more widely used material for shrinkage compensation in North America, it is limited in its use to specific applications such as grouts and flat slabs. Furthermore, it cannot be used for applications that require prestress and the degree of expansion is affected by higher temperatures. CSAs offer flexibility in use and the effects of adverse ambient conditions can be compensated by the use of higher dosages [73]. CSAs are widely used throughout the South East and Far East Asian countries, Japan and Australia. In Japan the product has a 20-year track record in use in large scale construction. [74].

6.6.1 Chemical composition

The most widely used single component, calcium sulfoaluminate admixture, is composed of 30% C_4A_3S, 50% $CaSO_4$ and 20% CaO with small amounts of glassy phase. Particle size is coarser than that of Portland cement. Larger particle size ensures that the potential expansion due to hydration is extended over a period of time. Chemical and physical properties of the most widely used proprietary product, Denka CSA, are given in Table 6.10 [74]. Other CSAs include mixtures of $C_4 ASH_{12}$ and 2 CS (monosulfate and gypsum) and mixtures of Type I cement, high-alumina cement, $CaSO_4.2H_2O$, $Ca(OH)_2$ and CaO [75].

Anhydrous calcium sulfoaluminate is formed by calcination of lime, gypsum and bauxite. The active expansive ingredient, $C_4A_3S^-$ is formed by solid-state reaction from mixtures of compounds composed of calcium oxide, aluminum oxide, sulfur trioxide gas formed during the calcination of gypsum, and bauxite. Crystal growth of CSAs is encouraged to proceed at a slow rate to preserve the potential force of expansion for extended periods [76].

For use in special fields of construction such as structural grouting and oil-well cements, the anhydrous calcium sulfoaluminate is combined with two or more of the following admixtures: (1) a dispersing admixture. (2) a gas liberating agent, e.g. aluminum powder or fluidized coke particles, (3) a powdered acrylic latex to increase bond strength, and (4) mortar density-increasing or -decreasing ingredients such as barytes or bentonite.

6.6.2 Mode of action

The expansion mechanism associated with the formation of ettringite in cement is not clear and several hypotheses have been advanced [77]. A

Table 6.10 Chemical and physical properties of calcined Denka CSA (courtesy S. Matsumoto)

Chemical properties		Physical properties	
Oxides:		Free CaO	19.4%
SiO_2	1.4%	Specific gravity	2.93%
Al_2O_3	13.1%	Specific surface area	2280 $cm^2 g^{-1}$
Fe_2O_3	0.6%		
CaO	47.8%		
MgO	0.5%		
SO_3	32.2%		
Ignition loss	0.9%		
Insoluble component	1.4%		

mechanism advanced by the manufacturers of a CSA and supported by the work of other investigators [74, 75, 77] is as follows. On reacting with water, CSAs form ettringite and expand. Formation of ettringite is thought not to occur in the liquid phase of the cement. The $C_4A_3S^-$ compound and lime react to form a solid solution consisting of hexagonal plate-like crystals of monosulfate and calcium aluminate hydrate of type C_4AH_{13}. Subsequent reaction of the monosulfate with gypsum produces acicular crystals of ettringite. Monosulfate apparently does not contribute to expansion, whereas the formation of ettringite involves expansion. A recently issued patent [75] covers the use of prehydrated high-alumina cement (H-HAC), lime and gypsum mixtures. Particle type, size, thickness of protective coating, and presence of moisture determine the rate and extent of expansion [Fig. 6.11].

The deformation that accompanies expansion as strength increases induces compressive stresses in the concrete, reducing the tensile stresses induced by drying shrinkage. Consequently, both cracking and contraction that occur on drying are decreased. In chemical prestressing, the expansive force generated produces tension in steel to a degree corresponding to the expansion produced. Concrete is simultaneously subjected to an equivalent compressive stress because of the restriction imposed by the reinforcement.

Although the mechanism of expansion for multicomponent admixtures is the same as that of CSAs, the rate and extent of expansion of the former is

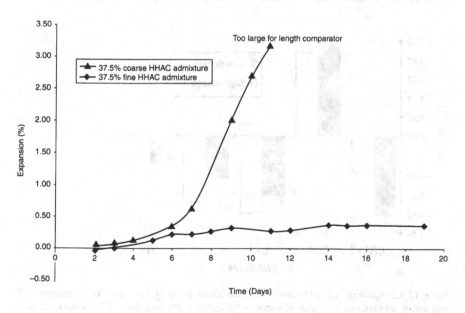

Fig. 6.11 Comparison of expansion for admixture containing coarse and fine grinds of hydrated high alumina cement (courtesy NRC).

also determined by the resulting modification produced by other components in the admixture. Formulation ingredients such as superplasticizers and silica fume significantly influence the extent of expansion in this regard [78, 79]. The 24 h early volume change values obtained for grout compositions containing two CSAs in conjunction with super plasticizer and silica fume is compared with that obtained for Type K cement in Fig. 6.12 [80].

6.6.3 Mix proportioning, mixing and curing

Admixture dosages vary from 8 to 11% by cement weight for shrinkage compensation and 12 to 17% for prestressing applications. Recommended dosage levels for the H-HAC-based admixtures vary from 10 to 20% depending on the desired level of expansion. Shrinkage compensation and crack prevention is attained at the lower dosage while for prestressing applications higher dosages are used.

When CSAs or H-HAC-based admixtures are used, cement is replaced in the mix and the proportions adjusted by reducing the sand by an amount corresponding to that of the admixture. Measurement of the powdered admixture quantities into the mix may be based on a given number of bags or a weighed amount of the material per weight of cement in the mix. In both cases the admixture is batched by weight and not by volume. The point

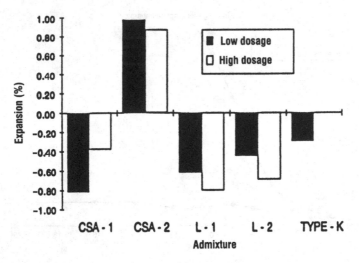

Fig. 6.12 Comparison of 24 h early volume change using low and high dosages of expansive admixtures vs Type K cement. (Courtesy Mailvaganam, N., Nunes, S. and Bhagrath R. (1993). Effectiveness of expansive admixtures in structural grout compositions, *Concrete International*, **15**(10), 38–43.

of addition in both precast and ready-mix applications is with the fine aggregate prior to the addition of the mix water.

The expansion achieved in concrete containing these admixtures is dependent on the type of aggregates used in the mixture. Thus the desired level of shrinkage compensation should take into account the modulus of the aggregate to be used on the job. The effect of cement and water contents on the extent of expansion should also be considered. Minimum cement content required to achieve desired expansion should be determined through mix trials. Compatibility with other admixtures may also need trials and the manufacturer's recommendation should be followed [80].

Satisfactory performance of the admixtures will depend to some extent on the degree of mixing. Slightly longer mixing times than those used for Portland cement mixes may be required, especially when small amounts of admixture are used. In hot weather, the sequence of charging the concrete ingredients into the ready mix truck can be changed. The truck drum is kept stationary until it reaches the destination when a 3–5 min mixing is carried out prior to discharge. This method minimizes slump loss and ensures that the expansive potential is not seriously affected. When mixing in cold weather using hot water, water should be added to the aggregate, followed by the cement and admixture; otherwise fast setting may occur.

Admixture dosage and post-placing curing conditions affect the properties of concrete, hence, selection of an appropriate dosage would depend on: (1) desired level of expansion, (2) degree of restraint of the structure and (3) curing conditions. In situations where adequate curing cannot be provided a slightly higher than normal dosage should be used to ensure effective shrinkage compensation. Since more than normal amount of water is required at the initial stages to produce effective expansion, water suction into the dry substrate should be minimized by thoroughly wetting the base or subgrade [78, 80, 81].

6.6.4 Factors influencing the reaction

The following factors affect the reaction which forms ettringite and the resultant expansion.

(a) Composition and fineness

Type and amount of aluminates, calcium sulfate and free lime present in the mixture governs the rate at which ettringite forms. The presence of lime is reported to be essential for both the initial and subsequent stages of ettringite formation because it maintains a solution phase saturated with calcium ions [81]. Crystal habit, particle size and range of particle size in CSAs determine its rate of hydration and duration of expansion (Fig. 6.11).

(b) Cement content

Higher expansion is attained in cement-rich mixes and this decreases as the cement content is lowered. In general, a minimum cement content of 280 $kg\,m^{-3}$ is required to obtain desired expansion values [78, 80].

(c) Ratio of water to cementing materials

At admixture dosage levels ranging from 9 to 13%, concretes with CSAs show significant increases in both total expansion and compressive strength at water–cement ratios less than 0.50 [81].

(d) Ratio of admixture to cement

The commonly used ratios of admixture to cement for the purpose of shrinkage compensation are 9–11 (admixture) to 91–89 (cement). At these ratios the properties of CSA concrete are similar to Portland cement concretes of similar mix proportions. At admixture dosages exceeding 11% however, concrete workability and strength decrease, while expansion and air entrainment increase. When expansion is unrestrained and exceeds 0.3%, strength is reduced [74].

(e) Curing conditions

At 8 to 11% dosage for CSA-containing concretes, larger expansions are produced when the concrete is water cured or at 100% RH than when sealed with curing compounds. When cured at 50% RH (air cured) little expansion (~0.05%) occurs and shrinkage occurs after 7 days [82]. Compressive strength for water-cured material is slightly lower than that of air-cured material. The effects produced above are more drastic in prestressed applications, since higher dosages are used. If no restraint is provided during moisture cure, disruptive expansion can occur.

(f) Temperature

Higher mix and ambient temperatures result in increased slump losses and reduced ultimate expansion. Reduced expansion also occurs at lower temperatures. In general, greater expansions are obtained at moderate (18–25°C) temperatures. At higher temperatures (>35°C) the rate of ettringite formation is accelerated and although a high rate of expansion occurs at early ages, the resistance offered by the concurrent acceleration of strength development results in a lower ultimate expansion value. At lower temperatures, and early ages the rate of ettringite formation is slower and the expansion produced is dissipated by higher creep.

(g) Degree of restraint

Adequate restraint during expansion must be provided to induce compressive stresses required for shrinkage compensation or for prestressing of steel. This is usually provided by the reinforcement, subgrade friction and forms. For a given admixture dosage, cement content and mix proportions, expansion decreases with increase in steel reinforcement. In the absence of restraint the level of compressive stresses necessary to offset shrinkage stresses is not achieved at lower dosages, while disruptive expansion results in self-stressing applications.

(h) Mixing time

Since the expansion produced is dependent on the uniformity of particle distribution, longer than normal mixing times are required. However, prolonged mixing will result in a significant reduction of the expansive potential, especially at higher temperatures.

(i) Compatibility with other admixtures

Superplasticizers and accelerators have been shown to reduce expansion significantly [79, 83]. The magnitude of the effect is dependent on whether the admixture retards or accelerates set. Retarders tend to increase ultimate expansion at normal temperatures [77, 80]. Under hot-weather conditions however, retarders offset the accelerating effects of high temperatures and allow the normal level of expansion to occur. Air-entraining agents do not influence the expansive reaction, although higher air contents may result when a CSA is used in air entrained concrete. The inclusion of silica fume in grout compositions made with Type K cement or CSA type expansive agents may reduce expansion. Silica fume is said to decrease the formation of ettringite by reducing the concentration of calcium (Ca^{++}) and hydroxyl (OH^-) ions involved in the formation of ettringite [78, 82–85].

(j) Aggregate type

Aggregate type influences expansion and shrinkage characteristics of concrete. Use of aggregates of high elastic modulus results in larger expansion [80].

(k) Age of admixture

CSA admixtures are more prone to loss of activity due to CO_2 and moisture pick up than are shrinkage compensating or Portland cements [74, 79, 85]. Consequently, exceeding the shelf life may seriously reduce the expansive potential. The materials are therefore packed in water-proof bags which should always be stored in a dry place. Once a bag is opened it should preferably be used the same day to maintain the activity. Shelf life is usually 9–12 months.

6.6.5. Effects on the plastic and hardened properties of mortar and concrete

The water requirement for equal slumps is generally higher for concretes containing these admixtures and they show significant reduction in bleeding. Both the high water demand and reduced bleeding are related to the avid water demand for ettringite formation [86]. When used at lower dosages (6–8%) CSA does not entrain any significant amount of air. At higher dosages however, there is a tendency for foaming to occur. Since CSA-containing mixes show increased cohesion and reduced bleeding, finishing operations should occur sooner than for Portland cement concretes. Due to the lack of bleed water, conditions that promote rapid moisture loss may cause plastic shrinkage. Precautionary procedures detailed in ACI 614-59 practice should be followed for satisfactory results.

Physical properties of CSA-containing concretes (compressive strength, creep, modulus of elasticity and durability) are comparable to those of corresponding Portland cement concretes, especially when CSA is used at dosages of 8–11%. When the dosage exceeds the ranges mentioned above and no restraint is provided either internally (by reinforcement) or

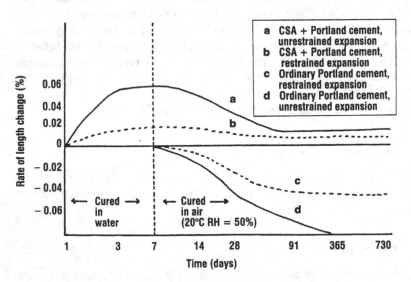

Fig. 6.13 Volume change at various ages for Portland cement concrete with and without CSA admixture and with and without restraint (Matusumoto [74]).

externally, a point is reached when the expansion will have a disruptive effect on mechanical properties. The admixture produces most of the expansion at early ages of wet curing. Subsequently on exposure to lower relative humidity levels as in air curing, a gradual decrease in the amount of expansion occurs with age. Depending on the extent of expansion achieved in the wet curing period and the rate of shrinkage soon after water curing ceases, significant contraction can occur (Fig. 6.13).

Under conditions of internal restraint, the expansion produced is proportional to the ratio of steel to concrete and the dosage of the admixture. Special care should be taken to ensure that reinforcement is located in its proper position during placement and consolidation so that adequate restraint and good bond to steel is obtained. Restrained expansion increases the density of the matrix and produces concrete or mortar with a lower coefficient of permeability than that of corresponding Portland cement concretes and mortar.

6.6.6 Applications

Shrinkage compensation is highly advantageous in many applications such as architectural precast and pneumatically applied concrete, water-retaining structures and for most horizontal slab applications such as floors, roofs and parking decks. The capacity for minimizing cracking, thereby allowing a reduction in construction joints, makes the admixture ideally suited for these applications. In structures where water leakage is a problem, use of these admixtures can result in the reduction of the number of cracks. Chemical prestressing is used in the production of pressure pipes, water tanks and tunnel linings. Multicomponent CSA are widely used in structural grouting of machine and column bases, repair of cracks and oil-well cementing operations.

6.7 Polymer-based admixtures

Of the several types of the polymer-modified mortars and concretes used for various construction applications, latex-modified mortar and concrete are by far the most widely used materials. Latex-modified mortar and concrete are prepared by mixing a latex, either in a dispersed liquid or as a redispersible powder form with fresh cement mortar and concrete mixtures. The polymers are usually added to the mixing water just as other chemical admixtures, at a dosage of 5–20% by weight of cement. Polymer latexes are stable dispersions of very small (0.05–5 μm in diameter) polymer particles in water and are produced by emulsion polymerization. Natural rubber latex and epoxy latex are exceptions in that the former is tapped from rubber trees and the latter is produced by emulsifying an epoxy resin in water by the use of surfactants [87].

6.7.1 Categories

Latex materials are classified into three types according to the type of surfactant used in the production of the latexes: cationic (positively charged), anionic (negatively charged) and nonionic (uncharged). The action of the emulsifier (surfactant) is due to its molecules having both hydrophilic and hydrophobic parts and the properties of the latex formed are very dependent on how the various constituents are put together.

Latexes are usually copolymer systems of two or more monomers, and their total solids content, including polymers, emulsifiers, stabilizers etc. is 40–50% by mass. Most commercially available polymer latexes are based on elastomeric and thermoplastic polymers which form continuous polymer films when dried [88]. The major types of latexes include styrene–butadiene rubber (SBR), ethylene vinyl acetate (EVA), polyacrylic ester (PAE) and epoxy resin (EP) which are available both as emulsions and redispersible powders. They are widely used for bridge deck overlays and patching, as adhesives, and integral waterproofers. A brief description of the main types in current use is as follows [87].

(a) Styrene–butadine rubber

Styrene–butadiene rubber (SBR) latexes which are compatible with cementitious compounds are copolymers. They show good stability in the presence of multivalent cations such as calcium (Ca^{++}) and aluminum (Al^{+++}) and are unaffected by the addition of relatively large amounts of electrolytes (e.g., $CaCl_2$). Outdoor exposure to sunlight tends to result in a gradual embrittlement of the cured latex, due to a lack of UV resistance. SBR latexes may coagulate if subjected to temperature extremes, or severe mechanical action for prolonged periods of time.

(b) Polyvinyl acetate latex

Commercial materials are copolymers manufactured by the emulsion polymerization process. Two main types of PVAs are used in construction:

1. Non-emulsifiable PVA, which forms a film that offers good water resistance, good light stability, and good aging characteristics. Because of its compatibility with cement, it is widely used as bonding agent and a binder for cementitious water-based paints and waterproofing coatings.
2. Emulsifiable PVA, which produces a film that can be softened with water, restoring adhesiveness and thereby permitting application of a film to a surface long before the subsequent application of a water-based overlay. Its use is limited to specific applications where the possible infiltration of moisture to the bond line is precluded.

(c) Acrylic latex

The resins used are polymers and copolymers of the esters of acrylic and methacrylic acids. They range in physical properties from soft elastomers to hard plastics, and are used in cementitious compounds in much the same manner as SBR latex. Acrylics are reported to have better UV stability than SBR latex and therefore remain flexible under exterior exposure conditions longer than SBR latex [88].

(d) Ethylene vinyl acetate

These polymers are thermoplastic materials usually supplied as redispersible powders. They are widely used in proprietary prepackaged products such as tile grouts, self-leveling underlay and patching materials.

(e) Epoxy latex emulsions

These are produced from liquid epoxy resin when mixed with the curing agent, which serves additional functions such as an emulsifying and wetting agent and as a surfactant [89–92]. The emulsions are stable and water dilutable from the time of mixing until gelation occurs. Pot life may be varied from 1 to 6 h depending upon the curing agent selected and by adding large amounts of water [89, 90]. Most epoxy emulsions are prepared just immediately prior to use on the job site rather than in the manufacturer's plant. This avoids the phase separation that occurs in previously prepared packaged emulsions. During mixing, equal parts of the epoxy are mixed with equal parts of the curing agent. The mixture is blended for 2–5 min and allowed to set for 15 min for polymerization to begin. Next, water is added slowly while the mixture is mechanically agitated to form the emulsion.

Cement hydration and epoxy polymerization occur simultaneously to form a structure that is similar to the latex-modified cementitious system. Epoxy systems develop high strength, adhesion and have low permeability, good water resistance and chemical resistance. A major advantage of this system is that it can be cured under moist or wet conditions. According to a recent study, the epoxy-modified mortars can be made without the hardeners with superior properties to those obtained with conventional epoxy mortars [89, 90].

(f) Redispersible polymer powders

These are used to produce prepackaged products such as decorative wall coatings, ceramic tile adhesives, self-leveling floor overlays and patching mortars for concrete structures. On site, addition of the product manufacturers specified amount of water produces a material with the desired properties. During the wet mixing, the redispersible polymer powders are re-emulsified and form continuous polymer films in the modified mortar and concrete, and behave in the same manner as the liquid latexes [87, 92].

6.7.2 Material parameters influencing performance

The properties of a latex depend on the nature of polymers in the latex, particularly the monomer ratio in copolymers and the type and amount of plasticizers. The monomer ratio affects the strengths of the latex modified mortars to the same extent as the polymer–cement ratio [87, 92]. Mechanical and chemical stability, bubbling and coalescence on drying all depend on the type and amount of surfactants and antifoamers and the size of dispersed polymer particles. It is important that the use of selected antifoamers and surfactants as stabilizers or emulsifiers produces no adverse effect on cement hydration.

Surfactants enable the polymer particles to disperse effectively without coagulation in the mortar and concrete. Thus, mechanical and chemical stabilities of latexes are improved with an increase in the content of the surfactants selected as stabilizers. An excess of surfactant, however, may have an adverse effect on the strength because of the reduced latex film strength, the delayed cement hydration and excess air entrainment. Consequently, the latexes used as cement modifiers should have an optimum surfactant content (from 5 to 30% of the weight of total solids) to provide adequate strength. Suitable antifoamers are usually added to the latexes to prevent excess air entrainment; increased dosages causes a drastic reduction in the air content and a concurrent increase in compressive strength [87, 92–94].

The molecular weight, glass transition temperature (T_g) and size of dispersed polymer particles in the latexes can affect the strength and chloride ion permeability of latex-modified mortar and concrete to a certain extent [87, 93] (Tables 6.11 and 6.12). SBR latexes with smaller particle size appear to initially provide lower chloride ion permeability to the mortars, but a difference in the permeability between the smaller and larger particle sizes eventually becomes insignificant as the concrete ages. The initial decrease in the permeability observed with smaller particles is attributed to the fact that smaller particle size coalesce into films faster than the larger particle sizes.

Polymer latexes used as admixtures in mortar and concrete should [87, 88, 93]:

- Have stability towards cations such as calcium (Ca^{++}) and aluminum (Al^{+++}) ions.
- Be stable to the high shear encountered in mixing and in transfer pumps.
- Have a low air-entraining capability during mixing when a suitable antifoaming agent is used in the emulsion or redispersible powder.

Table 6.11 Effect of glass transition temperature T_g on the properties of acrylic-latex-modified mortar (Ma and Brown)

Mortar identification	F	G
T_g by DSC (°C)	+12	−55
Flow, C 230 (%)	112	108
Wet density g ml^{-1}	2.19	2.15
Flexural strength, C 78 (CoV = 7.5%) (MPa)		
7 days	6.90	4.95
Compressive strength, C 109 (CoV = 5%) (MPa)		
7 days	31.90	17.80
28 days	46.00	29.85
Permeability, C1202 (CoV = 10) (C)		
9 days	3830	2800
15 days	2510	1680
28 days	1630	890
61 days	1060	670

Table 6.12 Effect of particle size on the properties of SBR-modified concrete

Concrete identification	A	B
Average particle size of SBR (nm)	90	190
Slump, C 143 (mm)	102	152
Air content, C 230 (%)	5.2	5.8
Permeability, C 1202 (CoV = 10%) (C)		
14 days	1150	2100
28 days	960	1420
56 days	810	800

- Have no adverse effect on the hydration of the cement.
- Have ability to form continuous films in mortar or concrete, due to a lower minimum film-forming temperature than the application temperature, which adhere well to the aggregates and cement hydrates and possess good alkali and water resistance.
- Possess good thermal stability in a wide range of temperatures during transportation and storage, particularly under cold- and hot-weather conditions, so that coagulation of the emulsion does not occur.

6.7.3 Modification of the cementitious matrix

Modification of mortar and concrete in the presence of a latex occurs by concurrent cement hydration and formation of a polymer film (coalescence of polymer particles and the polymerization of monomers). Cement hydration generally precedes the polymerization and as curing progresses, a comatrix phase consisting of cement hydrates and polymer films is formed (Fig. 6.15). The benefits of latex modification are best realized when both cement hydration and polymer film formation proceed well to yield a monolithic matrix network structure in which the hydrated cement phase and polymer phase interpenetrate. The superior properties of the polymer-modified mortar and concrete to conventional mortar and concrete are characterized by such a distinct structure. For example, improved bonding between the aggregates and cement hydrates which occurs with some reactive polymers such as polyacrylic esters (PAE) due to their reaction with calcium hydroxide or silicate surfaces increases flexural strengths; and the polymer films which provide a sealing effect around the aggregates and cement hydrates are able to bridge cracks and prevent crack propagation, increase waterproofing and watertight characteristics, chemical and freeze–thaw resistance [87, 93, 94].

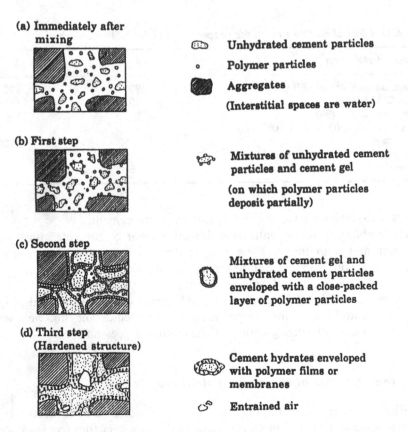

Fig. 6.14 Simplified model of formation of polymer–cement comatrix (Ohama [87]).

Such effects increase with an increase in the polymer content or the polymer–cement ratio (the weight ratio of total solids in a polymer latex to the amount of cement in a latex-modified mortar or concrete mixture). However, at levels exceeding 20% by weight of the cement in the mixture, excessive air entrainment and discontinuities form in the monolithic network structure, resulting in a reduction of compressive strength and modulus [87, 94, 95].

6.7.4 Mix proportioning

Although the mix design of latex-modified mortar and concrete is done in much the same way as that of ordinary mortar and concrete, properties such as workability, strength, extendibility, adhesion, watertightness and chemical resistance are controlled by the polymer–cement ratio rather

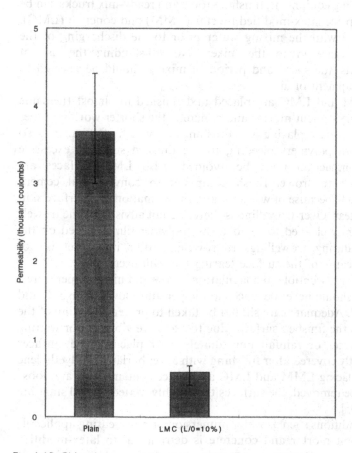

Fig. 6.15 Chloride permeability test results (Ohama [87]).

than by the water–cement ratio. Hence, the polymer–cement ratio should be based on the desired requirements.

Mix proportions of most latex-modified mortars are in the range of the cement–fine-aggregate ratio 1 : 2 to 1 : 3 (by weight), the polymer–cement ratio of 5–20% and the water–cement ratio of 0.3–0.6, depending on their required workability. Because of the inherent air-entraining characteristics of latexes, air-entraining agents cannot be used in latex-modified mortar and concrete. For cold-climate or winter applications, the use of high-early-strength cement, ultra-rapid-hardening cement and high-alumina cement should be considered. The ACI (American Concrete Institute) suggested guidelines for the mix proportions of latex-modified concretes [87, 93] and these are a useful resource for bridge deck and patching applications respectively.

6.7.5 Mixing, placing and curing

Conventional mixing equipment, finishing tools and ready-mix trucks can be used to mix and place latex-modified mortar (LMM) and concrete (LMC). The latex is added with the mixing water prior to the discharging of the cement and aggregate in to the mixer. Notwithstanding the use of antifoaming agents, the speed and period of mixing should be selected to avoid undue entrapment of air.

Although LMM and LMC are placed and finished in almost the same manner as ordinary cement mortar and concrete, the shorter working times of the former necessitate placing and finishing within 1 h after mixing. To prevent water with polymers bleeding to the finished surfaces, excessive vibration for compaction must be avoided. Also LMC surfaces are somewhat difficult to trowel finish compared to conventional cement mortar and concrete because of wet drag and the formation of a surface skin which will easily tear. Over-trowelling is therefore not advisable. The trowels should be frequently cleaned to remove thin polymer films formed on the trowel surfaces during trowelling. Retrowelling after initial set is not recommended because of the surface tearing that will occur.

Since the latex is susceptible to coagulation at low and high temperatures, LMM and LMC should never be placed at temperatures lower than 5°C and higher than 30°C. Adequate care should be taken to prevent bleeding of the polymer solids to the finished surfaces due to excessive vibration or wetting by sprayed-on water or rainfall immediately after placing. The surfaces should be promptly covered after finishing with a wet burlap or polyethylene sheets. Prior to placing LMM and LMC all surface contaminants and loose material should be removed, the surfaces thoroughly wetted and all standing water removed [95].

Wet curing conditions such as water immersion or moist curing applicable to ordinary cement mortar and concrete is detrimental to latex-modified mortar and concrete. Optimum strengths are obtained by providing a reasonable degree of wet conditions (for 1-3 days) for early cement hydration and subsequent dry conditions (at ambient temperature) to promote the coalescence of the polymer particles for film formation. The effects of these curing conditions on the strength of LMM and LMC are shown in Table 6.13 [87].

Generally, polymer latexes used as cement modifiers are not toxic, are safe materials to handle and require no special precautions. However, because they have an excellent adhesion to various materials, even to metals, all the equipment and tools such as mixers, trowels, and vibrators should be washed down or cleaned immediately with water after use. For concrete requiring formwork, it is advisable to use the most effective mold-release agents, e.g., silicone wax or grease [87, 96].

6.7.6 Properties of latex-modified mortar and concrete

The presence of the cement hydrate/polymer comatrix in LMM and LMC confers superior properties, such as high tensile and flexural strengths, excellent adhesion, high waterproofness, high abrasion resistance and good chemical resistance, when compared to ordinary cement mortar and concrete. The degree of these improvements however depends on polymer type, polymer–cement ratio, water–cement ratio, air content and curing conditions. Some of the properties affected by these factors are discussed below [87, 88, 93–95].

(a) Workability, setting, bleeding and segregation

The water content, required to produce a given slump, can be significantly reduced when a latex admixture is used and the extent of the reduction increases with an increase in the polymer–cement ratio. This effect is due to

Table 6.13 Effect of wet and dry curing of SBR-modified mortar E (Kuhlmann)

Flow C 230 (%)		115
Wet density (g ml^{-1})		2.21
Type of cure	Wet*	Dry*
Compressive strength, C 109 (CoV = 5%) (MPa)		
2 days cure	14.60	14.80
4 days cure	22.35	21.25
7 days cure	35.85	32.75
28 days cure	42.25	41.35
Chloride ion penetration, C 1202 (CoV = 10%) (Coulombs)		
28 days cure	3240	1320

*Wet cure was in saturated lime water until time 0 test; dry cure was storage in laboratory air (about 25°C and 50% relative humidity) until time of test.

the presence of polymer particles and attendant air entrainment and the dispersing action of the surfactant emulsifiers which provides a ball-bearing effect, resulting in an increase in workability. In general, the setting time of LMM and LMC is extended in comparison to normal mixes and the extent of the delay is dependent on the type of polymer and polymer–cement ratio of the mix. Setting delay is due to the presence of surfactants contained in latexes which mildly inhibit the hydration of cement [87, 94, 96, 97]. Due to the hydrophilic colloidal properties of the latex, the water reduction afforded by their use and the air entrained in the mixtures of mortar or concrete, there is little tendency for bleeding and segregation. This is in contrast to ordinary mixes which are apt to cause bleeding at similar mix consistencies.

(b) Air entrainment

Most latex emulsions cause heavy air entrainment in the absence of antifoaming agents which are either integral components of the emulsion or are added at the time of mixing. Air entrainment results from the action of the surfactants contained as emulsifiers and stabilizers in the polymer latexes. Today's commercial latexes usually contain the proper antifoaming agent, and unwanted air entrainment is therefore considerably reduced. The air content of latex-modified mortar can range from 5 to 20%, while that of latex-modified concrete, depending on the size of the coarse aggregate, can be less than 2%, much the same as ordinary concrete [97].

(c) Strength: tensile, flexural, compressive bond and abrasion

Overall improvement of the cement hydrate–aggregate bond and presence of interpenetrating polymer films in the comatrix of latex-modified mortar and concrete bestow a noticeable increase in tensile and flexural strength. There is however no improvement in compressive strength compared to ordinary concrete. The main factors affecting the mechanical properties of latex-modified mortar and concrete are: (1) the nature of the polymer latex, cement and aggregate used in the mix; (2) the controlling mix proportion parameters such as polymer–cement ratio, water–cement ratio, binder–void ratio and air content; (3) curing and testing methods [87, 98].

The nature of the polymer latex is determined by the monomer ratio in the copolymer and this property of the latex affects strength values in manner similar to that obtained with the polymer–cement ratio. The effects of polymer–cement ratio on strength are presented in Table 6.14 [87].

Most latex-modified mortars and concretes have good adhesion to most substrates (tile, stone, brick, steel and aged concrete) compared to conventional mortar and concrete. In general, bond strength in tension and flexure increases with an increase in the polymer–cement ratio,

Table 6.14 Effect of polymer–cement ratio on mortar properties (Ohama et al.)

Mortar identification	A	B	C	D
Styrene–butadiene				
polymer–cement ratio	0.00	0.05	0.10	0.15
Water–cement ratio	0.394	0.400	0.401	0.402
Flow, C 230 (%)	110	110	120	122
Wet density (g ml^{-1})	2.22	2.24	2.20	2.18
Water in mortar (%)	8.95	8.99	8.92	8.83
Average mass of flexural				
strength specimens (g)				
1 day	180.0	180.3	185.8	186.9
8 days	175.7	176.0	181.5	186.9
28 days	174.6	175.1	180.0	180.5
42 days	174.3	174.3	179.5	180.3
55 days	174.2	174.0	179.2	180.0
Loss of mass (%)				
55 days	3.2	3.5	3.6	3.7
Flexural strength, C 78				
(CoV = 7.5%) (MPa)				
55 days	7.60	9.80	12.60	13.70
Permeability, C 1202				
(CoV = 10%) (C)				
55 days	3840	3100	1320	880

irrespective of the type of polymer used [87, 94, 98]. The values can be affected by the porosity of the substrate, and its service conditions. An increase in abrasion resistance is dependent on the type of latex and polymer–cement ratio used. In general, abrasion resistance is considerably improved with an increase in polymer–cement ratio.

The combination of latex admixtures with other admixtures, such as superplasticizers and silica fume, provides significant improvements not readily attained by normal mortars. Extension of workability, superior bonding strength and increased ratio of bonding to compressive strength have been demonstrated in a polymer-modified mortar containing an acrylic latex and a superplasticizer [95–97, 98]. For example, the reference mortar containing no polymer had a bonding–compressive-strength ratio of 0.2, whereas the mortars containing 15% polymer, 15% polymer plus super-plasticizer and 15% polymer plus silica fume developed ratios of 0.38, 0.31 and 0.37 respectively.

Because of the temperature dependence of the polymers themselves, the mechanical properties of LMMs and LMCs are dependent on the nature of the polymer and the temperature the material will encounter in service. LMMs and LMCs generally show a rapid reduction in strength or deflection with increased temperature. The strength reduction is substantial at temperatures that are higher than the glass transition temperature of the polymers and at higher polymer ratios. The maximum temperature limit for retaining useful strength properties has been found to be 150°C [87, 93, 94].

(d) Deformability, elastic modulus and Poisson's ratio

The significant differences in the moduli of latexes and the cement hydrates (elastic moduli 0.001–10 GPa and 10–30 GPa respectively) causes most LMMs and LMCs to have a higher deformability and elasticity than ordinary cement mortar and concrete. Depending on the polymer type and polymer–cement ratio, the deformability and elastic modulus tends to initially increase with an increase in the polymer–cement ratio and subsequently decrease at higher ratios. Poisson's ratio however is only marginally affected [87, 94, 98].

(e) Drying shrinkage, creep and thermal expansion

Depending on the polymer type and the polymer–cement ratio, drying shrinkage of the LMM and LMC may be either higher or lower than ordinary mortar and concrete. Generally the drying shrinkage tends to decrease with increasing polymer–cement ratio. The type of curing has a significant effect on drying shrinkage; dry curing increases the drying shrinkage but it becomes nearly constant at 28 days, regardless of the polymer type and polymer–cement ratio [87, 98].

In general, at low latex dosage levels, the creep strain and creep coefficient of latex modified concrete and mortar are considerably smaller than those of ordinary cement cement, mortar and concrete [94, 98]. The low creep is probably due to the low polymer content which may not affect the elasticity, but increases the strength by improving the binding capacity of the matrix as well as providing better hydration through water retention in the mortar and concrete. The coefficient of thermal expansion at about $9-10 \times 10^{-6}$ is very similar to that of concrete, which is 10×10^{-6} [87, 94, 99].

(f) Water resistance, chloride ion penetration resistance, carbonation and chemical resistance

The large pores (ranging from 0.01 μm to 0.1 μm) are sealed by the continuous polymer film formed in the comatrix of LMM and LMC. Consequently, they show reduced water absorption, water permeability and water vapor transmission over ordinary cement mortar and concrete; and this effect increases with an increase in polymer content and polymer–cement ratio (Fig. 6.15). The improved water permeability also improves the resistance to chloride ion entry and hence corrosion mitigation [87].

In contrast to the increased strengths of normal mortar and concrete, most LMMs and LMCs tend to lose strength when immersed in water. This trend is more significant for the flexural strength [87, 98]. However, since the flexural strengths of most LMMs and LMCs are significantly higher than ordinary mortar and concrete and because they retain a large part of these strengths there is no problem in their practical applications. Polyvinyl-acetate-based mortars however have poor water resistance because they undergo alkaline hydrolysis when there is continuous exposure to moisture.

The carbonation resistance of latex-modified mortar and concrete is significantly improved because of the pore sealing effects of the latex. Figure 6.16 shows the degree of carbonation (depth of penetration) that occurs in both indoor and outdoor exposure. The magnitude of the resistance is dependent on the type of polymer, polymer–cement ratio used in the mix and carbon dioxide exposure conditions. The chemical resistance is generally rated as good to fats and oils but poor to organic solvents. Although they resist alkalis and most salts (except sulfates), their resistance to mineral acids is only moderately improved over ordinary concrete at the usual latex dosages used. Higher chemical resistance is obtained by selection of a more chemically resistant polymer and a increased polymer–cement ratio [87, 98].

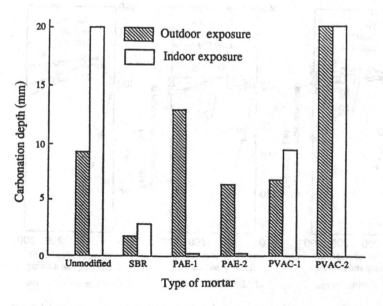

Fig. 6.16 Carbonation depth of latex-modified mortars after 10-year outdoor and indoor exposure (Soroushian and Tlili [91]).

(g) Freeze–thaw durability

The reduction in porosity, decreased water content, and air entrainment that results when latexes are used in mortar and concrete mixes make them much more resistant to freezing and thawing conditions than conventional mortar and concrete. Figure 6.17 presents the freeze–thaw durability in water (−18 to 4°C) of combined water- and dry-cured SBR-, PAE- and EVA-modified mortars [98]. The frost resistance of mortars made with these latexes is markedly improved even at polymer–cement ratios of 5%. However, an increase in the polymer–cement ratio does not necessarily produce further improvement in freeze–thaw resistance. LMM and LMC, when exposed to outdoor conditions involving freeze–thaw, UV radiation and carbonation show better weatherability when compared with conventional mortar and concrete.

6.7.7 Applications

Latex-modified mortars and concretes have become promising materials for preventing chloride-induced corrosion and for repairing damaged reinforced concrete structures. In Japan and the USA, latex-modified mortar is widely used as a construction material in bridge deck overlays and patching compounds, and for finishing and repairs [99]. Polymer–cement hydrate-

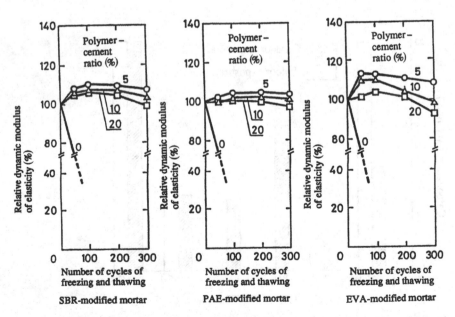

Fig. 6.17 Number of cycles of freezing and thawing vs relative dynamic modulus of elasticity of latex-modified mortars (Ohama [87]).

type flexible waterproofing materials with very high polymer–cement ratios of 50% have been used. Some of the advantages that these materials provide include excellent elongation, flexibility and crack resistance, and good waterproofing and chemical resistance characteristics.

LMC is used in shotcrete applications requiring a quick seal against flowing water. Two types LMC are used [100]: a shotcrete system containing an ambient-temperature-polymerizable monomer which reacts with normal Portland cement, and another system using ultra-rapid-hardening cement. The former is prepared by mixing concrete with magnesium acrylate monomer, where the setting time can be controlled to within a few seconds. The latter is produced by modifying ultra-rapid-hardening cement concrete with SBR and it is used for repair purposes [100].

LMC is used in underwater concrete for both new construction and repair. The important requirements to obtain antiwashout capability, such as segregation resistance, flowability, self-leveling characteristics and lower bleeding are provided by the addition of viscosity-enhancing polymeric admixtures at polymer–cement ratios of 0.2–2.0%. These admixtures are water-soluble polymers, and classified under two groups, viz., cellulose types such as methyl cellulose and hydroxy ethyl cellulose and polyacrylamide types such as polyacrylamide and polyacrylamide–sodium acrylate [101].

6.7.8 Standards and specifications

Current test methods for mortars and concretes having latex modifiers are generally ASTM methods with minor modifications to accommodate the properties unique to these materials. The variations in the tests typically relate to items such as proportioning, mixing sequence and curing. Standardization work on test methods and quality requirements that will address the changes needed in the appropriate existing standards and practices has been in progress in the USA, Japan, UK and Germany, and RILEM specifications and guidelines for polymer-modified mortars and concretes are now available.

6.8 Admixtures for recycling concrete waste

Every year millions of cubic meters of ready-mixed concrete are returned for disposal in dump sites. There are now increasing environmental concerns and restrictions regulating the disposal of returned plastic concrete. In the last decade, environmental protection agencies in Europe and North America have classified both returned plastic concrete and truck wash water as hazardous waste. Consequently, the conventional methods of disposal in slurry tanks and landfill sites have become redundant, and disposal of plastic concrete is now a major problem for the ready-mixed concrete industry [102].

In an effort to assist concrete producers to comply with new environmental regulations, admixture manufacturers developed a cost-effective admixture-based alternative to disposal in the late 1980s. An important application is the overnight or weekend stabilization of returned plastic concrete and wash water for reuse as part of the mix batched the next day. The technique keeps concrete plastic for longer periods making possible its storage and later use. Two non-chloride admixtures (stabilizer and activator), which alternately suspend and reactivate cement hydration, are used in the system. The stabilizer is readily adsorbed, coating the cement particles and interacts with hydration products that are in solution to suspend hydration. Depending on the dosage, the mix can be maintained in a deactivated state for hours or days. After the desired period of retardation an activator is added to the mix. The activator reactivates hydration, removing the adsorbed layers of the stabilizer, allowing hydration to proceed. The admixture system may also be used to maintain workability during long hauls, prevent setting during the long waiting periods as a result of pump, truck or equipment breakdown, so permitting the reuse of aged plastic concrete [102].

In practice, concrete returned to a ready-mixed concrete plant is stabilized (cement hydration stopped) in the ready-mix truck, or collection containers. The treated concrete remains in the transit mix truck or designated containers until needed. In most cases the reclaimed concrete is stabilized for 1–3 days. Figure 6.18 shows the retardation obtained with incremental dosage of the stabilizer. When ready for use, the activator admixture is added, and the stabilized/activated concrete is blended with fresh concrete. Figure 6.19 shows the reactivation of hydration with the accelerator. The system generally blends 15–35% reclaimed concrete with 65–85% of fresh concrete. It is generally used by authorized ready-mixed concrete suppliers fully trained (by the admixture manufacturers) to determine quantities of stabilizer and activator for each mix. A team of technical specialists visits each plant and develops either a computer program or a set of charts to optimize the dosage rates for the customer according to the variables mentioned below.

The stabilizer is most often used at dosages ranging from 0.325 to 8.47 l per 100 kg of cementitious materials. Dosage is influenced by a number of factors, including, (1) mix proportioning, (2) elapsed time between batching and stabilizing, (3) average ambient temperatures, (4) quantity of reclaimed concrete and (5) desired time of stabilization. The activator dosage range is 0.65–0.98 liters per 100 kg of cementitious material and is also governed by most of the factors influencing the stabilizer dosage. Unlike conventional retarders, the stabilizer can be used at high dosages without the attendant adverse effects such as flash set and poor strength development characteristics resulting from the use of normal retarders [103].

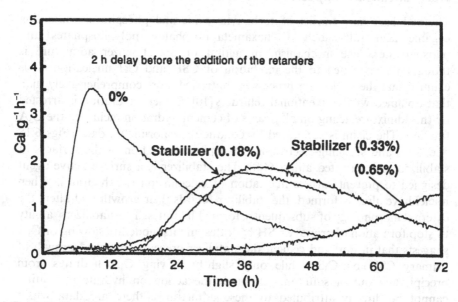

Fig. 6.18 The effect of varied concentrations of the stabilizer on the rate of cement hydration.

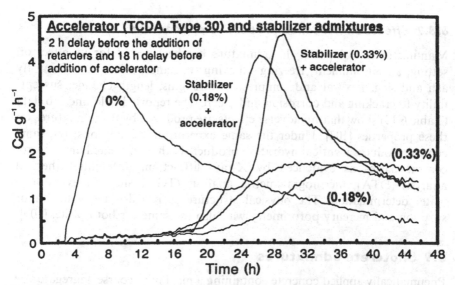

Fig. 6.19 The effect of an accelerator and stabilizer admixtures on the rate of cement hydration (courtesy NRC).

6.8.1 Chemical composition and mechanism of action

The stabilizer may consist of carboxylic acids and phosphorus-containing organic acid salts such as hexametaphosphates, polyphosphates and phosphonates. The mechanism of action of the stabilizer admixture is thought to be related to the inhibition of CSH and CH nucleation. It is claimed that the nucleation process is controlled more comprehensively than that obtained with conventional retarders[10]. Cement hydration is arrested by the admixture acting on all phases of cement hydration including the C_3A fraction. The claim is supported by conduction calorimetric data (Figs 6.18 and 6.19) and scanning-electron-microscopic examination of the surfaces of stabilized and activated alites (C_3S). The stabilizer is a surface active agent designed to prevent surface nucleation of calcium ion rich hydrates. When nuclei have already formed, the stabilizer retards their growth and alters the external morphology of subsequently formed hydrates. The stabilizers' ability to stop formation of primary CSH hydrates and its moderate slowing of C_3A suggests that in Portland cement the stabilizer prevents epitactic* growth of primary CSH on C_3S, while only slightly slowing C_3A hydrates from precipitating out of solution. Although the action on hydrate nucleation cannot be directly attributed to these admixtures, there are data which indirectly illustrate these effects. For example, the alteration of ettringite morphology on alite, and the more open texture of ettringite on C_3A suggests that these admixtures strongly influence how the hydrates are formed[104].

6.8.2 Effects on hardened properties of concrete

Manufacturers claim that the admixture has been tested for its effect on setting, air entrainment, freezing–thawing resistance, mixes containing fly ash and slag, flexural and compressive strengths, length change, susceptibility to cracking and corrosion and creep. The reported field and lab data (Table 6.15) show that no adverse effects are produced by the admixtures on these properties [103]. Under the same experimental conditions, the same cement produces identical hydration products both with treated and control cement pastes as evidenced by X-ray diffraction, differential thermal analysis (DTA), thermogravimetric analysis (TGA) and non-evaporable water determination. The physical structure as revealed by water vapor sorption and mercury porosimetry was also the same for both pastes [105].

6.9 Shotcrete admixtures

Pneumatically applied concrete containing a maximum coarse aggregate size of 10 mm is usually termed shotcrete. The process is particularly suited to

---

* Growth on a crystalline substrate of another crystalline substrate that mimics the orientation of the first substrate.

restoration work that requires the replacement of concrete that has been lost or cut away, and to insure against future damage by adding a further layer of concrete [106] (Fig. 6.20). Since low water–cement ratios are used, the 'no-slump' characteristic affords it to be placed in layers of limited thickness on vertical and overhead surfaces. The advantages of shotcrete include ease

Table 6.15 Field trial data (courtesy Cormix Inc., UK)

	Mix reference					
	1	2	3	4	5	6
Load volume (m³)	6.0	6.0	5.0	5.0	6.0	—
Concrete grade	C20P	C20P	—	—	—	—
Wash water (l)	300	—	300	—	300	—
Coretard (l)	1.5	—	1.5	—	1.5	—
OP cement (kg)	1860	1860	937.5	937.5	648	648
GXT slag (kg)	—	—	937.5	937.5	642	642
Sand (kg)	5520	5520	4070	4070	6162	6162
20–25 mm (kg)	6240	6240	5625	5625	6162	6162
Added water (l)	700	700	600	570	372	372
Plasticizer (l)	—	—	7.75	7.75	—	—
Slump (mm)	70	90	190	130	5	10
Compacting factor	0.97	0.99	—	—	0.84	0.88
Plastic density (kg m⁻³)	—	—	—	—	2324	2303
Temperature (°C)	5.0	5.0	—	—	8.0	9.0
Yield (m³)	—	—	—	—	1.003	1.012
Compressive strength (N mm⁻²) 1 day*	Concrete set but too weak to measure		—	—	3.0	3.0
2 days	15.0	17.5	10.3	12.5	—	—
	15.5	17.1	12.0	13.0	—	—
Mean	15.3	17.3	11.1	12.8	—	—
4 days	27.3	29.3	—	—	15.0	12.5
7 days	34.6	38.3	37.8	36.5	21.5	21.0
	33.1	36.6	34.2	38.8	24.0	21.5
Mean	33.9	37.5	36.0	37.7	22.8	21.3
28 days	42.0	50.2	53.5	55.0	38.0	33.0
	42.4	47.7	53.0	52.5	40.5	34.0
Mean	42.2	48.9	53.5	53.8	39.3	33.5
Initial set ⎫ (Proctor	—	—	—	—	3.4 h	2.8 h
Final set ⎭ Needle)	—	—	—	—	6.3 h	6.3 h

*Stored overnight at 5–8 °C.

Fig. 6.20 Application of shotcrete.

of placing, excellent bonding and relatively low shrinkage. One of the disadvantages and limitations of the process is that it requires specially trained workers and is difficult to control. Hence, quality is dependent on the workers employed. The method can be used in a variety of applications except in areas of poor access and is unsuitable for use where congested reinforcement is present.

Shotcrete can be applied by either a 'dry-mix' or 'wet-mix' process. The dry-mix process involves premixing of the cement and sand, and transfer through a hose in a stream of compressed air. Water is injected at the end of the hose and mixed with the material as it exits at high velocity; water content is restricted to approximately that required for proper hydration. Set-accelerating admixtures, normally in powder form, are introduced into the premix, whereas a liquid accelerator is added to the water at the discharge nozzle, or as a separate injection at the nozzle. Steel and other fibers are usually incorporated in the premix[107]. In the wet process, a predetermined ratio of cement, aggregate and water is batched, mixed and transferred to a pump. The concrete is pumped along a flexible hose to a discharge nozzle from whence it is projected at high velocity on to the surface to be coated. A rapid-setting admixture like sodium aluminate or metasilicate solution is commonly added at the nozzle to enable build-up of thick layers [108]. In the following pages the use of admixtures that are powerful accelerators of cement hydration and the hydraulically reactive solid silica fume are described.

6.9.1 Types of admixtures and mode of action

A variety of admixtures and additives are used to accelerate strength development, cohesiveness, bond, freeze–thaw and abrasion resistance, and to reduce rebound. Most of the accelerating admixtures used probably act by precipitating as insoluble hydroxides or other salts – a form of false set. Conduction calorimetric studies show that their main effect on early strength is due to the action on the C_3A fraction of the cement. The reaction of C_3A is intensified in the presence of these admixtures and a sharp peak occurs at an early age.

Admixtures used either in the wet or dry shotcrete process are required to comply with the following specifications [109]:

1. Initial and final set should occur within 3 and 12 min respectively.
2. The admixture should have a capability to form maximum layer thickness by increasing the cohesion of the mix and hence resistance to sloughing off.
3. It should reduce loss of material by rebound.
4. It should increase rate of strength development so that strength values of the order of 3.5–7 MPa (500–1000 psi) are rapidly attained.
5. It should provide an ultimate strength (28 days) that is not less than 30% of the non-accelerated mix.

The cost, lower ultimate strength and safety hazards are often cited as reasons in seeking to limit the use of shotcrete accelerating admixtures. However, when quick-setting characteristics are required for rapid build-up of layers and early strength development for immediate support and to seal off water leakage are essential, the use of an accelerator is indispensable. Consequently, accelerators are generally used on all overhead work, on some vertical walls and in locations of high ground water flow. Their use in shotcrete prevents sagging and sloughing off the shotcrete and reduces rebound of the material.

Materials used are usually strong alkalis, and organic bases. Proprietary materials generally consist of various proportions of these ingredients and may be marketed in both liquid and powder form. In some admixtures, single materials such as sodium aluminate may be present as the sole active ingredient. More commonly, the main accelerating ingredient is combined with one or more admixtures such as wetting agents, thickening admixtures, and stabilizing agents to obtain other desired modifications and are therefore multicomponent admixtures. The main accelerator ingredient may consist of sodium aluminate, sodium and potassium hydroxide, or carbonates, triethanolamine, ferric sulfate [110] and sodium fluoride [111]. Sodium salts of hydroxycarboxylic acids such as gluconates and tartrates may be used as wetting agents. Polyhydroxy compounds are used to stabilize solutions of sodium aluminate [112]. Low-causticity materials, which significantly reduce the potential for injury during application, have now been introduced into the market. The compositions are closely guarded trade secrets and therefore little published information is available. Currently used materials include the following.

(a) Carbonates and hydroxides of alkaline earth metals

These admixtures are used exclusively in the dry-mix process. Normal dosage range is from 2.5–6% by weight of cement. They mainly accelerate C_3S hydration and their reactivity is strongly influenced by cement composition and fineness, the presence of mineral additions and ambient temperature. A characteristic of this accelerator is the drastic decrease in final strength. Compared to plain concrete, the 28-day strength can be reduced significantly (typical values are 30–40%) and values up to 50% have been recorded in certain cases [113].

(b) Alkaline silicates (water glass)

Sodium- and potassium-silicate-based accelerators are mainly used in wet-mix shotcrete. They are added to the mix in liquid form and used normally at high dosages (>10% by weight of cement). Quick setting is provided by precipitating as calcium silicates. When used in large amounts, they produce a decrease in bond and strength and increased drying shrinkage [114]. The advantages of this admixture are that it is compatible with all types of cements, the decrease in final strength is less than with aluminate-based accelerators when used at normal dosages (4–6%) and it is not so aggressive to the skin. However, these admixtures promote rapid loss of consistency, do not provide good early strengths and are usually not specified for applications where this requirement is warranted [114, 115].

The use of sodium-silicate-based admixtures has been restricted in Germany and Austria for the following reasons [116]:

● Compressive strength decreases of 40–60% in comparison with concrete without the admixture.
● Decrease in the E modulus with time.
● Decrease in the waterproofing characteristics due to the extraction of lime when the concrete is subjected to a continuous exposure to moisture.
● High viscosity at low temperature which requires heating of the material.

Sodium silicate is permitted to be used only if it is shown that the specified strength can be attained (through laboratory trial mixes) even with a 50% decrease in strength and rigorous on-site quality control is provided to ensure that the proper admixture dosage and water content are used in the application.

(c) Sodium and potassium aluminates

These admixtures, used in liquid form at dosages ranging from 2.5 to 5.5%, are the most widely used materials for dry- and wet-mix shotcrete. Although their efficiency is less influenced by cement composition, fineness and presence of mineral additions, than that seen in carbonate-based admixtures, they nevertheless produce a latter-day strength decrease of 20–25% in relation to plain shotcrete. Potassium-aluminate-based accelerators generally perform better than the sodium-based material. The former however are more expensive. Their main characteristic is to promote a rather quick setting by acting directly on the cement hydration, through combination with gypsum, thus preventing ettringite formation around the cement grains. This allows almost instantaneous reaction of the C_3A, conferring the initial stiffness desired in most shotcrete applications [108].

The aluminate-based admixtures undoubtedly provide the best wet-mix shotcrete, making it possible to build thick linings even in overhead work. The high alkali content and the consequent health hazard are the main constraints to their more widespread use. Furthermore, there is concern that the admixture may promote the alkali–aggregate reaction in concretes containing reactive aggregates, as well as sulfate attack [117].

(d) Non-alkaline powder accelerators

The use of these materials commenced in the early 1990s. Generally these are calcium-aluminate-based admixtures used in dosages from 6 to 12%. Their chemical function differs from that observed in alkali-based admixtures [113]. Setting acceleration is provided by the reaction of the accelerator with water, without any direct interference in the hydration of the cement [109]. Dosages exceeding 7% are necessary to provide the required early strength, particularly for overhead work. Although their use also results in a strength decrease compared to plain shotcrete, the decrease is less than that observed for alkali-based aluminates. A disadvantage of their use is the susceptibility to humidity, which necessitates the installation of a dryer with the dosing device [108].

Other alkali-free accelerators used include aluminum hydroxide or a combination of this material with aluminum sulfate. They develop early strength even in low amounts (4%) and there is no decrease in later strengths up to 8%. However, a significant loss in strength has been observed with amounts above 10% [111]. In spite of the advantages, there is some reluctance to the use of non-alkaline accelerators in powder form because of the required special adaptations to available equipment and the 10–15% greater rebound compared to alkaline materials [108, 109].

(e) Non-alkaline liquid accelerators

Non-alkaline liquid accelerators are fairly new to the international market, therefore the bibliography on their use is still scarce. They were conceived to solve some classical problems stemming from the use of alkaline accelerators, such as caustic alkalis, hazardous conditions in underground work, risk of alkali–aggregate reaction, risk of handling conventional accelerators with extremely high pH level, and reduction of latter-age strength.

The chemical composition of these accelerators is neither described in the pertinent bibliography nor divulged by their producers. They have a pH between 3 and 5.5 and an alkali content of less than 0.3% and are used in amounts between 3 and 10% by weight of cement. When used in adequate amounts, they allow overhead build-up in linings of up to 300 mm thick. An important characteristic of these accelerators is that the latter-age strengths are not reduced when compared with a reference material (without accelerator) [108]. For example, in the construction of the tunnels of the south freeway (Via Express Sul) in Florianopolis, Brazil, comparative tests were made with an alkali–free liquid accelerator and the sodium-aluminate-based accelerator. The aluminate-based accelerator was added in an amount

Table 6.16 Shotcrete strength obtained at different ages (Prudencio)

Age	Compressive strength (MPa)			
	Reference	Aluminate 4%	Alkali free 4%	Alkali free 6%
15 min	—	0.44	—	0.51
30 min	—	0.54	—	0.55
I h	—	0.75	—	0.58
I h 30 min	—	0.93	—	0.72
2 h 30 min	—	—	—	0.84
3 h 40 min	—	—	0.74	—
4 h	—	4.15	—	—
5 h	0.76	—	—	—
6 h	—	7.03	—	4.38
6 h 40 min	—	—	3.05	—
8 h	5.97	17.12	5.80	8.52
I day	28.3	20.9	24.8	23.9
5 days	31.8	25.2	34.2	36.4
7 days	40.3	36.5	42.7	48.0
28 days	50.1	41.7	43.5	53.8

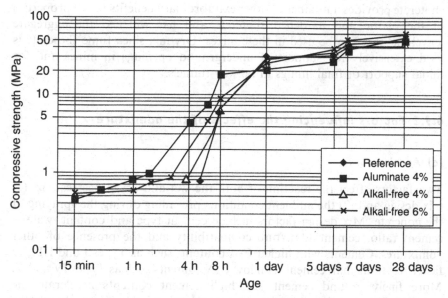

Fig. 6.21 Comparison between the performance of shotcrete with alkali-free and sodium-aluminate-based accelerators (Prudencio [108]).

of 4%, while amounts of 4 and 6% of alkali–free accelerator were added. the results obtained are summarized in Table 6.16 and Fig. 6.21 [108].

The data in Table 16.6 show that the aluminate-based accelerator performed better at early ages than the alkali–free accelerator. However, while the dosage of 4% of alkali–free accelerator gave an unsatisfactory strength performance at early ages, a 6% dosage amply satisfied the work design requirements (8 h: 4.7 MPa; 24 h: 11.8 MPa; 7 days: MPa; 28 days: 21.2 MPa). Furthermore, the shotcrete with alkali–free accelerator in the amount of 6% at 28 days showed a markedly superior performance to that of the shotcrete with aluminate-based accelerator and even to the performance of plain shotcrete. Similar results were found in other field investigations with this type of accelerator [108] supplied by another producer. Studies on shotcrete microstructure done on samples of shotcrete containing this type of accelerator did not reveal any alterations in the morphology of cement hydration products. The high cost of these materials has been a hindrance to their more frequent use.

(f) Silica fume

Silica fume has been used as a cohesion inducing agent to improve adhesion and sag and washout resistance. Incorporation of silica fume in dry-mix shotcrete provides a number of other extraordinary benefits to the properties of the plastic and hardened shotcrete and has widened the aggregate 'envelope' that can be used in these mixes. Dry-mix silica fume shotcrete is used extensively for support of underground openings in mines and for initial support or final linings in railway tunnels.

6.9.2 Factors influencing the effects of the admixture

(a) Accelerators

Two factors affect the reaction of both the wet and dry processes, viz. (1) mix design and (2) the ambient conditions prevailing during the spraying of the concrete. Mix design factors include cement type and content, water–cement ratio, cement-admixture compatibility and the presence of other admixtures. Cements with higher C_3A contents such as Types I and III give faster reactions than cement with low C_3A contents such as Types IV and V. More finely ground cement and high cement contents accelerate the reaction. Batch to batch variations in cement and admixture change the compatibility of the mixture, resulting in lower early strength development and strength variation. Conventional admixtures used for pumping (in the wet mix process) often retard the cement accelerator reaction.

Other mix parameters that affect the reaction during the spraying of the concrete are admixture dosage, mix and ambient temperature, water–cement ratio, prehydration of the cement and a delay in addition of the admixture. High admixture dosage produces flash set resulting in a mix that is too dry to stay on the sprayed surface. Low admixture dosages, on the other hand, may result in delayed set time and sloughing off of the mix. Consequently, in either case build-up of layers may not be possible. High and low temperatures ($>30\,^{\circ}C$ and $<10\,^{\circ}C$) accelerate set and extend set time respectively and produce results similar to that obtained with high and low admixture dosages. Prehydration of the cement due to the presence of moisture in the aggregate and a delay in the addition of the admixture result in lower early strength values.

The influence of accelerators on both setting time and early strength development is highly specific, depending not only on the particular accelerator and cement type used, but also on the specific brand and plant from which the cement came. Many of the more caustic accelerators, while very effective in accelerating set, can have a detrimental influence on the properties of the shotcrete at later ages. Thus, preconstruction compatibility testing of all shotcrete ingredients proposed for use on a given project is considered mandatory. Laboratory testing of cement–admixture combinations for setting time and compressive strength testing of laboratory-made mortar cubes can give misleading results. Such testing is therefore best performed by shooting test panels using the materials, equipment, and shotcrete crew proposed for use on the project [107, 109].

(b) Silica fume

Because of its extreme fineness and high glass content, silica fume is a very efficient pozzolanic material, i.e. it is able to react very efficiently with the products of hydration of Portland cement to create secondary cementing materials in hydrating concrete and shotcrete. The calcium hydroxide produced by the hydrating Portland cement is largely consumed in the ensuing pozzolanic reactions. This affects the microstructure and results in a product with very low permeability and absorption characteristics that is resistant to deterioration in a variety of chemically aggressive environments. The extreme fineness of the material increases the water demand of the mix and in order to realize the full potential (increased thixotropy of the projected concrete, good adhesion and quick setting) of silica fume in the mix, a superplasticizer must be used in combination with the mineral additive [109, 117–119].

6.9.3 Effects on the plastic and hardened properties of concrete and mortar

(a) Accelerators

The method of placement makes shotcrete potentially more variable and more anisotropic than normal concrete. Consequently, the development of the physical and mechanical properties such as strength, shrinkage, permeability and durability is drastically different from those of normal concrete. The effects are particularly noticeable during the very early and later ages [107, 109, 117].

- **Consistency of the mix:** high cement contents and optimum accelerator dosages (3–6%) usually produce cohesive mixes which enable the building of layers. However, depending on the dosage of the admixture, the mix may exhibit good cohesion or produce a dry gritty mix.
- **Initial and final set:** at optimum dosage levels, initial setting time may be reduced to 3–4 min while final set time can vary between 12 and 15 min. Higher dosages reduce the time even further.
- **Compressive strength and modulus of elasticity:** strength values are usually lower (15–20%) than for corresponding mixes containing no admixture for both wet- and dry-mix processes at optimum dosages of 2–4%. Drastic strength reduction occurs as the dosage exceeds 4%. The stipulated specification requirement that strength should not be reduced by more than 30% of the control concrete is not possible at dosages

above 6%. However, under special conditions such as stoppage of water leakage, such reductions are acceptable. In general, factors that tend to improve early strength of concrete seem to have a detrimental effect on ultimate strength. The relationship between compressive strength and modulus of elasticity is similar to that of regular concrete, in that if adequate curing is provided, values increase with age.

- **Tensile and flexural strengths:** values are usually lower than what would be expected from compressive strength values. This probably reflects the effect of laminations and other defects in the shotcrete on tensile strength. Flexural strength values follow a similar relationship observed for regular concrete.

- **Bond strength:** high bond strength can be attained between layers of shotcrete and the rock surface for both wet and dry mix processes when admixtures are used. Bond strength increases initially with dosages up to 2% and thereafter decreases with further addition.

- **Shrinkage:** higher values are obtained for mixes containing the admixture in comparison with the corresponding mixes with no admixture. Shrinkage increases with higher admixture and water contents of the mix, for both wet and dry shotcrete.

- **Durability:** most proprietary accelerating admixtures adversely affect the concrete's resistance to freezing and thawing [114, 118]. More recently, the widespread use of silica fume has enabled the use of considerably low dosages of accelerators and also contributed to the improvement of porosity values, thereby dramatically improving durability of such concretes.

- **Permeability:** since the admixtures improve spraying conditions and minimize rebound, it is argued that permeability should be improved. Actual tests, however, show that the effect is marginal and the values obtained are similar to mixes with no admixture. Silica-fume-containing mixes, however, show improved permeability [118].

(b) Silica fume

The addition of silica fume to dry-mix shotcrete in proportions of 10–15% by mass of Portland cement substantially improves the adhesive and cohesive properties of the freshly applied shotcrete. Silica fume creates a very dense and sticky mix with an almost complete lack of bleeding. These characteristics of the material are attributed to the extreme densification achieved by packing of the ultrafine silica fume particles between the cement particles in the plastic (fresh) shotcrete. Practical ramifications of these characteristics include the observations which follow [107, 117–119].

- **Washout resistance:** Silica fume shotcrete is very resistant to washout by water. It is thus well suited to operations where freshly applied shotcrete will be exposed to running water or tidal flushing, e.g. rehabilitation of marine structures in the intertidal zone, construction of dikes, and creek channelization.

- **Application thickness:** It imparts the ability to apply the material overhead in a single pass at thicknesses that previously had been unattainable with conventional shotcrete. Vertical surfaces can be built out to any desired thickness in a single pass, without the use of accelerators and without sloughing or tearing of the plastic shotcrete.
- **Adhesion:** Adhesion of the plastic dry-mix silica fume shotcrete, particularly in wet areas, is substantially improved. For example, adhesion in areas such as locks, dry docks, tunnels, and even leaking structures, can only be achieved through the use of high concentrations of shotcrete accelerators, to create a flash-setting condition, a process that is detrimental to the long-term durability of the hardened shotcrete. Silica fume has been found to promote excellent adhesion in such conditions with minimal or no accelerator addition.
- **Rebound:** Because of marked improvements of adhesion and cohesion, the amount of rebound is significantly reduced. In controlled tests, rebound of a conventional dry-mix shotcrete, applied to a thickness of 50 mm (2 in) to the smooth overhead surface of concrete deck slabs, was measured at 40% by mass (the rebound was caught in tarpaulins and weighed). Rebound under the same circumstances was reduced to 25% by mass in dry-mix silica fume shotcrete.
- **Compressive strength:** Tests conducted on cores extracted from test panels shot during routine quality control show that silica fume shotcrete produce consistently higher compressive strengths and lower permeability in the hardened shotcrete than shotcrete where accelerators were used.

6.9.4 Guidelines for use

(a) Addition, dispensing and mixing, hazards and safety measures, supply and storage

Addition rates used for powder accelerators are generally in the range of 2–6% by weight of cement. Liquid accelerators are batched by volume ratios with the accelerator-water ratios ranging from 1:20 to 1:1. The different solids content of various liquid accelerators should be taken into consideration in metering the admixture by volume ratios. Optimum concentrations are not used in all applications and where fast set and high early strength are to be attained high dosages are used. However, specifications may stipulate the maximum strength reduction allowed for the accelerated mix and this then puts a limit to the amount of admixture.

Powder accelerating admixtures used in dry-mix shotcrete are usually added to the concrete or mortar using either auger corkscrew or gear-wheel-type dispenser. The powder is fed by gravity to the corkscrew helix, conveyed thorough a tube and then discharged into a hopper containing the mix. Addition rate is adjusted by increasing or decreasing the rotational speed of the corkscrew. The admixture dispenser is usually positioned above the conveyor charging the shotcrete machine. This arrangement provides a relatively uniform distribution of the admixture in the incoming materials. The delivery rate of the admixture must by synchronized with the rate of material delivery past the dispenser so that desired cement accelerator mix proportions are achieved. When liquid admixtures are used in the dry process they are injected from a pressurized tank into the water line at a separate injection port near the nozzle. Proper proportioning by volume of water is obtained by adjustment of the admixture feed control value. Predampening of the bagged dry-mix silica fume shotcrete is generally recommended. Not only does it reduce the amount of dust generated by the shotcrete, but it also improves the uniformity and adhesion of the applied shotcrete.

At present the only reliable method of dispensing powder shotcrete admixtures directly into the wet mix at a uniform rate at the nozzle is by the use of a log feeder. In this method the powder is compressed to a log which, when placed in the feeder, is gradually ground off and carried away in the air stream to the nozzle. The accelerator mixes with the other shotcrete ingredients at the nozzle and in transit to the sprayed surface. The amount of admixture added is controlled by the rate of advance of the log in the feeder. The method of dispensing liquid admixtures in a wet mix is essentially the same as that used in the dry-mix operation, except that admixture is injected into the air stream instead of the water line.

In order to minimize rebound, thorough blending of the ingredients should be done. The aggregate should be well coated with the cement in the dry-mix process. Since shotcrete admixtures promote shrinkage, it is essential that the sprayed surfaces be kept moist for at least 7 days. Ambient temperature conditions should be maintained at a level so that air in contact with the shotcrete surface is above freezing – at least for 7 days [106, 115].

Virtually all dry-mix shotcrete, including that containing silica fume, is supplied in a premixed bagged form, typically in 30 kg (66 lb) paper bags or 1600 kg (3520 lb) bulk bin bags. Dry-mix silica fume shotcretes have been supplied in both coarse and fine aggregate gradations, with or without accelerators and/or steel fibers in the bags. Precise weight batching and plant mixing of quality materials has enabled suppliers to provide owners and users with high-quality uniform products with good field performance.

Powder admixtures are moisture sensitive. They must, therefore, be stored in bags lined with plastic sheeting, in an elevated and dry location. Bags of the accelerator should be placed on pallets not closer than 5 cm to the ground surface. The material in ripped bags should be used immediately after the bags are damaged or should be discarded. Shelf life of the powder is limited to a period of 6–8 months under proper storage conditions. The activity usually begins to decrease significantly after 6 months [107].

Caustic materials contained in the admixtures can cause reactions on exposed skin, ranging from mild irritation to severe burns. Most of the ingredients are classed as primary irritants. With liquid admixtures, burns can occur immediately. Because of the serious hazard to exposed skin and eyes, personal equipment is required for all those in the vicinity of the shotcrete machine and nozzle. These include long-sleeved shirt or coveralls. In addition, prior to shooting, protective cream or lotion should be applied to skin areas likely to be exposed to the caustic chemicals. A suggested treatment is to wash exposed skin with water and rinse with vinegar followed by treatment with a protective cream. All exposures to the eye require the immediate attention of a physician; immediate action would be to flush the eye with copious amounts of water [120, 121].

(b) Specifications

Some shotcrete specifications may require the accelerating admixtures to conform to ASTM C 494, 'Standard Specification for Chemical Admixtures' (such as rapid set), but ASTM C 494 is only partially applicable. The specifications for uniformity, packaging, storage, sampling and inspection stipulated for type C accelerating admixtures are still pertinent to these rapid set accelerators, but testing methods for set and strength are modified. ACI Standard 'Recommended Practice for shotcreting' (ACI 506-66) describes the application of shotcrete for different types of construction. Application procedures and equipment requirements are given for both the dry- and wet-mix processes. The testing of shotcrete is also covered in some detail.

In general the required performance criteria and product uniformity parameters stipulated in shotcrete specifications are as follows:

1. Type and composition.
2. Percentage active ingredient concentration.
3. Chloride content.
4. Maximum percentage of accelerator to be used.
5. Compatibility with the cement to be used as defined by the requirements of initial and final set times, early and ultimate strengths.
6. Variation of dosage according to location, type of structure or strength requirement.

6.10 Shrinkage-reducing admixtures

One of the most exasperating problems when building with concrete is its normal shrinkage cracking. Some of the structures where cracks due to drying shrinkage often occur, and where the consequences can be serious, are bridge decks, parking garages, marine structures, primary and secondary containment structures and industrial floors.

The susceptibility of concrete to cracking due to drying shrinkage depends on whether the concrete is restrained or unrestrained. If the concrete is unrestrained, it can shrink freely and change volume without cracking. However, since most concrete is restrained – by subgrades, foundations, reinforcement or connecting members – it often develops tensile stresses high enough to cause it to crack. Several additional factors affect cracking and these include the rate of shrinkage, creep characteristics, modulus of elasticity and tensile strength of the concrete. According to the leading theory (capillary tension theory) explaining the mechanism of drying shrinkage, one of the main causes of drying shrinkage is the surface tension developed in the small pores of the cement paste of concrete. When these pores lose moisture, a meniscus forms at the air/water interface. Surface tension in this meniscus pulls the pore walls inward, and the concrete responds to these internal forces by shrinking (Fig. 6.22).

This shrinkage mechanism occurs only in pores within a fixed range of sizes. In pores larger than 50 nm the tensile force in the water is too small to cause appreciable shrinkage and in pores smaller than 2.5 nm a meniscus cannot form [122]. The amount of cement-paste shrinkage caused by surface tension depends primarily on the water–cement ratio, but it is also affected by cement type and fineness and by other ingredients (such as admixtures, and supplementary cementing materials) which affect pore size distribution

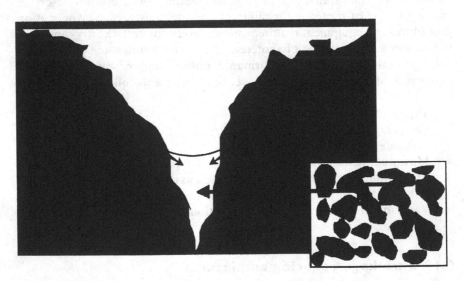

Fig. 6.22 Effect of pore meniscus and surface tension on the drying shrinkage of mortar (Balogh [122]).

in the hardened paste. Figure 6.23 shows the relationship between surface tension (measured as a filtered liquid) and drying shrinkage of the cement mortar and the effect produced by a number of surfactants [122].

Some of the measures taken to reduce shrinkage cracking have been the use of high-range water reducers to attain very low water–cement ratios, the use of expansive admixture and shrinkage compensating Type-K cements. A new liquid admixture, called shrinkage-reducing admixture (SRA), introduced to the North American market in 1995, is said to chemically alter the shrinkage mechanism without expansion. According to published patents the general composition of the SRA is indicated as $R_1 O(AO)n R_2$, where A is an alkalene polymer having carbon–carbon chain number of 2–4 or two different types of alkaline groups. R_1 and R_2 are selected from the hydrogen, hydroxyl, alkyl, phenyl, cycloalkyl groups. The symbol n is an integer between 2 and 10 indicating the degree of polymerization [122]. The commercially available SRA is a glycol ether blend [123].

The SRA reduces shrinkage by reducing the surface tension of water in the pores between 2.5 and 50 nm in diameter. It is added to the mix in the gauge water and dispersed in the concrete during mixing. After the concrete hardens the admixture remains in the pore system, where it continues to reduce the surface tension effects that contribute to drying shrinkage. In both laboratory and field tests of various concretes containing the admixture, the product seemed to be most effective when added at a rate of 1.5–2.0% by weight of cement [122, 123]. But dosages of 1–2.5% can be used to obtain the desired level of shrinkage without adverse side effects. Although the primary impact of the SRA is to reduced drying shrinkage, the admixture also affects other fresh and hardened properties of the concrete. These are discussed below.

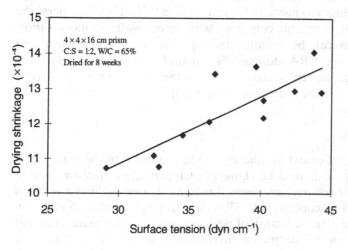

Fig. 6.23 Relationship between surface tension and drying shrinkage (Berke et al. [124]).

6.10.1 Effects on the fresh concrete properties

When added at dosage of 2% by weight of cement to a concrete mixture with 460 kg m^{-3} of cement without adjustment for the volume of the water introduced by the admixture, the concrete's slump and porosity are increased. However, when substituted for an equal volume of water, the SRA has little or no effect on concrete slump. It does have a slight retarding effect on the rate of hydration and may extend the setting time up to about an hour. The admixture also affects the air content of fresh concrete and therefore when used in air-entrained concrete, the air-entraining admixture dosage must be increased to achieve a specified air content.

6.10.2 Effects on the hardened properties of concrete

(a) Shrinkage cracking

The extent of shrinkage reduction obtained is dependent on the water–cement ratio (W/C) of the mixture. Generally, the lower the ratio the greater the percentage shrinkage reduction that could be achieved. For mixes with W/C < 0.60, reductions in 28-day shrinkage of 80% or greater and reductions in 56-day shrinkage of 70% were achieved at a admixture dosage rate of 1.5% by weight of cement. In mixes with W/C = 0.68 the level of shrinkage reduction at 28 days and 56 days was 37% and 36% respectively [121].

(b) Compressive strength

Previous work has shown that adding the SRA to concrete at a 2% dosage can reduce the strength as much as 15% at 28 days [121–124]. In general the strength reduction is less in concretes with lower water–cement ratios. Strength reduction can be counteracted by using a superplasticizer and slightly reducing the SRA dosage. The manufacturer recommends the addition of enough superplasticizer to reduce the mixing water by 7–10% while keeping the cement content constant [123].

(c) Thermal cracking

A large percentage of cracks in concrete can be attributed to thermal effects. Two basic causes of thermal cracking are temperature differences within massive sections of concrete and overall volume change of thinner sections of concrete caused by cooling [124]. SRAs due to their retardation effect on hydration and attendant reduction of peak temperatures can reduce thermal cracking because of the decreased thermal contraction on cooling. The addition of the SRA results in some strength reduction, which needs to be counteracted either by a decrease in water–cement ratio through the use of a superplasticizer or an increase in the cement content.

(d) Curling

Curling of concrete slabs is caused by the top surface drying and shrinking faster than the core concrete. In laboratory testing, with 8 ft × 4 ft (2.4 m × 1.2 m) slab specimens, SRA showed a significant tendency to reduce curling of the slab. This property of the admixture was confirmed in a field trial conducted by the Virginia DOT, USA, where 1.5 gal (US) (5.7 liters) of SRA were used in a test overlay for the Lesner Bridge [123].

(e) Air-void parameters and freeze–thaw resistance

Concretes that have air contents in excess of 6% generally, give adequate spacing factors (0.1–0.3 mm) However, some surfactants used as SRAs, even though yielding proper spacing factors, cause the concrete to readily attain saturation, rendering it more susceptible to freeze–thaw attack. It is probable that these surfactants are acting as wetting agents and promoting the imbibition of water into the concrete [125, 126].

6.10.3 Factors affecting the performance of shrinkage-reducing admixtures

(a) Effect of water–cement

The effect of water–cement ratio of the concrete mix on the degree of shrinkage reduction achieved was discussed previously. The reduction realized is greater for lower water–cement ratios.

(b) Effect of cement type and blended cementitious materials

Tests done on mortar mixes containing a variety of cementitious materials (including seven different cements and blends of Portland cement/fly ash and Portland cement/slag) gave long-term shrinkage reductions that ranged from 25 to 38% [124]. These tests stress the importance of determining the SRA's shrinkage reduction levels with local materials.

(c) Effect of wet curing

Longer wet curing increases the effectiveness of the admixture especially in early age concrete. Ultimate levels of shrinkage are also reduced by extended periods of wet curing [1]. Figure 6.24 presents a comparison of the shrinkage achieved at two dosages and with different curing regimes.

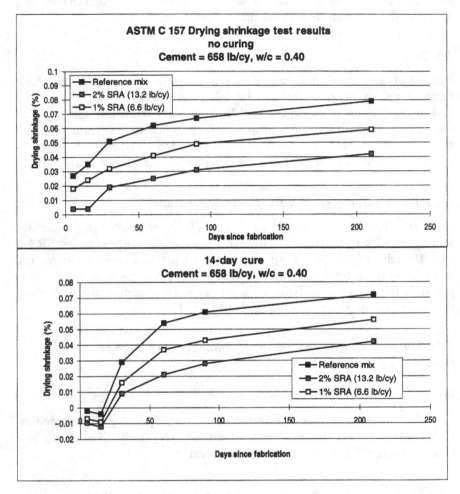

Fig. 6.24 Comparison of the shrinkage achieved at two dosages with different curing regimes (Fujiwara *et al.* [126]).

Although no ASTM designation yet exists for the SRA, the results obtained from tests done according to ASTM C 494-92 indicate that the admixture containing mix gave values that closely matched those of the non-admixtured concrete and therefore meets all requirements of the standard.

The application of these chemicals is not restricted to use as admixtures. Japanese investigators have carried out both laboratory and field trials in which the SRA was applied to the surface of the hardened concrete and obtained similar effects to that observed when it used as an admixture [122].

References

1 Hobbs, D.W. (1988). *Alkali Silica Reaction in Concrete*, Thomas Telford Ltd., London.
2 Durand, B. and Chen, H. (1991). *Petrography and Alkali-Aggregate Reactivity*, Course Manual, Ottawa, March 12–14, 399–489.
3 Nomachi, H., Takada, M. and Nishibiyashi, S. (1989). *8th International Conference on Alkali-Aggregate Reaction*, Kyoto, Japan, 211–15.
4 Wang, H. and Gillott, J.E. (1992). *9th International Conference on Alkali Silica Reaction*, London, 1090–99.
5 Hansen, W.C. (1960). *Journal of the American Concrete Institute*, **31**, 881–3.
6 Mehta, P.K. (1978). *4th International Conference on the Effect of Alkali in Cement and Concrete*, London, 229–33.
7 Stark, D., Morgan, B., Okamoto, P. and Diamond, S. (1993). *Eliminating or Minimizing Alkali Silica Reactivity*, Strategic Highway Research Program, Washington, SHRPC-343, 266.
8 Sakaguchi, Y., Takakura, M., Kitagawa *et al.* (1989). *8th International Conference on Alkali-Aggregate Reaction*, Kyoto, Japan, 229–34.
9 Ohama, Y. *et al.* (1989). *8th International Conference on Alkali-Aggregate Reaction*, Kyoto, Japan, 253–8.
10 Vivian, H.E. (1947). *CSIRO Bulletin*, **229**, 55–6.
11 Ohama, Y., Demura, K. and Wada, I. (1992). *9th International Conference on Alkali-Aggregate Reaction in Concrete*, London, 420–31.
12 Nakajima, M., Nomachi, H., Takada, M. and Nishibiyashi, S. (1992). *9th International Conference on Alkali-Aggregate Reaction in Concrete*, London, 690–7.
13 Gillot, J.E. and Wang, H. (1993). *Cement and Concrete Research*, **23**, 973–80.
14 Gudmundsson, G. and Asgeirsson, H. (1983). *Proceedings of the 6th International Conference Alkalis in Concrete*, Copenhagen, 217–21.
15 Hooton, R.D. (1987). *International Workshop on Condensed Silica Fumes in Concrete*, Montreal, Canada, 30.
16 Oberholster, R.E. and Westra, W.B. (1981). *Proceedings 5th International Conference on Alkali-Aggregate Reaction in Concrete*, Capetown, S. Africa, 10.
17 Perry, C and Gillott, J.E. (1985). *Durability of Materials*, **3**, 133–46.
18 Farbiarz, J., Schuman, D.C., Carrasquillo, R.L. and Snow, P.G. (1989). *Proceedings 8th International Conference on Alkali-Aggregate Reactions in Concrete*, Olso, Norway. 241–6.
19 Aitcin, P.C., Regourd, M. and Fortin, R. (1986). *Division of American Ceramics Society, Abstracts*, **63**, 470.
20 Wang, H. and Gillott, J.E. (1992). *Proceedings 9th International Conference on Alkali-Aggregate Reactions in Concrete*, London, 1100–6.
21 Hudec, P. (1993). *Cement and Concrete Composites*, **15**, 21–6.
22 Diamond, S. and Ong, S. (1992). *Proceedings 9th International Conference on Alkali-Aggregate Reactions in Concrete*, London, 269–78.
23 Jensen, D., Chatterji, S., Christensen, P. and Thalow, P. (1984). *Cement and Concrete*, **14**, 311–15.
24 Ratinov, V.B. and Rosenberg, T.I. (1994). *Concrete Admixtures Handbook*, Noyes Publications, New Jersey, 626.

25 Brook, J.W. and Ryan, R.J. (1989). *Third International Conference*, Ottawa, Canada SP–119, American Concrete Institute, Detroit, 535.

26 Sakai, K., Watanabe, H., Nomachi, H. and Hamabe, K. (1991). *Concrete International*, **13**(3), 26–30.

27 Korhonen, C. J. and Smith, C.E. (1991). *New Admixtures for Cold Weather Concreting*, CRREL Special Report 90–32, 200–9.

28 Korhonen, C. J. (1990). *Antifreezing Admixtures for Cold Regions Concreting, a Literature Review*, Special Report, CRREL 90–32, 14.

29 Korhonen, C.J., Cortez, E.R. and Charest, B.A. (1992). *Materials Engineering Congress*, ASCE, Atlanta, USA, 382–97.

30 Lagoyda, A.V. (1984). *Perfecting Technology of Concrete by Use of New Chemical Admixtures*, Moscow, Znanie, MDNTP, 94–102.

31 Mazur, I.I. (1984). *Perfecting Technology of Concrete by Use of New Chemical Admixtures*, Moscow, Znanie, 125–8.

32 Koroleva, O.E., Lagoyda, A.V. and Zhdanova, R.E. (1987). *Perfecting Technology of Concrete by Use of New Chemical Admixtures*, Moscow, Znanie, 135–8.

33 Korhonen, C.J. and Cortez, E.R. (1991). *Concrete International*, **13**, 38–41.

34 Ratinov, V.B. and Rozenberg, I.I. (1989). *Concrete Admixtures*, Strojizdat.

35 Scanlon, J.M. and Ryan, R.J. (1989). *The Construction Specifier*, Dec., 58–65.

36 Korhonen, C.J., Cortez, E.R. and Charest B.A. (1992). *Material Engineering Congress*, ASCE, Atlanta, GA, 382–97.

37 Krasnovskij, B.N., Dolgopolov, N.N., Zagrekov, V.V. *et al.* (1991). *Concrete and Reinforced Concrete*, No. 2, 17–8.

38 Nagataki, S. (1989). RILEM Committee TC 84-AAC, May, 8.

39 Kwai, T. (1987). *Proceedings 5th International congress on Polymers in Concrete*, Brighton Polytechnic, Brighton, Sept., 385–90.

40 Mailvaganam, N.P. (1994). *Concrete Admixtures Handbook*, Noyes Publications, 1000–6.

41 Rakitsky, W.G. (1993). *Conchem International Exhibition and Conference*, Karlshrue, Germany, Nov., 155–81.

42 Sakuta, M., Yoshioka, Y. and Kaya, T. (1985). *Use of Acryyl- Type Polymer as Admixture for Underwater Concrete*, ACI SP-89, ACI, Detroit, 261–78.

43 Khayat, K. (1996). *ACI Materials Journal*, **93**, 134–8.

44 Kawakami, M. *et al.* (1989). *Proceedings 3rd International Conference on Superplasticizers*, ACI SP-119, Ottawa, Canada, 493–516.

45 Ohama, Y. *et al.* (1990). *Underwater Adhesion of Polymethyl Methacrylate Mortars*, ACI SP-137, 93–108.

46 Saucier, K.L. and Neeley, B.D. (1987). *Concrete International*, **9**, 42–7.

47 Anderson, J.M. (1983). *Concrete*, **17**, 12–15.

48 Griffin, D.F. (1975). *Corrosion Inhibitors for Reinforced Concrete, Corrosion of Metals in Concrete*, ACI SP-49, American Concrete Institute, Detroit, 95–102.

49 Craig, R.J. and Wood, L.E. (1970). *Highway Reserve*, Record No. 328, Washington, 77–88.

50 Berke, N.S. and Weil, T.G. (1992). *Advances in Concrete Technology, Energy, Mines and Services*, Canada, 899–924.

51 Desai, M.N., Shaw, V.K. and Ghandi, M.H. (1974). *Anti-Corrosion Methods and Materials*, **21**, 10–2.

52 Mailvaganam, N.P. (1984). *Concrete Admixtures Handbook*, Noyes Publications, 562–4.
53 Treadaway, K.W.J. and Russell, A.D. (1983). *Highway and Public Works*, **63**, 19–21.
54 Slater, J.E. (1983). *Corrosion of Metals in Association with Concrete*, STP 818, ASTM, Philadelphia, 3.
55 Berke, N.S. (1989). *Corrosion Inhibitors in Concrete*, Corrosion 89, Paper no. 445, National Association of Corrosion Engineers, Houston, TX.
56 Andrade, C., Alonso, C. and Gonzalez, J.A. (1986). *Cement and Concrete Aggregates*, **8**, 110–16.
57 Berke, N.S. (1991). *Concrete International*, **13**, 24–7.
58 Berke, N.S. (1987). *NACE Corrosion-85, Paper No. 273*, (Part 1) National Association of Corrosion Engineers, Houston, Texas, USA, 1985. Berke, N.S., Ibid, Part 2, 134–44.
59 Berke, N.S. and Rosenberg, A. (1989). *Transportation Research Record*, No. 1211, 18–27.
60 Chin, D. (1987). *A Calcium Nitrite-Based Non-Corrosive, Non-Chloride Accelerator*, ACI SP-102, 49–77.
61 Gaidis, J.M., Rosenberg, A.M. and Saleh, I. (1980). *Improved Methods for Determining Corrosion Inhibitors by Calcium Nitrite in Concrete*, ASTM, STP 713, 64–74.
62 Virmani, Y.P., Clear, K.C. and Pasko, T.J. (1983). *Time to Corrosion of Reinforced Steel in Concrete Slabs, Vol. 5, Calcium Nitrite Admixtures or Epoxy Coated Reinforced Bars as Corrosion Protective Systems*, FHWA-RD-83-012, FHWA, US Department of Transportation, 71.
63 Berke, N.S. and Weil, T.G. (1988). *2nd International Conference on the Performance of Concrete in Marine Environment*, St. Andrews, New Brunswick, Canada, August 21–6.
64 Hope, B.B. and Ip, A.K.C. (1990). *Admixtures for Concrete: Improvement of Properties*, Chapman & Hall, London, 299–306.
65 Treadaway, K.W.J. and Russell, A.D. (1968). *BRS Current Papers, Highway and Public Works*, **36**, 19–21; 40–1.
66 Lewis, J.L.M., Mason, C.E. and Brereton, D. (1956). *Civil Engineering and Public Works Review*, **51**, 602, 881.
67 Gouda, V.K. and Halaka, W.Y. (1970). *British Corrosion Journal*, **5**, 204–8.
68 Andrade, C., Alonso, C., Acha, M. and Malric, B. (1992). *Cement and Concrete Research*, **22**, 869–81.
69 Nmai, C.K., Farrington, S.A. and Bobrowski, G.S. (1992). *Concrete International*, **14**(4), 45–51.
70 Yau, S.S. and Hartt, W.H. (1980). *Corrosion of Steel in Concrete*, ASTM STP-713, 51–63.
71 Gouda, V.K. and Shater, M.A. (1975). *Corrosion Science*, **15**, 199–204.
72 Hime, W.C. (1993). *Concrete International*, **15**, 54–7.
73 Mailvaganam, N.P., Nunes, S. and Bhagarath, R.J. (1993). *Concrete International*, **15**(10), 38–43.
74 Matusomoto, S. (1970). *CEER*, **6**, May.
75 Sheik, S.A., Fu, Y. and O'Neill, W.M. (1993). *ACI Materials Journal*, **9**(6), 74–83.

76 Masaichi, O., Kondo, R., Mugruma, H. and Ono, Y. (1980). *4th International Symposium on the Chemistry of Cements*, Tokyo, Japan, Supplementary Paper IV–86.

77 Rosetti, A., Chiocchio, G. and Paolini, A.E. (1982). *Cement and Concrete Research*, **12**, 577–85.

78 *Denka CSA, Manual*, Introduction Edition (1982). Denki Kagaku Kogyo, Kabushiki, Kaisha Company, Tokyo, Japan.

79 Xie, P. and Beaudoin, J.J. (1992). *Cement and Concrete Research*, **22**, 845–54.

80 Mailvaganam, N.P. (1981). Report DIMI-*/11/81*, Research and Development Laboratories, Sternson Ltd., Canada, 21.

81 Klein, A., Darby, T. and Povilka, M. (1961). *Journal of ACI, Proceedings*, **58**, 59–79.

82 *Denka CSA, Manual*, Data Edition (1982). Denki Kagaku Kogyo, Kabushiki, Kaisha Company, Tokyo, Japan.

83 Bayasi, Z. and Abifaher, R. (1992). *Concrete International*, **24**(4), 35–7.

84 Chen, M.D., Olek, J. and Mather, B. (1991). *Concrete International*, **13**(3), 31–7.

85 Mailvaganam, N.P. (1994). *Concrete Admixtures Handbook*, Noyes Publications, Boca-Raton, 862–3.

86 Hoff, G. (1972). CTIAC *Report 8*, US Army Engineer Waterways Experiment Station, Vicksburg, Mississippi.

87 Ohama, Y. (1995). *Handbook of Polymer Modified Concrete and Mortar – Properties and Process Technology*, Noyes Publications, 157–9, 161–3.

88 Ma, W. and Brown, P.W. (1992). *9th International Congress of Cement Chemistry*, Delhi, 424–9.

89 Popovics, S. (1985). *Polymer Concrete, Uses, Materials and Properties*, ACI, Detroit, SP-89, 207–29.

90 Ohama, Y., Demura, K. and Endo, T. (1993). STP 1176, ASTM, *Polymer-Modified Hydraulic Cement Mixtures*, 90–103.

91 Soroushian, P. and Tlili, A. (1993). STP 1176, ASTM, *Polymer-Modified Hydraulic Cement Mixtures*, 104–19.

92 Tsai, M.C., Burch, M.J. and Lavelle, J.A. (1993). STP 1176, ASTM, *Polymer-Modified Hydraulic Cement Mixtures*, 63–75.

93 Walters, D.G. (1993). STP 1176, ASTM, *Polymer-Modified Hydraulic Cement Mixtures*, 6–18.

94 Bright, R.P., Miraz, T.J. and Vassallo, J.C. (1993). STP 1176, ASTM, *Polymer-Modified Hydraulic Cement Mixtures*, 44–62.

95 Ohama, T., Demura, K., Satoh, Y., *et al.* (1989). *Proceedings 3rd International Conference on Superplasticizers and Other Admixtures in Concrete*, Ottawa, Canada, ACI, SP-119, 321–39.

96 Ohama, Y. (1989). *International Meeting on Advanced Materials*, Materials Research Society, **13**, 79–97.

97 Chandra, S. and Flodin, P. (1987). *Cement Concrete Research*, **17**, 875–90.

98 Ohama, T., Demura, K., Satoh, Y., *et al.* (1990). *RILEM Symposium on Admixtures for Concrete*, Barcelona, Chapman & Hall, London, 317–24.

99 Kuhlmann, L.A. (1993). STP 1176, ASTM, *Polymer-Modified Hydraulic Cement Mixtures*, 125–40.

100 Amano, T., Ohama, Y., Takemoto, T. and Takeuchi, Y. (1989). *Proceedings 1st Japan International SAMPE Symposium*, Chiba, 1564–69.

101 Atzeni, C., Mantegazza, Massida, L. and Sanna, U. (1989). *Proceedings 3rd Conference Superplasticizers and Other Admixtures in Concrete*, ACI SP-119, 457–70.

102 Anon. (1988). *Concrete Construction*, **33**, 316–19.

103 Anon. (1988). *'Delvo System', Research and Development*, Dept. Master Builders Technologies, 24.

104 Kinney, F.D. (1989). *Proceedings 3rd International Conference on Superplasticizers and Other Admixtures in Concrete*, ACI SP–119, 19–40.

105 Dolch, W.L. (1989). *An Investigation into the Nature of the Cement Paste Treated with the Delvo System*, Consultant's Report, 21.

106 ACI Standard 506–566. *Recommended Practice for Shotcreting*, Reaffirmed 1972.

107 Morgan, D.R. (1985). *Geotechnical News (Vancouver)*, Mar., 24–5.

108 Prudencio, L.R. (1998). *Journal of Cement, Concrete and Composites*, Feb., 213–2.

109 Mailvaganam, N.P. (1994). *Concrete Admixtures Handbook*, 2nd Edn, Noyes Public., 1009–17.

110 Mahar, J.W., Parker, H.W. and Weller, W.W. (1975). *Shotcrete Practice in Underground Construction*, Report No. FRA-OR & D 75–90, Department of Civil Engineering, University of Illinois, USA, 5–23.

111 Singh, M.N., Seymour, A. and Bortz, H. (1973). *Use of Shotcrete for Structural Support*, ASCE and ACI SP-45, July.

112 Litvin, A. and Shideler, J.J. (1966). *Shocreting*, ACI SP-14, 165–84.

113 Blank, J.A. (1975). *Use of Shotcrete for Underground Structural Support*, ASCE and ACI SP-45, 320–29.

114 US Patent 3 656 955; assigned to Progil (1972).

115 Reading, T.J. *Concrete International*, **3**(1), 27–33.

116 Morgan, D.R., Private communication.

117 Mehta, P.K. (1985). *Cement and Concrete Research*, **15**(6), 969–78.

118 Gilbride, P., Morgan, D. R. and Bremner, T.W. (1996). *Third CANMET/ACI International Conference on Performance of Concrete in Marine Environment*, Aug. 4–9, 1996, St. Andrews-The-Sea, Canada, 163–73.

119 Aitcin, P.C. (1983). Condensed Silica Fume, Faculte des Sciences Appliques, Université de Sherbrooke.

120 Malhotra, V.M. and Carette, G.G. (1983). *Concrete International: Design and Construction*, **5**(5), 40–6.

121 Bates, R.C. (1975). *Use of Shotcrete for Underground Structural Support*, ASCE and ACI, SP-45, 130–42.

122 Balogh, A. (1996). *Concrete Construction*, July, 546–51.

123 Tomita, R. (1992). *Concrete Library of JSCE*, No. 19, **6**, 233–45.

124 Berke, N. S., Dallaire, M.P., Hicks, M.C. and Kerkar, A. (1994). *Proceedings ACI International Conference on High Performance Concrete*, Singapore, 326–33.

125 Shoya, M., Sugita, S. and Sugawara, T. (1990). *Proceedings of the International RILEM Symposium*, 484–95.

126 Fujiwara, H., Tomita, R. and Shimoyama, Y. (1994). *A Study of the Frost Resistence of Concrete Using an Organic Shrinkage Reducing Agent*, ACI SP-145–34, 643–55.

Chapter 7

Application of admixtures

7.1 Introduction

Concrete is no longer a rationalized mixture of cement, sand and stone, but has matured to an engineered material containing a mixture of admixtures. Admixtures modify the structure–property relationship of normal concrete by chemically altering the rate of cement hydration and or, the nature of the hydration products. The last decade has seen a wide interest in many aspects of admixtures and research on the chemistry of the aqueous phase of the cement–water–admixture system has demonstrated that admixtures can control the type of products formed so that many properties can be designed into concrete. The correct combination of admixtures can produce concrete that is custom-made for the particular job at hand.

In addition to the well-established applications (water reduction, retardation, acceleration and air entrainment), chemical admixtures are used for a wide range of special purposes. Included in this category are both conventional admixtures used in unusual ways and special admixtures designed for specific applications. In this chapter both the established uses and the special applications are described to enable the practicing engineer and architect to take a new look at the potential of these materials which are now integral components of the concrete. The diverse applications of admixtures discussed here as well as in Chapter 6 give an account of the scope, function, mechanism and attributes of each admixture. They present the reader with objective information on the theory and practice relating to admixture use.

The urgent need to repair structures that have failed has made durability the most pressing construction problem of the day. Factors that control the durability of concrete are the selection of suitable materials and their proportioning followed by effective mixing, handling, placing, compaction, finishing and curing. Many innovative applications of super-plasticizers enable the placing of highly fluid concretes that have excellent filling capabilities and also provide a homogeneous concrete. The dramatic improvements to freeze–thaw and chemical resistance through the use of air-entraining and water-reducing agents, and super-plasticizers, are presented in a discussion that details the enhancement of workability, durability and water tightness. Specific problems with concreting in cold and hot environments, practical implications, and the manner in which admixtures can be used to overcome the limitations are dealt through a basic understanding of the effects produced on the cement and concrete.

Environmental protection agencies in many countries now classify fresh concrete as a hazardous material. Consequently, the dumping of plastic concrete waste and wash water from ready-mix trucks in landfill sites is banned. Admixture manufacturers responded to this constraint by developing a chemically-based concrete recycling system that stops hydration for a desired period and allows reactivation at a given time. Additional innovative uses for this admixture system, as well as the manner in which it enables the recycling of fresh concrete, are also discussed. With the exception of situations where the chemically induced change may be essential for a given application, admixture selection is based on techno-economic considerations. A summary of the factors that should be taken into account when considering the economic aspects of admixture use is presented.

As more chemicals are added to the concrete mix, compatibility becomes the central parameter governing selection. Side effects or reactions between chemicals due to the sequence of addition, cement type, temperature change and batching equipment can all affect performance. The discussion on admixture cement interaction discussed in the section on admixture problems presents some of the adverse effects that arise from the wide variability in the dispersing action of superplasticizers with different brands of cement. Field problems that result due to overdosage and incompatibility of admixtures are also described. The benefits obtained in using admixtures are also dependent on dosage, for overdosing can be detrimental. The aspects of storage, safety and method and efficiency of dispensing are therefore important; the section on guidelines for use, batching and dispensing equipment gives detail specifications and equipment requirements necessary to ensure that the concrete mix contains the stipulated admixture dosage.

The need for careful selection of aggregates, optimum proportioning, effective batching, mixing and curing is not lessened by the use of an admixture. It is prudent to remember that admixtures are tools that should be used judiciously, and not for the purpose of covering up deficiencies in the mix. Notwithstanding the significant beneficial effects and expanded applications provided by admixtures, it is important to be cognizant of the potential for serious problems presented by the use of a mixture of chemicals in concrete. Compatibility becomes the key issue and verification of field performance through trial batches with materials to be used on the job under field conditions is crucial.

7.1.1 Reasons for use of admixtures

Admixtures are generally used for the following reasons:

1. To enable the concrete to meet requirements of job specifications, namely, permitted maximum water–cement ratio, minimum early and ultimate strengths, and retention of workability when the available raw materials are of poor quality.
2. To provide modifications which improve the quality of plastic concrete, e.g. increased workability or water reduction at given consistencies, retard or accelerate time of initial setting, retard or reduce heat evolution during early hardening, improve finishing qualities, control bleeding and segregation, improve pumpability and reduce the rate of slump toss.
3. To reduce the cost of concreting operations by effecting a reduction of the overall cost of concrete ingredients, permitting rapid mold turnover and ease of placing and finishing.
4. To improve the quality of hardened concrete such as increased early and long-term strengths and modulus of elasticity, decrease permeability (and hence inhibit corrosion of embedded steel reinforcement) and absorption, increase abrasion resistance and increase bond with reinforcement.
5. Control expansion caused by the reaction of alkalis with certain aggregates.

In many instances, an admixture may be the only feasible means of achieving the desired result. However, in other instances, certain desired objectives may be best achieved with greater economy by changes in composition or proportions of the concrete than by using an admixture.

7.2 Air-entraining admixtures

A concrete's susceptibility to damage by freeze–thaw cycles arises when its moisture content reaches the critical saturation point, with the severity of the deterioration depending on the degree of exposure. The general method of counteracting freeze–thaw damage of concrete is to provide an adequate air-void system through the use of air-entraining admixtures. Air entrainment has provided great benefit in many applications but nowhere more dramatically than in situations where the structure is exposed to freeze–thaw conditions.

It has sometimes been argued that concrete is seldom damaged by frost action if low water–cement ratio mixes are used. However, the paramount effect of air entrainment in improving freeze–thaw resistance was clearly demonstrated in a study of the freeze–thaw resistance of both air-entrained and non-air-entrained superplasticized concrete with low water–cement ratio (<0.35). The investigators concluded that the effect of providing an adequate air void system in concrete far overshadowed any improvement in freeze–thaw resistance as a result of lowering the water–cement ratio [1].

In Russia, Scandinavia and North America not only is the concrete exposed to severe freeze–thaw cycles, but it is also subjected to the damaging effects of ice removal salts. Highway bridges often span water courses; abutments and piers are situated in areas where wet and dry conditions alternate. Certain parts of hydraulic structures such as dams, spillways, tunnel inlets and canal structures are exposed to fluctuating water levels or spray. Concrete in these structures is often saturated and quite vulnerable to freeze–thaw damage [2]. Consequently, a large portion of the concrete placed in North America is air entrained. In the United Kingdom the application of air-entrained concrete is largely confined to the production of concrete carriage ways, aircraft runways, lightweight concrete and precast concrete cladding panels, otherwise concrete mixes are not usually designed for frost resistance.

In order to provide adequate protection against freeze–thaw action most North American specifications for durable concrete always include the requirement of low water–cement ratio and the proper amount of entrained air consistent with the maximum aggregate size used. Figures 7.1 and 7.2 show the effect of entrained air on the resistance of concrete to freezing and thawing conditions [3]. The air-entraining agents used can be conventional neutralized wood resins, synthetic detergents (alkyl-aryl sulfonates), salts of petroleum acids (sodium salts of naphthenic acid), or the more recent types which include salts of fatty acids, and organic salts.

The recent increased use of pozzolanic and supplementary cementing materials such as fly ash and slag in the ready-mixed industry requires close control of air-entrained concrete. This is due to the difficulty of entraining the required air content when these materials are present in the concrete mix. Problems arise because of the adsorption of the surface active agent on the high-surface-area carbon particles [4]. For concrete mixes containing fly ash with carbon content less than 5% and at cement replacement levels under 25%, little difficulty is encountered in entraining air contents of 4–5%. Slight increase in the dosage of the admixture will usually adequately compensate for the air-detraining effects of such concretes. However, when 6–7% air contents are required, not only must the carbon content and cement replacement levels be below 5% and 25%, respectively, but also special care must be used to ensure uniformity. Slight changes in the consistency of the mix may cause large changes in air content.

Manufacturers of a multicomponent mixture of fatty acids, salts of sulfonic acid and stabilizing polar compounds (such as sodium octonate) claim that it provides concrete extra protection by creating ultra-stable air bubbles that are small and closely spaced [5]. In addition to many advantages, it can be used in fly ash concrete containing large amounts of

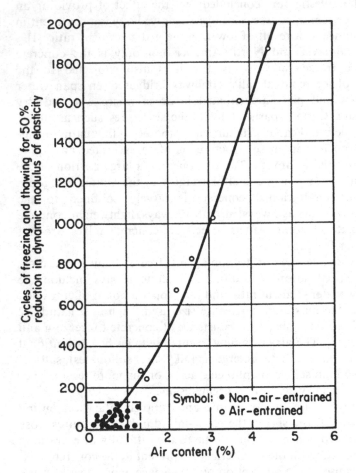

Fig. 7.1 Effect of entrained air on the resistance to freezing and thawing in laboratory tests (courtesy Portland Cement Association, Skokie, Illinois, USA [2]).

carbon, concrete containing a high-alkali cement and in concrete requiring extended mixing times.

Other fine materials which tend to inhibit air entrainment include pigments, particularly carbon black. This is of concern to the ready-mixed operator supplying integrally colored concrete for exterior exposure [6]. Accelerating admixtures which are used to reduce down time and in cold weather concreting can be used successfully in air-entrained concrete, but should be added separately and in solution form to the mix. Direct contact of these admixtures with some types of air-entraining agents mixed in the same water phase may adversely affect both admixtures.

Fig. 7.2 Effect of freeze–thaw cycles on concrete with and without air entrainment. (Left) Concrete without air-entrainment local aggregate. This sample failed at 100 cycles of freezing and thawing. Expansion 0.6%, weight loss 25%. Since 25% loss is considered equivalent to failure, further cycles of freezing and thawing were discontinued on this sample. (Right) Air-entrained concrete 4% air local aggregate. After 300 cycles of freezing and thawing, expansion was 0.2% and weight loss 4%. This sample shows the direct advantages that occur through use of air entrainment in concrete (courtesy, Protex Industries Inc. Denver, Colorado, USA).

7.2.1 Control measures used to ensure proper air entrainment

It has been shown that the air content of a concrete mix is usually indicative of the adequacy of the air-void system when the admixture used meets pertinent specifications of the national standards (ASTM C 260, CSA A.266.2 and BS 5075). However, factors such as variability in the material, mixing, placing methods, temperature and the use of blended cements, may make it difficult to adjust the required amount of air with the right bubble size and spacing. It is also not easy to adjust the air in superplasticized concrete and high-strength concretes. Twenty-three factors that affect air entrainment have been listed [7].

An understanding of the effect of variables on air entrainment is best shown by a consideration of the data collected during the construction of some dams in Japan [8]. The following relationship was found to exist where the concrete was air entrained either by an air-entraining agent alone or an air-entraining agent with a water-reducing admixture:

$$\Delta\sigma = -3\Delta s - 5.9\Delta a$$

where $\Delta\sigma$ is the variation in compressive strength (MPa) at 91 days, Δs the variation in slump (mm) and Δa the variation in air content.

An increase in slump of 25 mm reduced compressive strength by 3 MPa. Thus, the control of air content is necessary to obtain reasonable standard deviation or coefficients of variation in compressive strength. In the study cited the compressive strengths at 3 months shown in Table 7.1 were obtained.

Achievement of the benefits of air entrainment in a consistent manner therefore requires close control of the proportioning and batching of the concrete and a good understanding of the factors which affect air-entraining characteristics. It is important that the concrete producer ensures that the properties of the concrete materials, mix proportioning and all aspects of the mixing, handling and placing procedures, are maintained as constant as possible. Some control measures that should be used particularly in ready-mix concrete include the following [9]:

Table 7.1 Variability of concrete for dams produced containing air-entraining agents, water-reducing agents, or a combination of both

Dam	Type of admixture	Coefficient of variation (%) of compressive strength at 3 months
Okutadami Tagokura Hatanagi No. 1	AEA only	12–15
Kurobe No. 4 Hitotsuse Moromaki Arimine	AEA and WRA	8–10.5
Sakamoto	WRA only	5

1. Air-entraining admixtures of proven performance should be used with an accurate dispensing system.
2. Special attention should be paid to proper batching technique and maintenance of fins in the mixers to assure formation of a proper air-void system.
3. Concrete should arrive at the job site with the proper slump. Excess water added at the job site (retempering) could cause the air bubbles to cluster thereby reducing durability.
4. Air content should be checked both at the plant and at the job site.
5. Air loss from plant to job site should be established and admixture dosage requirements adjusted at the plant to compensate for the loss.
6. Other parameters of the air-void system in the hardened concrete should also be determined to confirm that the air-void spacing factor is within the limits specified in the standards.

Air entrainment plays a significant role in providing the required plasticity for lightweight aggregate mixes. The texture of some lightweight aggregates tend to make the concrete harsh, requiring an increased fine aggregate content. Consequently, the density of the resultant concrete is increased. Air entrainment enables the percentage of fine aggregate to be kept low, helps to prevent the coarse aggregate particles from floating in the mortar fraction, and facilitates pumping of such mixes. A major application of air-entraining agents in the precast industry is in the production of lightweight cladding panels. High air contents are incorporated in the panels to provide proper coarse aggregate distribution and enhance durability.

7.2.2 Methods of placing

The method of placing air-entrained concrete can vary from simple manual placing between retaining side forms to more complex slip-form operations where the consecutive procedure of concrete placing, compaction, forming, finishing and insertion of joints is carried out by one or more machines. Figure 7.3 shows the placing of a concrete sidewalk by the slip-form technique.

Air-entrained concrete can also be placed by pumping. Concretes with air contents of 4–6% can be satisfactorily pumped, though some trouble may be experienced at the high end of the range since the elastic compression of air on each stroke reduces the efficiency of the pump. This is particularly likely to occur with long pipelines. Notwithstanding this limitation, it has been shown that air entrainment assists pumping in more cases than otherwise [10]. The air content of the placed concrete is not significantly affected by pumping, although small losses (1–2%) have been noted. Air entrainment is important especially if the concrete without air is deficient in fines and therefore inclined to be harsh

Fig. 7.3 Placing of a concrete sidewalk using a slip-form paver (courtesy, Barber Greene, USA).

7.2.3 Air-entraining admixture/superplasticizer compatibility

The use of superplasticizers in air-entrained concrete has caused much debate. Two main problems are associated with superplasticized air-entrained concrete: (1) a decrease in air content by 1–3% when slump is increased from 75 mm to 220 mm after the addition of the superplasticizer to create flowing concrete, and (2) a change in the air void system to less desirable values. However, most investigators [10–11, 12] have shown that, although the air-void spacing factor required for adequate frost resistance is altered, the change did not necessarily affect the freeze–thaw durability of concrete. The decrease in air content in flowing concrete is due to the release of entrapped air as a result of reduced paste viscosity or the coalescence of the bubbles. The reduction is usually offset by an increase in the dosage of the air-entraining agent [10, 12] to compensate for the lost air content. Addition of the air-entraining agent after the blending of the superplasticizer in the mix has been suggested as a suitable means of minimizing increases in the air-void spacing [13]. Problems in entraining the proper amount of air and suitable air-void structure are discussed in more detail in the section on admixture compatibility (pp. 393, 398).

7.2.4 Composite air-entraining–water-reducing admixtures

Air-entraining admixtures confer significant benefits to concrete produced in hot climates such as those of Central and South America and the Middle East. The admixture is widely used to upgrade poor-quality sands by compensating for deficient fine material. Consequently, bleeding and segregation are reduced, e.g. air entrainment eliminates the considerable separation of the mix (into an upper supernatant layer and lower aggregate layer) that occurs in cast *in situ* columns when sands deficient in particles less than 150 mm diameter are used in the mix. Figure 7.4 [14] illustrates this.

The variability in strength of air-entrained concrete is greater than that of plain concrete. It is, therefore, prudent to use the air-entraining agent in combination with a water-reducing admixture, or to use a single composite air-entraining and water-reducing admixture, since it leads to a lower standard deviation than a plain air-entraining admixture. Also, as the mean strength at an equivalent cement content is higher, the use of the admixture will result in a more consistent product. This is illustrated in Table 7.2 where the mean strengths and standard deviations of batches from one plant using plain concrete, air-entrained concrete with neutralized wood resins, and composite air-entraining and water-reducing agent, are presented.

Because of the essential role that air-entrained concrete has played in reducing freeze–thaw susceptibility, its use in North America has been predominantly associated with this property. Consequently, other favorable modifications have often been ignored. In Europe, Australia and Africa where freeze–thaw action seldom or rarely affects the concrete, the secondary modifications produced by air-entraining admixtures have been realized to their full potential, when used as a composite water-reducing and air-entraining admixture. The increased use of this type of admixture is due to the following reasons:

1. Frequently specified maximum water–cement ratio of 0.55 with the required mix cohesion and minimum strengths are readily attained at reasonable cement contents.

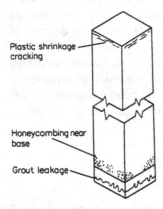

Plastic shrinkage cracking

Honeycombing near base

Grout leakage

Fig. 7.4 Diagramatic representation of problems arising from poor aggregate qualities.

Table 7.2 Air-entrained concrete has a higher standard deviation that plain concrete, but this can be minimized by the use of an air-entraining, water-reducing agent

Type of admixture in 270 kg m⁻³ OPC mix	Mean strength ($N\ mm^{-2}$)	Standard deviation ($N\ mm^{-2}$)
None (plain mix)	31.8	5.2
Neutralized wood resins	31.6	8.6
Air-entraining water-reducing agent	33.4	7.9

2. The lower water–cement ratio and higher cohesion produced at the desired compacting factor values in the mixes provide a more 'structured' concrete under conditions of no applied force. This is of particular importance in slip-formed concrete where reduced edge slump is required.
3. The higher mean strength obtained results in a concrete with a lower rejection rate than corresponding concretes containing conventional air-entraining admixtures only. Typical results of comparative mixes containing no air-entraining agent, a normal neutralized wood resin, and a composite water-reducing and air-entraining agent, are shown in Table 7.3, where the benefits obtained are clearly illustrated.

In North America almost all air-entrained concrete contains a water-reducing admixture. There is some reluctance to use a composite type of admixture because of the concern that it reduces flexibility in use due to the variation in concrete materials and job site conditions. Therefore, the use of

Table 7.3 Comparative porperties of plain concrete and air-entrained concrete produced with neutralized wood resins and a water-reducing air-entraining agent

		Control	Mix I	Mix 2
Admixture		None	Traditional neutralized wood resin	Water-reducing air-entraining agent
Mix design	Aggregate type	40–10 mm rounded gravel	40–10 mm rounded gravel	40–10 mm rounded gravel
	Sand,	Zone 2	Zone 2	Zone 2
	Air–cement ratio	6.0	6.0	6.0
	% fines	33	33	33
	Water–cement ratio	0.59	0.56	0.54
Properties of plastic concrete	Slump (mm)	45 mm	40 mm	45 mm
	Compacting factor	0.89	0.91	0.93
	VeBe (s)	5.0	3.5	2.0
	Air content	1.0	3.9	4.1
Properties of hardened concrete	Average compressive strength N mm⁻² (lbf/in²) 7 day	30.1 (4300)	28.0 (4000)	30.8 (4400)
	28 day	37.3 (5335)	32.5 (4650)	37.2 (5320)

a composite water-reducing and air-entraining admixture for the purpose of improving product consistency finds little favor in North America.

7.3. Normal-setting water-reducing admixtures

Durable concrete is typically characterized by low porosity because the fundamental porosity of concrete influences all of its material properties. For this reason, most of our standard practices for the construction of concrete structures have as their objective the minimization of paste porosity, which consequently increases both strength and durability. Although low water–cement ratio (W/C) is responsible for improved mechanical properties and enhanced durability, attaining a low W/C necessitates either a sacrifice in workability, or the use of high cement content, neither a desirable consequence. A more advantagious alternative is the use of water-reducing admixtures (WRAs).

Normal range WRAs have been available for over 30 years, and superplasticizers for the last 20 years. The newest product is the mid-range WRA, occupying a vital position in the spectrum of water-reducing capability (Table 7.4). Mid-range WRAs are reported to provide water-reducing effectiveness without the attendant inconsistent setting time produced by high dosages of normal WRAs. Normal, mid- and high-range WRAs are used for the purpose of increasing the durability of concrete primarily by means of decreasing the permeability and improving mechanical properties (Fig. 7.5). Their effectiveness will depend on the dosage used, temperature, cement composition, fineness and other mixture characteristics.

The relationship of the concentration of the admixture to the water reduction produced has been found to be linear (Fig. 7.6). Two conclusions were reached from the study which included several brands of cement [15]: (1) it is the addition rate of the admixture which is important, and (2) the chemical nature of the admixture does not play a part in water reduction, but is an important factor in determining the retention (or loss) of slump in concrete with age. For example, the extended workability of the more

Table 7.4 Water-reducing admixtures (Hover [119])

Appropriate ranges of water reduction effectiveness (%) *	Nomenclature
5–7	Normal water reducer
7–15	Mid-range water reducer
15–30	High-range water reducer or 'superplasticizer'

*Effectiveness varies with a large number of factors, especially admixture dosage.

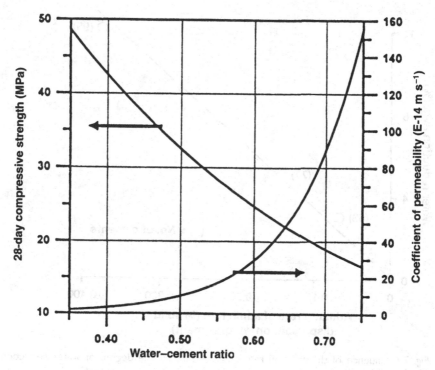

Fig. 7.5 Impact of water–cement ratio on both compressive strength and permeability based on data from ACI 211.1 (ACI 221, 1991, and Powers et al., 1954b).

recently developed acrylic-ester type superplasticizers is attributed to a steric hindrance mechanism due to the presence of side chains in the molecule. Water reduction is more noticeable in cements of low alkali or C_3A contents. Dodson [16] states that the ratio of the percentage C_3A to percentage SO_3 (as reported in the mill certificate) must be less than 2.5 and preferably close to 2.0 in order to realize the maximum performance of chemical admixtures.

The use of WRAs (normal, mid-range and high-range) in ready-mixed and precast concrete to reduce water content and increase workability is described in the following pages.

7.3.1 Ready-mixed concrete

Ready-mixed concrete is now widely used in many sophisticated ways including low-slump concrete for highway paving, heavyweight concrete for thermonuclear plants, low-heat concrete for prestress work, lightweight concrete for high-rise buildings, and gap graded concrete for exposed

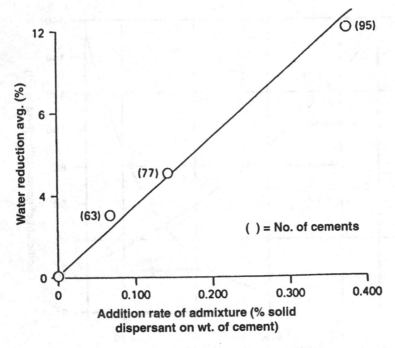

Fig. 7.6 Influence of the addition rate of dispersant on the degree of water reduction (Dodson [15]).

aggregate architectural treatments. A variety of admixture types are used for this purpose; often two or more admixtures are used in combination in the same concrete mix.

Most of the ready-mixed concrete used in Canada, Europe and the United States contains an admixture [17]. The ready-mixed concrete producer in both continents usually supplies concrete-containing admixtures either on request from the client to provide a specific material or as a means of providing a specific type of concrete, e.g. watertight concrete or as a means of achieving the most economic use of mix ingredients. In North America, since freeze–thaw resistance plays a significant role in the durability of the concrete, the combined use of water-reducing and air-entraining agents is common. In Europe, the use of air-entraining agents is less frequent.

When concrete is specified by compressive strength and workability, normal-setting and retarding water-reducing admixtures (both normal and mid-range types) are widely used as a means of attaining the required properties at lower cement contents than mixes which contain no admixture. Although the use of admixtures by the ready-mix sector of the industry is generally similar to that of site-batched concrete, there are several unique elements and problems in the control of concrete quality in the former sector. For example, in ready-mixed concrete, reducing variation in concrete properties in both the plastic and hardened state from batch to batch is of considerable importance in minimizing rejection levels in field batches. In this context, admixtures play a critical role in the maintenance of uniformity. This is illustrated in the following examples.

Table 7.5 presents a comparison of the results obtained from concrete batches produced on the same plant with and without admixtures. The table summarizes data collected over a 6-month period for two concretes of differing slump values (50 mm and 75 mm). It can be seen that the hydroxycarboxylic-acid-based normal water-reducing admixture produced no effect on the standard deviation for the 50 mm slump mixes, whilst an increase is noted for the higher-workability mixes.

The effect produced by the incorporation of a lignosulfonate-based water-reducing agent is shown in Table 7.6. The results were obtained from a series of mixes over an 8-month period by a ready-mix plant used in the production of concrete piles. Since the standard deviation of this particular plant was 5.0 MPa for mixes produced without the use of admixtures, it is evident that the use of the admixture resulted in reduced variability.

These results indicate that in high workability mixes with cement contents in the median range, the admixture may cause an increase in the standard deviation. Thus in redesigning the mix to have a lower cement content in this class of concrete, adequate consideration should be given to this difference in standard deviation. Increased uniformity can be attained in this instance

Table 7.5 Changes in standard deviations of a ready-mix concrete plant using a hydroxycarboxylic water-reducing agent

Mix	Slump (mm)	Admixture	No. of results	Average 28-day strength (N mm^{-2})	Standard deviation (N mm^{-2})
I	75	No	59	46.0	4.3
	75	Yes	61	52.0	5.8
2	50	No	386	44.0	5.0
	50	Yes	43	48.0	4.9

Table 7.6 Seven-day and 28-day strengths and standard deviations for concrete containing a lignosulfonate water-reducing agent produced on a ready-mix concrete plant

No. of mixes	Mean strength (N mm^{-2})		Standard deviation (N mm^{-2})	
	7 day	28 day	7 day	28 day
53	55.4	66.4	4.6	4.2

if the increase in standard deviation is compensated for by not utilizing the full potential cement reductions indicated by the mean 28-day strengths.

Figure 7.7 shows statistical data for strength tests from a ready-mixed concrete plant [18]. The coefficient of variation of 13.7% indicates an operation with a fair degree of control. However, these were random strength tests taken during the placing of foundation, sidewalks, driveways and miscellaneous construction typical of the small user of ready-mixed concrete. The results represent concrete mixes where there was wide variation in slump, sand gradation, moisture content, mixing time and where a high coefficient of variation (20%) is usually anticipated. The significant difference between this and the usual concrete delivered to the small consumer is that a water-reducing admixture was used throughout.

Another example of the effects produced by admixtures in ready-mixed concrete is also shown in Fig. 7.8, in which data from another ready-mix plant is presented. The slope lines show that a change in slump from 75 to 175 mm without a water-reducing admixture required an increase in water–cement ratio of 0.08. With the admixture, the same variation in slump required an increase in water–cement ratio of only 0.05, indicating that such concretes permitted variations in slump with less than the usual variation in water demand and water–cement ratio.

7.3.2 High-strength/high-performance concrete

Although the more modern high-performance concretes are being increasingly used in typical applications where high-strength concrete was used, high-strength concrete continues to be routinely produced in precast and prestressed applications through the use of low-slump cement-rich mixes which require prolonged vibration or shock methods for consolidation. However, site-batched concrete which utilizes less robust forms than those

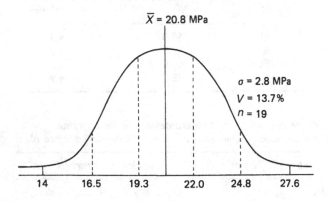

Fig. 7.7 Frequency distribution of control tests from a typical ready-mixed concrete plant (Howard et al. [18])

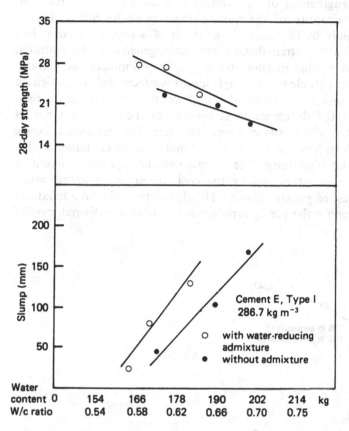

Fig. 7.8 Influence of water-reducing admixture on water content, slump and water—cement ratio (Howard *et al.* [18]).

used in precast concrete and therefore cannot use the same compaction procedures is particularly suited to the use of high-performance concretes. The term *high-performance concrete* refers to concrete mixtures which possess high workability, high-strength, high-dimensional stability and durability. This type of concrete finds application in heavily reinforced structural elements in high-rise buildings, offshore platforms, superspan bridges and heavy-duty pavements. The microstructure principles underlying the composition and properties of high-performance concrete are related to improved homogeneity, particularly in the porous and weak transition zone which exists at the paste—aggregate interface. By densification and strengthening of the transition zone, the properties of the concrete can be improved and the risk for easy microcracking (and consequential increase in permeability) can be reduced [19].

To achieve strengthening of the transition zone, firstly, a substantial reduction in the quantity of the mixing water must be made. This is accomplished mainly by the use of large doses of a superplasticizer. For additional densification, strengthening and homogeneity of the hydrated cement paste and mortar in concrete, a number of mineral admixtures possessing very fine particle size and high specific surface, such as condensed silica fume, rice husk ash, fly ash, slag and metakaolin, have been used. Figure 7.9 shows the difference in strengths obtained in two types of concretes fabricated with Portland cement concrete. The reference concrete contained a WRA with no silica fume; the second concrete contained both a superplasticizer and silica fume. The selection of the type and amount of mineral admixture is determined by the cost, reactivity, required workability, and the desired packing density. Highly reactive mineral admixtures like silica fume reduce the curing time needed to obtain a desired level of

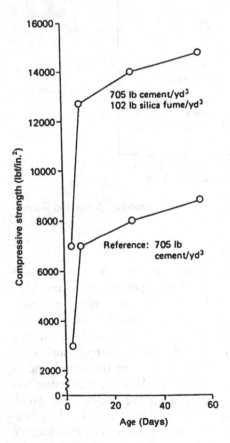

Fig. 7.9 High-strength concrete through the use of silica fume and a high-range water-reducing admixture (Dodson [15]).

strength, while the less reactive materials improve the thermal resistance by lowering the peak temperatures during hydration [20].

High-strength concrete continues to be produced in precast operations because of the relatively high material cost of high-performance concrete (HPC). As in HPC, aggregate–cement bond and matrix strength plays a significant role in determining the strength of high-strength concrete. However, the high cement contents that are generally required for such mixes are often counterproductive. High shrinkage stresses produced cause loss of aggregate–cement bond or cracking of the cement paste due to the restraint induced by the aggregate particles. Matrix strength is primarily dependent on matrix porosity, which is governed by the water–cement ratio [19]. Furthermore, plastic and workable concretes are necessary to avoid segregation or honeycombing [20]. The increased plasticity and reduced water and cement contents required to achieve high strength can be attained using normal and retarding water reducing or superplasticizing admixtures. In general, water-reducing retarders (based on hydroxycarboxylic acids) are more effective in such mixes than normal water reducers. Depending on the type of admixture used, two approaches are feasible [16]:

- **Method A.** Using a normal or retarding water reducer, the water–cement ratio is reduced 6–10% at the same cement content and slump.
- **Method B.** Using a superplasticizer, a lower water–cement ratio is produced at a lower cement content with the same or increased workability.

Table 7.7 shows the results obtained with a hydroxycarboxylic acid using Method A, while Table 7.8 shows the results obtained with superplasticizers using Method B.

Table 7.7 High-strength concrete produced by a hydoxycarboxylic acid water-reducing agent

Mix details	Control	Test mix 1	Test mix 2
Rapid hardening cement (kg)	500	500	500
Zone 2 sand (kg)	785	785	785
Crushed rock (10–5 mm)(kg)	1320	1320	1320
Hydroxycarboxylic acid water-reducing agent (ml/50 kg cement)	0	70	140
Water–cement ratio	0.46	0.42	0.41
Properties of concrete			
Slump (mm)	35	35	45
Air content (%)	0.8	0.8	0.7
Compressive strength at: 1 day	19.3	21.0	20.7
7 days	40.5	50.5	51.6
28 days	49.3	59.5	62.8
Density (kg m^{-3})	2415	2432	2437

Table 7.8 Cement reduction and change to type I with superplasticizers (La Fraugh [54])

Cement		Water–cement ratio	Admixture		Slump (mm)	VeBe (s)	Compressive strength (MPa)				Remarks
Type	Content (kg m^{-3})		Type	Dosage			18 h	60°C cure	28 day		
							21°C cure		21°C cure		
III	415	0.37	Zeecon	228 ml	57	3.8	24.24	40.34	75.50		Control
III	386	0.36	Mighty 150	1.2%	51	3.9	29.23	44.75	77.16		
III	386	0.33	Sikament	2.5%	38	5.8	29.10	49.71	82.46		
I	386	0.34	Mighty 150	1.2%	57	3.8	17.72	39.65	72.33		
I	386	0.31	Sikament	2.5%	114	2.1	22.62	41.99	81.02		

* Zeecon in millimeters per 50 kg cement.
Superplasticizers in percent of liquid by weight of cement.

Admixture systems which use high dosages of two or more admixtures have been promoted for high-strength mixes requiring high workability. The system usually consists of the use of a combination of admixtures based on the concept of (1) complementary and (2) compensatory action between two materials with similar functions; for example, in the production of high early strength, the increased rate of strength gain achieved by water reduction through use of a normal setting water reducer can be augmented by accelerated hydration produced by the use of an accelerator. Additionally, the adverse side effect (excessive set retardation) that results from the use of high dosages of the water reducer can be offset by the set-acceleration effect produced by an accelerating admixture [14].

7.3.3 High-workability mixes

The term *high-workability concrete* as applied to medium-cement-content concrete utilizing conventional lignosulfonate, hydroxycarboxylic and carbohydrate polymer based water reducers refers to mixes with slumps above 100 mm that are readily consolidated by vibration. Normal and mid-range WRAs usually produce slumps of up to 150 mm with little adverse effect, provided the fines content of the mix is increased to maintain cohesion. Such mixes find use in areas of congested reinforcement and thin sections with poor access, such as piling and diaphragm walling. These admixtures offer an alternative to the traditional methods which required increase in both cement and sand contents of the mix. Combinations of a water reducer and retarder or water reducer and a low dosage of superplasticizer may be used to effect economy in the production of flowing concrete (see later).

7.3.4. Pumping

Normal concrete mixes resist being moved under pressure due to segregation of the cement paste that occurs when pressure is applied. Basically a pumpable concrete will have a suitable aggregate–void system and a cement paste consistency that will flow adequately through the void channels. Both these properties of the mix are required to meet the primary demands of pumpability which are:

1. The pump pressure must be transmitted to the solids.
2. A continuous annular grout film must be formed adjacent to the pipe wall to ensure sliding of the concrete core.

Such requirements can often be met by specially designed mixes for individual jobs, almost always involving a compromise in sand–aggregate ratio and water content. However, even with good mix design the incidence of blockage is high. Variations in concreting materials (aggregate shape, gradation and moisture content) account for the major pipeline blockages. Furthermore, today many specifiers will not allow alterations to mix proportions merely to accommodate pumping because of adverse side effects on the properties of hardened concrete. A more viable approach to the design of mixes to be pumped is offered by the use of admixtures. Chemical admixtures broaden the envelope of aggregate gradations which may be used in the mix, enable concrete to be placed under a wider range of job conditions, and enhance the physical properties while making the mix more pumpable.

Three broad classes of pumpable concrete usually used are:

1. Low-cement-content mixes (210 kg m^{-3}).
2. Medium-cement-content mixes ($200-300$ kg m^{-3}).
3. High-cement-content mixes (> 300 kg m^{-3}).

Mixes in both low- and high-cement-content classes are more prone to problems than the medium range. In low-cement-content mixes, poor cohesion results in segregation and in high-cement-content mixes thixotropy causes pipeline friction. Admixtures will modify the flow characteristics of the paste, helping to achieve and maintain optimum flow characteristics. Because pumped concrete must not only meet specified job performance criteria (e.g. strength, freeze–thaw resistance) but should also remain stable under a variety of job conditions, particularly in hot and cold weather, it is common to find that concrete to be pumped often contains two or more types of admixtures.

In low-cement-content mixes, WRAs impart water retentivity to the cement paste under forces tending to separate the mix water. Special admixtures called pumping aids are available for assisting pumping of low-cement-content mixes. Pumping aids are admixtures that are generally used for the purpose of improving the plastic properties of concrete in situations where strength is not of primary concern. Chemicals that have been used include water-soluble synthetic and natural organic polymers, organic water-soluble flocculants, emulsions and inorganic materials of high specific surface area. Most of the organic materials act by increasing the viscosity of the cement paste, while the inorganic materials act as pore fillers. More information on these admixtures can be found in Chapter 6 and by consulting the literature [21, 22].

Concretes in the medium-strength range, although having satisfactory paste flow properties, often run into problems due to a lack of supply of consistent quality aggregates. Common problems are decreased cohesion of the cement pastes for mixes in the lower-cement-content range and increased friction to flow in mixes in the higher-cement-content range. In both instances, the use of either normal or retarding water reducers in combination with an air-entraining agent will alleviate these problems. Air entrainment increases the cohesion of the cement paste, while the retarding water reducer enables the release of water to reduce the friction that develops in a thixotropic paste. Mixes of high cement content tend to have thixotropic pastes. Consequently, flow through the aggregate–void channels is inhibited and the mobility of the peripheral grout layer decreases. Admixtures used in this class of concrete are of the dispersing agent type which induce lubrication by an increase in the free water content of the mix. Commonly used materials are calcium lignosulfonates and sodium salts of hydroxycarboxylic acid.

Pumping of lightweight concrete is another area where admixtures play a significant role in improving pumping characteristics. Such concretes are inherently more susceptible to segregation and absorption of water under pressure than normal concretes. The use of admixtures (air-entraining agent, superplasticizer or thickener) imparts increased viscosity and plasticity to the mix, resulting in improved pumpability.

7.3.5 'Watertight' concrete

Concrete which is to be subjected to water under a hydrostatic pressure is often specified as being 'watertight' or 'waterproofed' concrete. As explained earlier, hydrophobic-type waterproofing admixtures are ineffective in preventing the ingress of water under these conditions. By far the best method of producing concrete suitable for these applications is to utilize a good quality conventional water–reducing agent or a superplasticizer. Water reductions ranging from 10–20% are afforded by the use of these admixtures resulting in water–cement ratios (W/C) significantly lower (< 10%) than that obtained in corresponding mixes without admixtures; therefore, a considerable reduction in concrete permeability is obtained. This is illustrated in Fig. 7.5. Concrete of low W/C is essential to resist water ingress and so the concrete must be fully compacted and designed to have sufficient workability, yet have as low a W/C as possible. In this regard, water-reducing admixtures can be used to produce the dual requirement of good workability and low W/C when:

1. The water-reducing admixture is used to produce half the potential water reduction attainable (say 5%) and the slump is increased to a higher values (say 125 mm) than a mix with no admixture.
2. When high dosages of normal water-reducing and accelerating admixtures are used, to produce high water reduction and high workability.
3. When high dosages of superplasticizer are used to effect low W/C ratio and high workability.

The significantly lowered W/C and the efficient compaction that is achieved by the use of these admixtures is the best means of producing good-quality impermeable concrete.

7.3.6 Piling

Cast-in-place piling has similar mix design requirements to pumping mixes, e.g. high workability with adequate cohesion to minimize segregation and bleeding during placing. The maximum coarse aggregate size is usually small (< 40 mm) and a high sand content is used. Although they are not subject to weathering exposure, such as may occur with precast piles, cast-in-place piles may be subject to aggressive water in the soil. Consequently, the water–cement ratio must be kept to a minimum [12, 23]. The higher water demand of the mix (high fines content, small coarse aggregate) coupled with workability and durability requirements makes the use of an admixture essential. Conventional water-reducing admixtures and superplasticizers can readily provide lower water–cement ratios at the desired high workability.

7.4 Set-retarding and water-reducing admixtures

ASTM recognizes three types of retarding admixtures: type B, which simply retards the hydration of Portland cement; type D, which not only provides set retardation but also water reduction; and type G, which is a retarding supeplasticizer. The materials that are generally used in these admixtures include:

1. Lignosulfonic acid or modifications and derivatives of lignosulfonic acid and its salts.
2. Hydroxylated caboxylic acid or modifications and derivatives of hydroxylated acid and its salts.
3. Salts of high-molecular-weight condensation product of naphthalene sulfonic acid.
4. Carbohydrates, polysaccharides and sugar acids.
5. Inorganic salts such as borates and phosphates.

These materials may be used singly or in combination with other organic compounds. Type B admixtures have now been replaced by the bifunctional type D water-reducing set retarders.

The primary role of set-retarding admixtures is to offset the accelerating effect of high ambient temperature and to maintain the workability of the concrete during placing, thereby eliminating form deflecting cracks. Extension of the set time is particularly valuable to prevent cracking of concrete beams and bridge decks, caused by form deflections. Set retarders are also used to keep the concrete workable long enough so that succeeding pours can be placed without development of cold joints or discontinuities in the structural unit. The effectiveness of retarders in reducing the rate of slump loss, however, depends on the particular combination of materials used.

7.4.1 Retarded concrete for large pours

Aspects pertaining to the use of set-retarding admixtures are essentially the same for ready-mixed concrete as site-mixed concrete. However, when ready-mixed concrete is to be used for slip-forming or mass pour operations, greater care and good judgment should be exercised in determining the dosage required to obtain the desired retardation time, because of the time involved in transportation from plant to site. In the case of large pours this requires continuous planning, to provide delivery rates that meet the demands of the job site.

The placing of large volumes of concrete in successive layers to form a monolithic structure is carried out in the construction of dams, bridge piers, bridge decks and caissons for jetties. Projects such as dams which require substantial amounts of concrete are commonly located in remote areas of rough terrain where the cost of cement is high. Water-reducing retarders can contribute to the quality, economy and ease of placing of mass concrete by affording water and cement reduction, improved workability and resistance to segregation of the mix during transportation. The lower cement content made possible may also decrease the heat generation and temperature rise in the structures (see later). Since the thermal properties of the concrete play a significant role in the differential volume change of mass concrete, mixes for the construction of dams have mainly low cement contents [24].

Construction of bridge piers and bridge decks involve pours as large as $3000-4000$ m^3 of high-strength concrete [23]. Consequently, the potential for development of damaging internal stresses in reinforcement or cracks in the structure is increased. However, special techniques developed to offset such stresses allow pours of large magnitude to be placed without adverse effects on the concrete [25, 26]. Among the techniques used in such pours is the use of water-reducing retarders which can make the following contributions [25, 27]:

1. It is recognized that success in reducing thermal stresses is dependent on minimizing the temperature differential between two points in the concrete; a maximum temperature differential of 20°C has been suggested [25]. Therefore, any reduction in temperature rise will augment the measures taken to reduce this differential. Although water-reducing retarders themselves will not lower the total heat evolved, the lower cement content made possible by the use of admixtures decreases heat generation and temperature rise. In addition,

admixtures delay the liberation of heat and reduce the peak temperature attained during hydration [24, 27]. Figure 7.10 gives results for the heat output for mixes containing 385 kg m^{-3} cement in a control concrete containing no admixture and 330 kg m^{-3} cement in a concrete containing an admixture [25]. The reduction in maximum temperature is a favorable effect which supplements other measures taken to minimize the temperature differential.

2. In bridge construction, the use of retarding admixtures produces uniformity in the rate of setting and lessens the risk of deflection in partially hardened concrete that may occur in continuously reinforced structures such as spandrel beams. The retarded concrete poured over the supports remains plastic until the final pours are placed at mid-span.

3. Continuous girder bridges are another type of structure where set control plays an important role. The weight of the concrete as it is successively placed on the deck deflects the girders causing the partially hardened concrete to crack. Conventional concreting procedures consist of placing alternate panels so that a positive moment results (Fig. 7.11) [28]. The method is time intensive and costly to the contractor, since finishing operations are intermittent and also involve periods of waiting for the concrete to achieve adequate strength. The use of retarded concrete provides a situation where deflection occurs when the concrete is still plastic and capable of deforming. No cracking occurs and the finishing operations can follow the pour sequence.

4. Structures with a high degree of reinforcement are susceptible to loss of bond between steel and concrete. This is particularly true in situations where the vibration required to consolidate the concrete is transmitted along the reinforcement and could cause loss of bond in previously placed partially hardened concrete. The adjustment of admixture dosage as placement proceeds ensures that the concrete remains plastic throughout the entire pouring period.

5. Retarding admixtures compensate for adverse ambient conditions, particularly in hot-weather concreting where they help to overcome the damaging accelerating effects of high temperature. The admixtures will lengthen the permissible period for vibration between batching and placing by an extension of the vibrational limit of the concrete (see later). Thus, as an operational procedure it is an advantage in such pours to use concrete retarders, since inevitable delays will not necessarily result in the loss of concrete that is being produced by rapid consecutive batching.

6. Retarding admixtures also help to eliminate cold joints and other discontinuities when concrete is placed in layers by enabling adjacent layers to be vibrated into each other. Since setting time governs the optimum time at which concrete can be revibrated, the slower the set,

the more effectively can concrete be revibrated at later ages without loss of strength [28]. Figure 7.12 illustrates this effect.

7. If a pour is to be halted either by operational problems or by design, the last layer placed before the interruption can be further retarded by the use of larger dosages of the admixture. This can eliminate the necessity for construction joints. The heavily retarded concrete can be horizontal in the placing of layers or vertical when a sloping layer of concrete should be left.

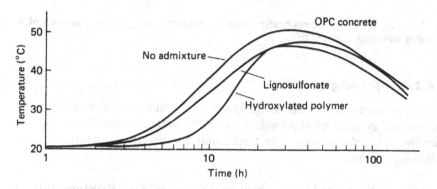

Fig. 7.10 Water-reducing admixtures can be used to reduce cement contents whilst maintaining the required compressive strength. In large pours reduced temperature rise is beneficial in minimizing the risk of cracking due to thermal expansion under restrained condition (Browne [24]).

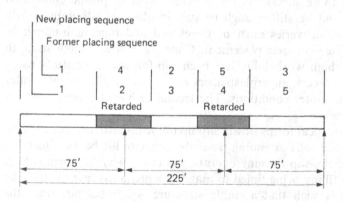

Fig. 7.11 Concreting sequence for continuous bridge deck (Shutz [28]).

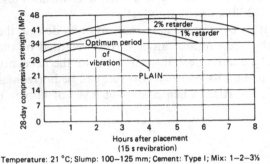

Temperature: 21 °C; Slump: 100–125 mm; Cement: Type I; Mix: 1–2–3½

Fig. 7.12 The extension of revibration time of concrete using various dosages of a retarding admixture (Shutz [28]).

7.4.2 Slip-forming

While pavements and canal linings are slip-formed in the horizontal plane, the normal concept of slip-forming is in the vertical plane. Vertical slip-forming is usually chosen over traditional concreting procedures for the following reasons:

1. The method produces a high-quality monolithic structure with a minimum number of joints.
2. It is fast and enables other phases of the job to be scheduled and completed more rapidly and efficiently.

For slip-forming to be successful, the concrete must be plastic enough to consolidate well and be stiff enough to stay in place. Continuous tightly scheduled material deliveries must be developed and then maintained to ensure uninterrupted concrete placement. Concrete of low to high strength and from low to high workability has been slip-formed successfully using retarding water reducers, superplasticizers and air-entraining agents under both summer and winter conditions. The technique has also been used for lightweight concrete.

In this system, vertical forms slowly slip up the semi-hardened concrete on jacks riding on steel rods extending from the concrete lift below. Once the form is filled and the slip-forming operation is under way the form is kept filled, the rate of filling being timed to match the predicted rate of rise. The lifting rate used is such that a stable structure is left behind, since the primary feature of the slip-form process is the ability of the concrete to support itself at an early age. Typical exposure times for the concrete after placing could range from 1.5 to 6 h. The requirements for setting time of the concrete are dictated by the time and distance between batching and placing and the time taken for concrete to emerge from the slip-form. In warm weather, retarding water-reducing admixtures, cooled mix water or fly ash are used to delay set so that sufficient time elapses to allow the transport and slipping of concrete. Conversely, when slip-form operations are conducted under winter conditions, the desired slipping times may be attained by the use of Type III cement (RHPC), silica fume, heating of the concrete to 22°C in the forms, use of protective enclosures, and high dosages of an accelerating admixture. Air-entraining admixtures are also used in slip-forming, not only to enhance durability characteristics but also to minimize the staining caused by excess bleed water running down the sides of the structure [29]. Table 7.9 shows mixes where both fly ash and admixtures were used to achieve the desired slipping times in the construction of the perimeter wall of a nuclear reactor building.

Mix proportions and set-controlling methods are usually determined from preliminary trial mixes conducted at the job site using actual construction materials to provide a guide for the slipping and set times required during the slip-forming process. The required data are usually obtained by preparing penetration resistance charts (using the Proctor penetration resistance test, ASTM C 403–20) which shows set-time variation with different admixture dosages (Fig. 7.13).

Variation of admixture dosage or delayed addition of the retarder during the course of the slip-forming process is determined by changes in temperature conditions. Each dosage change corresponds to the amount required to compensate for the effect of a change in temperature of 8.2°C, averaged between the concrete mix temperature and the air temperature [29]. The changes made should be incorporated in a gradual manner to avoid drastic changes in concrete setting time. This is particularly true for delayed addition which can extend the set time by an additional 2–3 h, increase the slump and air content of the treated concrete (Table 7.10) [16].

The Proctor penetration test is widely accepted for both initial laboratory trial and on-site testing to determine correlations between setting characteristics of the concrete and the rate of slipping of forms. Penetration resistance values in the range of 0.7–3.0 MPa have been found optimal for successful slip-forming. When using the Proctor penetration test as a means of estimating slipping times, care must be taken to ensure that storage conditions for the mortar containers are strictly comparable with those of the concrete in the slip-forms. The effects of ambient temperature and high winds, heat generation, etc., should be taken into account [30].

Vertical slip-forming is used in the construction of cooling towers, tapered chimneys, service cores for buildings, bridge piers, shaft linings and, more recently, in reactors for nuclear power stations. Two typical structures slip-formed from mixes containing hydroxycarboxylic acid, carbohydrate polymer type water-reducing and air-entraining admixtures are shown in Figs 7.14(a) and (b).

Table 7.9 Monthly statistical analysis and concrete mix design summary sheet
Area: Reactor building
Class: 3.5–3.7

Period ending: April 1976

			Mix designation 3.5	Mix designation 3.6	Mix designation 3.7
Period in use		From: To:	April 13, 1976 April 28, 1976	April 15, 1976 April 28, 1976	April 16, 1976 April 27, 1976
Design requirements:	Strength	MPa	34.5 (39.7)	34.5 (39.7)	34.5 (39.7)
	W/C		0.395	0.394	0.397
	Slump	±12 mm	100	100	100
	Air	±1%	4.5	4.5	4.5
	Sand	% of total agg. by vol.	35	35	35
Mix design proportions:	Cement	kg m^{-3}	368	380	380
	Fly ash	kg m^{-3}	83.1	83.1	83.1
	Water	kg	178	182	184
	Sand		575.5	575.5	575.5
	Stone		1074	1074	1074
	AEA MBVR	ml mm 0.19	280	280	315
	WRA Pozz	ml mm 0.12	680 (100 × R)	440 (100 × R)	215 (100 × R)

		3.5 Month	3.5 To date	3.6 Month	3.6 To date	3.7 Month	3.7 To date
Statistical analysis:	No. of samples, n	19	19	15	15	31	31
	Average strength, $\bar{x}$ (MPa)	43.6	43.6	40.7	40.7	38.8	38.8
	Standard deviation, σ MPa	3.38	3.38	2.62	2.6	2.3	2.3
	Coefficient of variation						
	Overall, V	7.8	7.8	6.4	6.4	6.0	6.0
	Within, v	2.2	2.2	2.9	2.9	4.4	4.4

Courtesy, New Brunswick Power Commission, Canada.

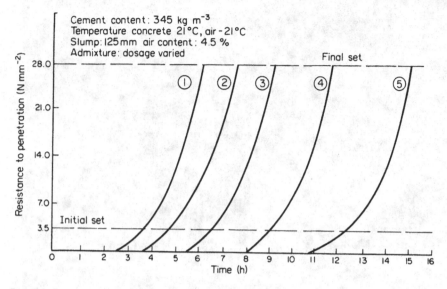

Fig. 7.13 Proctor needle penetration resistance for mortar sieved from concretes containing varying retarder levels.

Table 7.10 Physical properties of concrete with and without the delayed addition of water-reducing set-retarding admixture (Dodson)

Cement no. *	Method addition†	Water–cement ratio	Slump, mm (in)	Air, %	Delay in time of setting, min	
					Initial	Final
A	1‡	0.54	100 (4)	1.0	—	—
	2	0.48	100 (4)	3.6	60	85
	3	0.48	200 (8)	5.5	235	240
B	1	0.59	100 (4)	2.2	—	—
	2	0.55	89 (3½)	5.1	90	105
	3	0.51	83 (1¼)	6.9	240	270
C	1	0.53	95 (3¾)	1.0	—	—
	2	0.48	95 (3¾)	3.1	134	140
	3	0.48	100 (4)	3.3	250	255
D	1	0.58	114 (4½)	2.5	—	—
	2	0.54	89 (3½)	4.6	65	115
	3	0.50	114 (4½)	6.7	215	285
E	1	0.53	100 (4)	1.2	—	—
	2	0.49	89 (3½)	3.5	60	60
	3	0.49	178 (7)	5.9	180	170

* Cement factor = 256 kg (564 lb) ([20], Table 5.3).
† Addition rate of calcium lignosulfate = 0.225% solids on weight of cement.
‡ 1: none; 2: addition with mix water; 3: added after 2 min of mixing.

(a)

Fig. 7.14 Slip-formed bridge columns: (a) Sehoe Bridge, Seoul, Korea (Mr C. Chun); (b) CN Tower, Toronto, Canada (courtesy J.I.R. Mailvaganam).

(b)

Fig. 7.14 Continued

7.4.3. Marine structures

Concrete exposed to a marine environment will be subjected to wave action which imposes dynamic load, shock, impact stresses, erosion by abrasion, freezing and thawing, wetting and drying and the chemical reaction of chlorides and sulfates. Normal concrete which typically has a compressive strength of 20–40 MPa is relatively permeable to aqueous solutions and has relatively low tensile and flexural strengths; the material therefore cracks easily under tensile stress. Consequently the durability of the concrete is adversely affected particularly when exposed to severe environments. The rate of ingress of injurious ions and subsequent chemical attack depends on the permeability and abrasion resistance of the concrete. The durability of concrete under these conditions is therefore correlated with low permeability and high strength.

A number of recommendations pertaining to the requirements of concrete to be used in marine structures have been made [31–33]. These include specification of cement type and cement contents of 400–500 kg m^{-3}, a water–cement ratio of less than 0.45 and an air content of 4–7% depending on the maximum aggregate size. The concrete must be fully compacted and mix designed to have sufficient workability, yet have the lowest possible water–cement ratio. Such parameters usually involve mix modifications which include an admixture.

Conventional water-reducing admixtures used in marine construction are normally of the set-retarding type. The use of air-entraining agents in

Table 7.11 Characteristics of concrete used in a marine concrete structure

	Control	Mix 1	Mix 2
Mix details			
Ordinary Portland cement (kg m^{-3})	425	425	425
Graded PFA (kg m^{-3})	105	105	105
Sand zone 2 (kg m^{-3})	611	611	611
20/10 mm coarse aggregate	1053	1053	1053
Lignosulfonate WRA (ml 50 kg^{-1})	0	140	140
Hydroxycarboxylicacid WRA* (ml 50 kg^{-1})	0	140	400
Properties of concrete			
Slump (mm)	105	100	110
Proctor needle penetration resistance (hrs to 0.5 N mm^{-2})	8.5	12.5	27.0
Air content (%)	1.4	1.8	1.6
Compressive strength (N mm^{-2}) at 7 days	39.0	53.4	58.7
28 days	52.7	60.2	64.8
Density (kg m^{-3})	2410	2420	2460

* Level varied according to the slip-forming requirements of retardation.

conjunction with water reducers in concrete located at the splash zone confers improved resistance of freeze–thaw cycling and sulfate attack. Typical basic mixes containing a water-reducing admixture for use in a marine environment for gravity platform construction are shown in Table 7.11. Superplasticizers have largely replaced conventional admixtures in the construction of offshore structures due to the significant water reduction ($> 20\%$) and workability (slump of 200 mm) produced in the concrete enabling ready placing of low water–cement ratio (< 0.45) mixes in heavily reinforced areas that are characteristic of these structures. The water reduction afforded by the use of superplasticizers helps to achieve the desired low permeability and high strength that are so important to ensure durability and longevity in service under marine conditions.

The construction of offshore structures is the most complex and demanding of all concreting operations. Working conditions are difficult and delays and alterations in procedures occur more frequently. A variety of mixes incorporating two or even three admixtures are used. Notwithstanding these constraints, it is possible that with proper dispensing equipment and a good degree of control of the operation, a high level of consistency in concrete properties can be obtained. This is illustrated by the histogram in Fig. 7.15 for the production of concrete for an oil production gravity structure.

In the area of tidal concreting for sea defense work, e.g. between tides, the concrete is required to develop strength rapidly even under winter conditions. The advent of viscosity-enhancing admixtures (VEAs) which

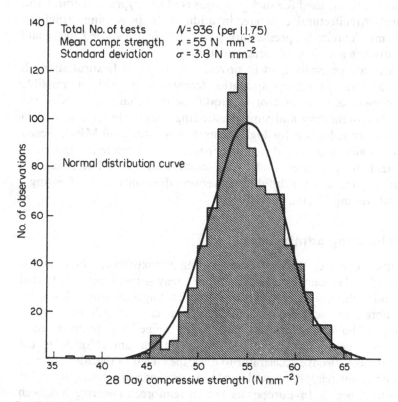

Fig. 7.15 The distribution of concrete cube tests for an oil production platform containing various admixtures.

enable the production of highly flowable concrete that is resistant to cement washout (see later) has been a boon to tidal construction. VEAs when used in conjunction with superplasticizers, silica fume, and accelerators such as calcium nitrite, offer significant advantages (high flow, complete filing and self-leveling, antiwashout due to high cohesion) in this application [34]. The combined accelerating effects of silica fume, the superplasticizer and the accelerator produce the required strengths even at low temperatures (5°C). In addition, superplasticized silica fume concrete improves other hardened properties of concrete such as density, abrasion, resistance to and freeze-thaw and sulfate attack, which are important for concrete exposed to sea water.

7.4.4 Tilt-up construction

The technique of horizontal on-site precasting of wall panels (often up to heights of 12.2 mm) is commonly referred to as tilt-up construction [34, 35].

Ready-mixed concrete used for such purposes is of two types, structural and architectural. Architectural concrete, in addition to possessing structural properties, may require a specific mix parameter (e.g. exposed aggregate finish) to produce a variety of surface textures.

Admixtures are generally used to provide the required flexural strength (2.1 MPa) and design tilt-up strengths (compressive) with permissible minimum cement content, and/or to modify setting characteristics of the mix. Normal set-retarding and superplasticizing agents are used to obtain the desired water reduction for high-strength concrete (> 50 MPa). Accelerators are usually used in cool weather or when the concrete is exposed to high ambient temperatures and wind conditions. The latter use of accelerators is particularly helpful in preventing durability under freezing–thawing and wetting–drying conditions.

7.5 Accelerating admixtures

The objectives sought by the use of accelerating admixtures can be obtained by the use of higher cement contents, more finely ground cement, heated materials and higher curing temperatures. Accelerating admixtures, however, provide a more convenient and economical way to achieve the desired results. A wide range of both inorganic salts and organic amine-based materials have been used as accelerating admixtures. Although calcium chloride is the cheapest and most effective accelerator to date, the effects of the admixture on the corrosion of the reinforcing steel has led to an increasing restriction of its use in North America. In Europe, its use in reinforced concrete has been banned and ACI 201 stipulates limits for chloride ion levels for concretes subjected to different exposure conditions [36]

7.5.1 Purpose and advantages resulting from the use of accelerators

The benefits provided by an increase in strength gain and shortening of the setting time are presented below.

1. Offsetting of the decreased rate of hydration of cement at low temperatures.
2. Reduction of the required curing and protection period and allowance of earlier removal of forms, enabling the contractor to meet specified construction schedules.
3. Reduction of the downtime of the structure (hence minimizing productivity losses) and shutdown for repair and renovation and advancement of the time when the structure can be put to use.
4. Reduction of the hydraulic pressure on forms.
5. More effective plugging of leaks against hydraulic pressure [37].

7.5.2 Non-chloride admixtures

The banning of calcium chloride during the last decade provided the impetus for the development of alternative materials which accelerated the hydration of cement without the potential for corrosion. A number of inorganic and organic compounds including aluminates, sulfates, formates, thiosulfates, nitrates, silicates, alkali hydroxides, carbonates, nitrites and calcium salts have been evaluated. Commercialization and field experience, however, is limited to only a few of these materials.

Calcium formate is usually sold as a powder and used at a dosage of 1–2% by weight of cement. Due to its very low solubility in water (16 g/100 g H_2O), the material is not commercially available in liquid form. The admixture accelerates the hydration and setting of all types of cement but beyond 2%, especially with respect to the first 24 h, the effect is not significant. In combination with sodium nitrite, however, early strength development is significant, producing strength increases of 125% of the control concrete at 24 h [38]. The rate of strength development in concretes containing formates may depend on the cement composition, particularly the CA_3/SO_3 ratio. It has been suggested that the benefit from the use of formates can best be realized in cements with CA_3/SO_3 ratios greater than 4 [39].

The best non-chloride admixture to date appears to be calcium nitrite; marketed as a 20% solids solution, it can be readily added to the mix using the usual dispensers. The admixture has therefore been more readily accepted by the ready-mixed concrete industry than the formate-based admixture. Strength development effects produced in concrete are reported to be comparable to those obtained with calcium chloride [40]. Table 7.12 shows a comparison of the results for concrete containing calcium chloride, calcium nitrite, and no admixture. In addition to the improvements in both compressive and tensile strengths, calcium nitrite is an effective inhibitor of chloride-induced corrosion. The material is now finding wide acceptance in bridge and parking deck repair. Other non-chloride accelerators used in Russia and special applications include the alkali-metal carbonates (Na_2CO_3, K_2CO_3, $LiCO_3$).

7.5.3 Accelerators for use in blended cement (fly ash or slag) mixtures

The significant increase in the use of supplementary cementing materials (such as fly ash and slag) in the last decade has dictated the need for an admixture that can offset the slowed hydration that results when such materials are incorporated in concrete. Strong basic salts such as sodium aluminate, alkali hydroxides, silicates, sulfates and thiosulfates have shown some promise. A number of proprietary admixtures which claim to catalyze the pozzalanic and thereby increase the rate of hydration are now marketed.

Table 7.12 Acceleration of strength development with calcium nitrate (Rosenberg et al. [40])

Cement brand	Admixture by weight of cement	3-day compressive strength		Setting time	
		MPa	lbf/in^2	Initial h:min	Final h:min
A	None	10.5	1525	8:45	12:21
A	Calcium nitrite 1%	10.8	1568	6:00	10:20
A	Calcium chloride 1%	14.8	2151	4:20	7:30
A	Calcium nitrite 2%	13.2	1924	3:05	6:55
A	Calcium chloride 2%	18.10	2624	2:10	4:55
B	None	10.11	1467	8:38	—
B	Calcium nitrite 1%	10.86	1576	5:24	9:05
B	Calcium chloride 1%	15.30	2220	3:16	5:00
B	Calcium nitrite 2%	14.30	2075	3:12	5:42
B	Calcium chloride 2%	17.66	2562	2:15	3:40

Water–cement ratio was 0.56 to 0.57; slump was 4.0 ± 0.5 in (100 ± 12 mm); air content was 1.95 ± 0.25%.

Because the chemical compositions of such admixtures are closely guarded trade secrets, there is a paucity of information relating to the mechanism by which these admixtures produce their effects.

7.6 Superplasticizers (high-range water reducers)

The dramatic modifications provided by superplasticizers are essentially marked extensions in performance of the basic function of conventional admixtures. The chemicals used as superplasticizers do not significantly affect the surface tension of water in the concrete mix, they can therefore be used at higher dosages without the attendant adverse side effects (of air entrainment, segregation and retardation) produced by conventional water-reducing admixtures. Since their introduction in the early 1970s, superplasticizers have evolved to a third generation which through their molecular structure have overcome the previous limitations associated with superplasticizer use. The three generations of superplasticizers can be identified as follows.

First-generation superplasticizers are primarily anionic materials. They create negative charges on the cement particles, causing them to repel each other, thereby reducing surface friction. They have little or no effect on the hydration process. Because of the short period of workability produced, first-generation superplasticizers are normally added at the job site. Second-generation superplasticizers are more adsorbed on the cement particles. This action not only produces plasticity of the cement paste, while allowing a low water–cement ratio, but also controls the hydration process. This hydration control enables second-generation materials to be used at lower concrete temperatures than was possible with first generation materials. The retarding versions extend the period of workability and some materials can be added at the batch plant. Third-generation superplasticizer chemicals are selected for the steric effects of their side chains. They are also adsorbed onto the cement particles but provide their effects more by a steric-hindrance mechanism than by electrostatic repulsion. Since they retain high consistencies for longer periods, they are more often added at the batch plant than on site. Thus, the primary difference between second- and third-generation materials is that a third-generation superplasticizer can maintain initial setting characteristics similar to normal concrete while producing a highly plastic mix at an extremely low water–cement ratio. The two main applications for superplasticizers are [41]:

1. The production of concretes with very high flowability where the admixture is added to the mix with no alteration in water–cement ratio to produce slumps in excess of 180 mm.
2. The production of concrete with low water–cement ratios to attain high early and ultimate strengths and much reduced permeability to moisture and salts.

With the advent of high-performance concrete and the development of third-generation (mostly acrylic polymer-based) superplasticizers which provide significantly higher water reduction at flowable consistencies, this demarcation has blurred. Through the use of a mixture of admixtures it is now possible to obtain highly durable low water–cement ratio concretes that are nearly self-leveling and yet quite cohesive.

7.6.1 Flowing concrete

Standard concrete mixes possessing good cohesion and slumps in excess of 180 mm are categorized as flowing concrete. Figure 7.16 (b) shows the fluid-like character that is achieved by the addition of a superplasticizer to a 50–75 mm slump (a). The unique characteristics of flowing concrete, in which high workability and high cohesion coexist, enable the coarse aggregate to be evenly distributed throughout the height of the placed concrete. This is clearly illustrated in the cores shown in Fig. 7.17, obtained after 3 years from the pour illustrated in Fig 7.18 and which gave the results presented in Table 7.13. Increased strengths shown for the cores taken from flowing concrete slabs can be correlated with the lack of voids (Fig. 7.17, core A).

Mix proportion aspects that should be considered in the design of highly flowable mixtures include: (1) cement content; (2) fines content; (3) type of superplasticizer; (4) the presence of other admixtures in the mix; (5) type of cement or cementitious material; (6) dosage of the admixture; (7) sequence

(a)   (b)

Fig. 7.16 (a) 50 mm slump prior to the addition of a superplasticizer; (b) collapsed slump of 100 mm generally used in flowing concrete (SKW, Trostberg, W. Germany).

Fig. 7.17 Cores taken from 3-year-old slabs poured with (A) flowing concrete; (B) 50 mm slump concrete.

(a)

(b)

Fig. 7.18 The placing of slabs (a) by the use of flowing concrete and (b) by the traditional method.

Table 7.13 Characteristics of the 3-year-old cores shown in Fig. 7.17

	Slab from flowing concrete	*50 mm slump concrete*
Visual examination		
Distribution of materials	Good	Good
Compaction of concrete	Good	Poor–fair
Voids: large	None	Few
medium		
small	Negligible	Considerable
Test results		
Corrected cylinder strength (N mm^{-2})	32.8	26.3
Estimated cube strength (N mm^{-3})	40.8	33.0
Density (kg m^{-3})	2320	2305

of addition; (8) coarse aggregate characteristics. Control of all these variables ensures that consecutive loads are similar in their handling, placing and finishing characteristics [42].

Optimum cement contents which provide flowing consistencies have been found be in the range 270–415 kgm^{-3} [43–48]. The fines content of a mix is important to prevent bleeding and segregation of the high-slump mix. However, the use of viscosity-enhancing admixtures which enable very highly flowable mixes to remain cohesive has made this parameter less critical than before. The use of a mixture of admixtures in flowing concrete makes compatibility of admixtures a key requirement. Many of the viscosity-inducing admixtures compete with the superplasticizer for adsorption sites on the cement particles and render the superplasticizer less effective. This is particularly true for naphthalene formaldehyde sulfonate type superplasticizers [48]. The newer types of superplasticizers produce their effects by a steric mechanism which induces higher flowability even at low water–cement ratios [44, 48]. It is therefore important to consider this effect when the cementitious material is cement types IV or V, blast furnace slag or fly ash, all of which require lower admixture dosage to produce a given slump. Furthermore, due to the varying solids content and hence the effective ingredient concentration in the different proprietary superplasticizers, particular attention should be paid to the manufacturer's recommended dosage for flowing concrete.

Factors which affect the dosage rate are concrete temperature, initial slump (i.e. slump before the addition of the superplasticizer), cement type and content, the presence of other admixtures in the mix prior to the addition of the superplasticizer and the sequence of addition to the mix. The latter factor is more applicable to the first-generation superplasticizers (melamine and naphthalene types) rather than the newer acrylic-polymer-based types which operate through the steric mechanism and is therefore not affected by reduced adsorption sites [48]. High concrete temperatures, finely ground cement with high C_3A contents, high cement contents (> 415 kg m^{-3}), the presence of a viscosity enhancing admixture and low initial slumps will require higher admixture dosages than the manufacturer's standard recommended dosage [48]. The presence of an air-entraining agent, retarder or water reducer and the use of acrylic type superplasticizers will produce a higher than anticipated slump increase due to the cumulative dispersing action of the two admixtures. Consequently, bleeding and segregation may occur in mixes with lower cement and fines content [45].

Ready-mixed concrete involves a delay between the mixing and placing of the concrete and on-site addition of first- and some second-generation superplasticizers. Consequently, the time-dependent properties of the plastic concrete (such as slump and setting time) and the manner in which the admixture is incorporated into the mix requires good control. For concrete which is to be produced in a highly workable form, it is necessary to deliver to the site concrete having a slump of 75 ± 5 mm and add the correct plasticizer dosage using a suitable dispenser. Manual dosing on site is sometimes used for centrally plant-mixed concrete with first- and second-generation superplasticizers. The superplasticizer dosage to be added for each load is calculated at the plant and carried in a container on the mixer truck. The required dosage is then fed under gravity through a tube into the mixing drum [43]. The calculation is generally based on charts prepared (prior to the pour) for the superplasticizer dosages required to produce non-segregaing flowing concrete at various slumps. This enables the required dosage adjustments to be readily made when difficulties in providing the requisite slump on site are encountered. The method usually requires sufficient control of consistency and relies on minimum periods of delay and, therefore, requires tight scheduling. After the addition of the superplasticizer, slump measurements are taken and recorded. With third-generation superplasticizers of the acrylic ester type, however, addition with the gauge water is done at the batch plant [48].

Many concrete producers use a combination of superplasticizer and a conventional water-reducing, set-modifying or air-entraining admixture to achieve the desired performance. The superplasticizer provides the major portion of the required water reduction and the conventional admixture is added to achieve one or more of the following objectives: (1) further water reduction; (2) admixture economy; (3) the desired air content; (4) increased workability; and (5) extension of set and workability. Commercial conventional admixture formulations used for this purpose are usually based on sodium lignosulfonates, hydroxycarboxylic acids or processed carbohydrates. Such combinations, besides reducing the dosage of the superplasticizer required, are reported to improve mix workability by reducing the high mix cohesiveness and consequently, the finishing effort required in some mixes [43]. More recently, field projects using super-plasticizers based on acrylic esters have shown that the desired fluidity, workability extension (up to 1 h), admixture economy, and greater quality control can be obtained by the sole use of this admixture [48].

Despite wide publication of the adverse effects that retempering causes, job site addition of extra water to compensate for slump loss is still a common practice, particularly in hot-weather conditions. Under these conditions, the use of both conventional and superplasticizing admixtures helps to minimize the amount of water required for re-tempering so that the loss of strength is minimized [46, 47]. This is shown in Table 7.14.

One of the significant limitations of the use of first- and second-generation superplasticizers in ready-mixed flowing concrete is the rapid decrease in the initially achieved high workability and this constitutes one of the chief constraints to their wider acceptance. Therefore, a number of major producers of admixtures have sponsored active research to improve the workability retention characteristics of their superplasticizers. Some recent developments [48] have shown promise, among these are materials based on acrylate polymers (AP). The AP-based materials are reported to be more effective than SNF- or SMF-based surplasticizers in terms of water reduction, slump increase and slump retention. Figure 7.19 shows the remarkable improvement in the retention of workability produced by the AP type superplasticizer compared to an SNF type.

The important features of this type of admixture is that such benefits are realized by addition of the admixture with the gauging water (like conventional admixtures and very unlike the SNF and SMF materials) thus eliminating some of the problems associated with the first- and second-generation superplasticizers. Table 7.15 [48] presents the effect of the time of

Table 7.14 Effect of water-reducing admixtures on retempering (Previte)

Cement B*		Time (min)		
		10	120	126
Reference	Slump, mm (in)	92 (3⅝)	35 (1½)	98 (3⅞)
0.18% modified	W/C	0.56	0.56	0.62
lignosulfonate	Slump, mm (in)	92 (3⅝)	41 (1⅝)	114 (4½)
0.75% hydroxylated	W/C	0.52	0.52	0.58
carboxylic acid	Slump, mm (in)	95 (3¼)	38 (1½)	95 (3¼)
1% s/s sulfonated	W/C	0.49	0.49	0.56
melamine formaldehyde	Slump, mm (in)	92 (3⅝)	35 (1⅜)	92 (3⅝)

*Temperature: 21°C (70°F).
Water-reducing admixtures allow a significant reduction in total water after retempering.

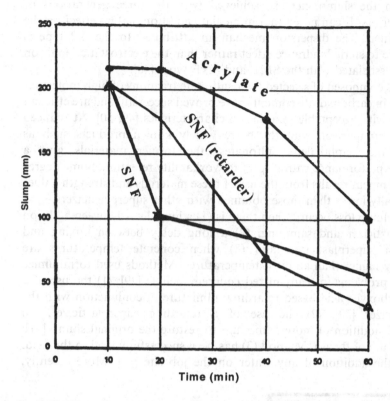

Fig. 7.19 Effect of superplasticizer on slump of concrete (Collepardi [48]).

Table 7.15 Effect of mode of addition of CAE, SMF and SNF superplasticizers on the slump of Portland cement concrete mixes (Collepardi)

Admixture			Concrete mixture	
Type	Dosage (%)	Mode of addition*	W/C	Slump (mm)
CAE	0.30	Immediate	0.39	230
CAE	0.30	Delayed	0.39	235
SMF	0.50	Immediate	0.41	100
SMF	0.50	Delayed	0.41	215
SNF	0.48	Immediate	0.40	100
SNF	0.48	Delayed	0.40	230

* Immediate: admixture with mixing water. Delayed: admixture after 1 min of mixing.

addition on the slump increase achieved with the three generations of superplasticizers. It can noted that immediate addition had no effect on the slump attained. The dispersion mechanism attributed to the AP type is related more to steric hindrance effect rather than the electrostatic repulsion [Fig. 7.20] associated with the SMF and SNF types [48].

Materials composed of selected lignosulfonate fractions of high molecular weight, low in carbohydrate content, have proved successful and are efficient dispersants with acceptable retardation characteristics [49, 50]. At reduced dosages in comparison with the more widely known products such as melamine and naphthalene sulfonates, the former materials give a comparable performance with respect to workability retention. Some degree of air entrainment results from the use of these materials and strength values are marginally lower than those obtained with other superplasticizers.

Greater slump loss occurs when there is (1) a long delay between addition of superplasticizer and sampling, (2) a long delay between mixing and addition of superplasticizer, and (3) when concrete temperatures are significantly higher than ambient temperatures. Methods used to minimize slump loss problems in ready-mixed concrete include [49]: (1) the use of a hydroxycarboxylic-acid-based retarding admixture in conjunction with the superplasticizer [50]; (2) the use of a retarding superplasticizer; (3) incremental addition of more admixture to restore the original slump [51]; and (4) the use of fly ash. Method (3) has been successfully used in the field, excluding the addition of any water on the job site [52]. More recently,

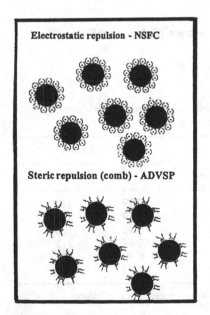

Fig. 7.20 Modes of dispersion of cement (Jolicoeur [125]).

products have become available which by their chemical nature allow considerable extension of the period of workability [48].

The range of application for which flowing concrete is used includes diaphragm walling, slabs on grade and casting heavily reinforced structural elements and the areas where the designer can use highly workable (flowing) ready-mixed concrete are as follows [53]:

1. For pumping of concrete to high elevations and longer distances.
2. In difficult pours with congested reinforcement and poor vibrator access.
3. In large pours where construction joints can be reduced thus speeding up construction.
4. In complex shuttering (forms).
5. In smaller structural elements where there is economy in design because of higher strengths attainable.

(a) Pumping

Flowing concrete and the small-bore pumping units revolutionized concrete placing and pumping techniques providing significant savings in manpower and costs. High workability and concomitant cohesion achieved through the use of superplasticizers enables concrete to be placed further, at faster rates and lower pumpline pressures regardless of elevation or site obstructions. Fig. 7.21 shows the dramatic effects produced by incorporating super-plasticizers in pumped concrete.

Successful pumping of concrete necessitates two requirements in the mix: (1) sufficient paste content to form an annular grout film against the pipe

Fig. 7.21 The use of a superplasticizer in pumped concrete (courtesy SKW, Trostberg, Germany).

wall to act as a slip surface, and (2) suitable grout consistency and interstitial void structure to offer good resistance to the forced bleeding of water from the cement paste so that dewatering of the mortar under the pressure of pumping can be prevented. Such requirements are met by superplasticized mixes in which the admixture increases the mortar fraction and provides a good void structure through water reduction so that pump pressures are transmitted to the solids. Superplasticizers also increase the viscosity and decrease the yield value of the mix. The former prevents segregation of the mix and bleeding of water, while the latter reduces the resistance of the mix to move under pressure, thus decreasing pumpline pressures.

(b) Flat-slab applications

Another area of application where the advantages resulting from the use of flowing concrete is obvious is in the construction of floor slabs, roof decks and concrete bay areas. The distribution of concrete over large areas and areas with congested reinforcement or constricted form configuration (which precludes the use of normal vibratory compaction due to poor vibrator access) is both time consuming and labor intensive. However, the use of a highly mobile mix (Table 7.16) such as flowing concrete will reduce the time and labor required for placement and consolidation.

The benefits that accrue from the use of flowing concrete are mainly realized by the contractor who recoups the expense for the admixture in savings in placing cost. There are also advantages to the ready-mixed concrete industry since the quicker placing allows faster truck turnaround time and shorter standing time on site. Flowing concrete is particularly significant in jobs requiring placing of concrete in foundations or footings where it is often necessary for the truck to make several maneuvers to complete placement. The greater radius of flow (about 5 m) of this type of concrete from the point of placing considerably reduces the number of maneuvers required. Wherever possible the chute should reach as far towards the middle of the placing area as possible. Extension tubes can be added to increase the discharge range accordingly. The use of such tubes

Table 7.16 Mix details and concrete characteristics for flowing concrete

Ordinary Portland cement (kg m^{-3})	=	350
10 mm crushed limestone (kg m^{-3})	=	1009
Zone 2 marine sand (kg m^{-3})	=	925
Water–cement ratio	=	0.52
Slump (before admixture addition) (mm)	=	50
Compressive strength (N mm^{-2}) at: 7 days	=	23.0
28 days	=	28.0

enables concrete to be placed further, as well as into inaccessible parts of the job. This is illustrated in Fig. 7.18, where concrete slabs are being placed by the use of traditional 50 mm concrete by hand raking and finishing, in comparison to a slab poured with flowing concrete.

(c) Precast applications

Flowing concrete can be utilized in the precast industry to aid production of complex-shaped precast concrete units. Often it is the only solution to a problem posed by such situations, e.g., in the construction of toroid-shaped dome units for the Ninian Platform in Scotland, the low water–cement ratio required for high early strength had to be rationalized with proper compaction in heavily reinforced components. The use of a high-strength flowing concrete produced with the help of a superplasticizer not only enabled proper consolidation but also provided the high early strength required for a fast turnover of molds. Figure 7.22 shows construction of the heavily reinforced dome units. Flowing concrete also offers significant advantages to architectural concrete where texture is required. Complex shapes are produced by casting flowing concrete on textured form liners to obtain the desired effect. The good compaction obtained reduces surface defects such as 'bug holes' and also permits early stripping of the unit.

7.6.2 High-range water-reduced concrete

Higher water reductions afforded by the use of superplasticizers enable the production of concrete having normal workability (75–90 mm slump) but

Fig. 7.22 Construction of the heavily reinforced toroid-shaped dome structures for the Ninian Platform in Scotland (courtesy SKW, Trostberg, Germany).

with significantly reduced water–cement ratios. First and second-generation superplasticizers provide water reductions in the region of 15–25% [53, 54] depending on the type of cement used, aggregate properties and concrete temperature. Manufacturers of third-generation superplasticizers claim further water reductions than those achieved with the first and second-generation materials because they can be used at higher dosages [48]. The water content is progressively reduced as the dosage of the superplasticizer is increased at a given workability particularly for mixes with cement contents in the range of 300–600 kg m^{-3} [55]. Such water-reducing and strength-increasing effects, however, vary with the type and brand of cement used; greater water reduction but not significantly higher strengths being achieved with some cements. The reason for this relative performance has not been fully determined. Some factors that may account for difference in behavior are the C_3A and SO_3 contents and the Blaine fineness of the cement [56, 57].

The dosage of the superplasticizer required for high-range water reduction applications will depend on the admixture type, the amount of water reduction required, the desired placing slump, the cement type and content, the concrete temperature and the presence of other conventional admixtures in the mix prior to the addition of the superplasticizer. The actual values used for initial and placing slump are usually governed by parameters such as desired early strength levels, durability criteria, the degree of vibrator compaction to be used and the finishing characteristics required for the application. In general, normal slumps at lower admixture dosage may be produced with a retarding superplasticizer, cements with low C_3A contents (< 5%) [53, 54], mid-range cement contents [56, 57], coarsely ground cements, and the presence of the other conventional admixtures in the mix. Low initial slump (< 50 mm) will usually require higher admixture dosage to produce conventional slumps (75–100 mm).

(a) Combined use of chemical, mineral and superplasticizing admixtures

Calcium-lignosulfonate-based water reducers produce complementary effects when used with superplasticizers. However, they should be added to the mix separately, the former with the gauging water and latter at the end of the mix cycle. The combined use of the two admixtures results in the following modifications in comparison with similar mixes which contain only the superplasticizer: (1) higher water reduction and extended initial set times for mixes with slumps less than 100 mm. For mixes with slumps greater than 200 mm, depending on the dosage of the normal water reducer used, significant set extension and some strength reduction; and (2) greater susceptibility to segregation on re-dosage of the superplasticizer [49]. The cumulative dispersing action is higher for naphthalene formaldehyde than the melamine formaldehyde. The use of acrylic-ester-based superplasticizers with other admixtures should be evaluated prior to field use. Because of the different mechanism of action, such superplasticizers may not require combination with other admixtures and in some instances the cumulative effects may be adverse [48].

When used with superplasticizers, hydroxycarboxylic-acid-based (HC) retarding admixtures confer benefits to concrete, particularly under high ambient temperature conditions [51] and in steam curing in precast operations [54]. Depending on the dosage of the HC admixture, the concrete temperature and placing slumps, an extension of 3–4 h in the setting time should be expected. Most currently marketed air-entraining agents (neutralized vinsol resin, sulfonated hydrocarbon and fatty-acid types) are compatible with all types of superplasticizers. The four main problems (lack of air entrainment at low water–cement ratios, loss of air due to very high slump, extension of the required bubble spacing and poor bubble size) encountered with air-entrained flowing concrete have also been identified for water-reduced superplasticized concrete. However, with the exception of reduced air content in low-slump mixes, the magnitude of the other problems is much reduced in comparison with flowing concrete. Reduced air content in low-slump mixes is caused by the lack of a minimum paste viscosity required for the air-entraining mechanism. Some producers have overcome this problem by adding the air-entraining agent after the superplasticizers so that the required viscosity is provided [58].

Previous work on superplasticized Portland cement concrete containing fly ash or blast furnace slag has shown that such mixes require 10% less admixture than reference Portland cement concrete to attain the same workability. Therefore, a given dosage may produce higher water reduction. The reason for the reduced admixture requirement has not been determined. It is probably due to the lowering (dilution) of the C_3A content of Portland cement that results when fly ash or slag replaces part of the cement in a concrete mix [59, 60].

The proportion of coarse to fine aggregate, the fineness modulus of the fine aggregate and the maximum size of the coarse aggregate affect properties such as percentage water reduction, early and ultimate strengths, placing and finishing characteristics of such concretes. Water-reduced cement-rich mixes containing superplasticizers produce mortar-rich cohesive concretes which respond slowly to applied vibration, often exerting a strong frictional drag on finishing tools. Vibrator effectiveness and reduced cohesion can be produced by increasing the coarse aggregate fraction and/or using a more coarsely graded (FM 2.8–3.1) fine aggregate [60, 61].

Common objectives sought in the water-reduced high-strength application include: (1) lowering of curing cost; (2) reduced cement content while maintaining same workability and strength levels; (3) the use of an alternative cement type, e.g., from Type III to Type I; (4) the production of more workable concrete with high early and ultimate strengths. These are discussed below.

(b) Lowering of curing cost, cement reduction and use of alternative cement

Heat curing is the most common method of accelerating strength development in concrete. Accelerated curing finds wide application in the precast industry for quick turnaround of forms and casting beds. Superplasticizers afford substantial reduction in the curing temperature and the length of the curing cycle normally used for accelerated curing of concrete products, e.g. when cement contents are not reduced, significant water reduction is effected, resulting in high early strengths under elevated curing temperatures. This advantage has permitted some producers to reduce the maximum curing temperature by 50–60°C and to reduce the curing period by 3–4 h, while maintaining strength levels comparable to concrete cured in a conventional manner [58, 62].

The use of auxiliary water reducers or retarders added in the gauge water can lead to higher early strengths than with the high-range water reducer alone. In general, water reducers should be added if lower curing temperatures are to be used, retarders if higher temperatures are expected. The choice of an auxiliary admixture and its addition rate will vary with the cement and cure temperature employed [56].

If a non-retarding superplasticizer is added to the mix at the normal recommended dosage, no modification of the set time results and the normal curing cycle may be used. However, a modification to this curing cycle may be necessary for higher superplasticizer dosages or when a retarding superplasticizer is used, or when the superplasticizer is used in combination with a normal water-reducing retarder. In these situations the final setting time may be lengthened, and the preset period may need to be increased accordingly. The length of the total curing cycle, however, may not necessarily be increased [57].

The lower water–cement ratio afforded by the use of a superplasticizer may be used to increase existing compressive strengths or to reduce cement content. Thus the use of superplasticizers may enable the precast producer to use lower cement content without reduction in mix workability and rate of strength development. The actual amount of cement reduction achieved will depend on the cement type used and the mix proportion used in the concrete. Previous work [63] indicates that even with low cement content (306 kg m^{-3}) a normal dose of superplasticizer can accelerate 3- and 28-day strengths by 90% and 55%, respectively, over levels attained with a plain mix. Cement reductions in the range of 11–20% have been achieved in mixes with a cement content of 415 kg m^{-3}, while maintaining desired strength levels [56]. An example of the use of a superplasticizer admixture to achieve cement reduction was presented in Table 7.8.

Most often precast producers use finely ground cement such as ASTM Type III to obtain desired production cycles. The high water reduction and concomitant rapid strength gain afforded by the use of superplasticizers will enable the precast producer to change to the more economical Type I cement (Table 7.8).

(c) High early strengths for early stripping and destressing of wire reinforcement

The concrete slump used in any particular prestressed and precast operation is usually determined by balancing the opposing requirements that it be readily placeable but also that it achieves maximum strengths. Superplasticizers are often used in water reduction application to obtain suitable 12–18 h release strengths. The substantial reduction in the time required to achieve stripping strength allows for early destressing of wire reinforcement and increased mold turnover for precast items. High strengths realized by the use of superplasticizers have been used advantageously in the production of high-strength precast columns in high-rise buildings and precast girders. Here, the admixture permits the maximum attainable strength by effecting high water reduction, increasing mix workability, and contributing to the more efficient use of available cement.

A typical application where the three desired objectives of high water reduction, high workability and high early and ultimate strengths are simultaneously required is in nuclear concrete. The concrete for the construction of prestressed concrete pressure vessels for nuclear power plants is required to provide high early and ultimate strengths and high workability at low water–cement ratios. Additionally, it must exhibit the lowest possible thermal expansion. Consequently, special concretes incorporating high water demand aggregate such as limestone are required. The high-water-reduction and substantial plasticizing action afforded by superplasticizers is often the most convenient manner of achieving these objectives.

7.6.3 High-performance concrete and mortar

The ready availability of superplasticizers and the nearly 30-year track record of their use has led to the development of innovative concretes and mortars for specialized applications both in new construction and the repair of deteriorated structures. Four such specialized applications are: (1) high-performance, high-strength concrete; (2) high-performance/high-volume slag or fly ash concrete; (3) self-leveling and self-compacting concrete and mortar; and (4) fibre reinforced shotcrete are discussed below.

High-performance concrete (HPC) is defined as an engineered material in which one or more specific characteristics have been enhanced by the careful selection and proportioning of its constituents and diligent curing of the post-hardened material [64]. It is characterized by a low water–binder ratio and although made from the same basic ingredients used in conventional concrete it results in a hardened material with a very dense microstructure having fine pores that are not interconnected. These two microstructural features produce different properties in terms of mechanical strength, durability and shrinkage, which makes HPC a significantly different material from ordinary concrete. Because of its dense microstructure, it drastically reduces the ingress of moisture and aggressive salts. Consequently, HPC is able to withstand higher operating loads and is more durable in aggressive environments than ordinary concrete. However, such microstrutural characteristics also make it quite vulnerable to autogenous shrinkage due to self-desiccation.

Self-desiccation occurs in low-water–cement-ratio/silica fume concrete when the cement paste receives little additional water during its curing. Chemical shrinkage then occurs making the intrinsic voids increase after the framework of hydrates is formed by setting. Macroscopic volume reduction of hardened concrete caused by cement hydration, and not by external load, moisture movement to and from the external environment, temperature change, or carbonation is known as autogenous shrinkage [64]. Earlier work by Davis [65] showed autogenous shrinkage to be much less than drying shrinkage and it was thought that it could be ignored for practical purposes. Several researchers [66–68], however, have demonstrated that autogenous shrinkage of cement paste with low water–cement ratio is quite large. These studies suggest that autogenous shrinkage is one of the most important properties which should be considered for crack control in high-strength concrete structures and emphasize importance of the timing and degree of water curing.

As HPC is relatively new to most ready-mixed concrete plants and finishers that have not worked with the concrete before it is important that all parties involved in a construction project be informed about the production, placing, finishing and curing of HPC. The following critical aspects of HPC use in the field are presented as a reference to the practitioner. They cover three main topics:

● Issues that should be addressed prior to commencing construction.
● Production and placement of HPC in the field.
● Post-placement concerns.

Since each field installation will use locally available materials, mix proportions may have to be adjusted for the character of the aggregate, cement and specific brand of the admixtures. A sufficient number of lab trial batches should be done in order to ensure that the desired performance will be achieved. Laboratory trial batching and specimen testing should be performed as closely as possible to the conditions expected in the field. Once the mix proportions are refined, field trials should be conducted at the batching plant to confirm mix proportions, the production process and for the evaluation of the performance of the concrete to insure compliance with the specifications. The plant trials must simulate truck mixing and haul time, and the concrete should placed as intended for full-scale use. Conducting field trials will avoid serious difficulties during placement and allow those involved in the project to become sufficiently acquainted with HPC.

The need for tight control of water content, including corrections for free moisture in the aggregate, should be emphasized to the batch plant operatives. In addition, more care must be taken in the batching process to ensure that the correct materials are batched in the proper sequence. Drivers must understand that transport delays can cause serious problems. They must also understand the need to fully discharge all wash water prior to charging the drum. The contractor's personnel should be informed of the sticky nature of HPC, particularly as the mix begins to lose slump. Experience has shown that adequate mixing of the constituent materials is vital to the quality of HPC. This is due in part to the low water–cement ratio of the mixtures. Mixing action in trucks is adversely affected by worn blades, fin build-up or an overloaded truck, and the concrete will no longer be uniform. It is therefore highly recommended that the batch sizes be limited to no more than two-thirds of the rated capacity of the truck, and that worn blades be replaced [69].

The concrete engineer should be cognizant of the effects of a mixture of admixtures – the incremental effects at higher dosages, the interactive, antithetical or synergistic effects, variation with different cement brands, ambient temperature and water–cement ratios – so that preventive or compensatory action can be taken. Some of the important points to be considered are as follows:

- The low initial water content requires the use of large doses of high-range water-reducing agents, which may retard the setting of the concrete. In some instances, a reduction in strength at very early ages particularly when ambient temperatures are moderate.
- The quantity of air-entraining agents required for these concretes is considerably higher than for conventional concretes.
- Corrosion inhibitors, like calcium nitrite, are accelerators and depending on their dosage may cause rapid slump loss and shorten the setting time when added at the batch plant.

Since HPC produces little or no bleed water, the use of an evaporation retarder such as misting of the concrete surface or the conventional method of applying a curing membrane immediately should be done to minimize plastic shrinkage cracks. Premature cracking due to delayed joint sawing (past 12 h) is a strong possibility due to thermal changes of the concrete at early ages as it undergoes significant thermal drops. Therefore proper curing and sawing of concrete must occur as early as practicable to avoid cracking of the concrete. Prior to commencing construction, a meeting should be held with the various parties involved and all aspects of the field trials should be discussed and potential problems addressed.

(a) Applications of high-performance/high-strength concrete: selected examples

Concretes that were previously (late 1970s) known as high-strength concretes are now referred to as high-performance concretes. This is because high-strength concrete embodies enhanced performance in other areas such as durability and abrasion resistance. Superplasticizers are essential ingredients in such concretes which are now being used increasingly for structures that

Table 7.17 Nothumberland Strait Bridge Project: characteristics of structural concrete (Malhotra [69])

Blended silica fume cement	450 kg m^{-3}
Classified sand	737 kg m^{-3}
Coarse aggregate	
20 mm–10 mm	570 kg m^{-3}
10 mm–5 mm	460 kg m^{-3}
Water	153 l m^{-3}
Air-entraining admixture	160 ml m^{-3}
Water-reducing admixture	1.71 lm^{-3}
Superplasticizer	3.0 l m^{-3}
Slump prior to addition of superplasticizer	40 mm
Slump after addition of superplasticizer	200 mm
Slump loss in 1 h	40 mm
Air content	6.1% (range 5.5–7.8%)
Spacing factor, $\bar{L}$	153 μm
Specific surface, α	25.8 mm^{-1}
Linear coeff. expansion	8.3 × 10^{-6} °C^{-1}
Compressive strength (100 × 200 mm cylinders)	
24 h	34.9 MPa
3 days	52.2 MPa
7 days	62.6 MPa
28 days	81.9 MPa
Modulus of elasticity, E (150 × 300 mm cylinders)	
28 days	40.0 GPa
91 days	41.0 GPa

have been designed for a 100-year life [64, 70]. Typical structures are large concrete bridges exposed to severe environments and offshore drilling platforms. The concretes mixtures used for such purposes incorporate silica fume and high dosages of superplasticizers and are proportioned for high strength and high durability. The following examples describe the use of this type of concrete in two large structures in Canada [69].

CONFEDERATION BRIDGE, CANADA

The bridge crossing the Northumberland Strait links Prince Edward Island with the Canadian mainland. At its narrowest point it is 12.9 km long and consists of 44 spans of 250 m each. Bridge piers are embedded in bedrock in a depth of 35 m of water [59]. Main span girders (weighing about 8000 tonnes) were match cast and post-tensioned on site. The bridge, which will be exposed to a very severe marine environment and more than 100 cycles of freezing and thawing annually, is designed for a 100-year service life [60]. Mixture proportions and properties of the HPC are shown in Table 7.17. A view of the completed main span girder is shown in Fig. 7.23.

HIBERNIA OFFSHORE CONCRETE PLATFORM

The Hibernia offshore concrete platform on the Grand Banks off the coast of Newfoundland, Canada, is a 111 m structure and required about 165 000 m^3 of superplaticized high-strength concrete. The typical mixture proportions and the properties of the concrete used for the 'skirt' elements of the structure are shown in Table 7.18 [70]. The precast 'skirts' are a series of reinforced concrete elements which were joined together to form continuous walls to support the base slab of the structure [69].

(c) High-Performance/high-volume fly ash and slag concrete

In high-performance/high-volume fly ash concrete (HP/HVFAC), 50–60% of the cementitious material is fly ash. The water content is kept very low and the high workability is obtained by the use of large dosages of a superplasticizer. Because of the very low water–cement ratio, this type of concrete has adequate early strength development, high latter-day strengths and excellent durability characteristics. The low cost, very low heat of hydration and high impermeability make this concrete very attractive for use in large pours for thick retaining walls, large columns and mass concrete. This type of concrete using ASTM Class C fly ash has been used successfully in a variety of structural applications [71].

The strength data in Table 7.19 show that using very low amounts of ASTM Type 1 (normal Portland) cement compressive strength of the order of 8 and 35 MPa can be obtained at 1 and 28 days respectively. The concrete's resistance

Fig. 7.23 Views of the completed main span girder for the Northumberland Strait Bridge, Canada (Malhotra [69]).

to chloride attack (indicated by the coulomb values) as determined by the rapid chloride penetration test (ASTM C 1202) has been shown to be high.

Like HP/HVFAC, high-performance/high-volume slag concrete (HP/HVSC) has a very low water–cement ratio ranging from 0.27–0.45, and large doses of a superplasticizer are used to obtain slumps exceeding 125 mm [69]. Table 7.20 shows the mixture proportions, and properties of the fresh and hardened concrete. It can be seen from Table 7.20 that 1-day compressive strength of slag concrete ranges from 2.5 to 6.9 MPa and 7-day

Table 7.20 Mechanical properties of high-volume slag concrete (Malhotra [69])

Mixture series	Batch no.	W/C+S	S/C+S	Density at 1 day (kg m⁻³)	Compressive strength* (MPa)				Flexure strength† (MPa)	Modulus of Elasticity* (GPa)	Chloride-ion penetration (C)
					1 day	7 days	28 days	91 days	14 days	28 days	28 days
1	A	0.45	0.60	2475	2.5	15.5	24.4	28.4	3.9	38.3	830
	B	0.45	0.60	2465			19.6				
2	A	0.34	0.70	2435	2.8	31.5	45.6	51.4	7.8	42.4	230
	B	0.34	0.70	2435			43.6				
3	A	0.29	0.75	2470	4.5	26.1	49.6	53.6	8.3	44.3	175
	B	0.29	0.75	2455			37.1				
4	A	0.45	0.50	2450	2.5	14.1	21.2	23.5	3.4	32.6	1160
	B	0.45	0.50	2450			19.9				
5	A	0.36	0.60	2435	4.3	25.6	45.6	49.9	7.0	42.3	325
	B	0.36	0.60	2420			50.8				
6	A	0.28	0.70	2460	6.3	40.1	57.3	66.0	8.6	45.1	215
	B	0.28	0.70	2480			51.2				
7	A	0.38	0.50	2455	6.9	34.8	52.3	55.2	7.0	44.3	385
	B	0.38	0.50	2440			53.6				
8	A	0.30	0.60	2430	6.7	40.9	54.6	58.3	9.0	42.5	275
	B	0.30	0.60	2440			54.4				
9	A	0.27	0.65	2445	4.9	44.7	63.2	69.5	8.7	43.8	320
	B	0.27	0.65	2465			57.4				
10	Control	0.39	0	2455	16.4	27.8	34.6	40.3	4.5	40.3‡	2985
11	Control	0.31	0	2475	27.8	43.3	55.1	66.4	6.3	45.1‡	1285
12	Control	0.27	0	2485	37.2	51.2	61.3	71.8	7.2	46.2‡	1305

* Testing carried out on 152 × 305 mm cylinders.
† Testing carried out on 76 × 102 × 406 mm prisms.
‡ Results are at the age of 91 days.

Table 7.18 Hibernia offshore platform: mixture proportions and strength test results of concrete for the precast skirt elements (Malhotra [69])

Constituent	Amount
Blended silica fume cement	450 kg m^{-3}
Coarse aggregate	976 kg m^{-3}
Fine aggregate	730 kg m^{-3}
Water	162 kg m^{-3}
Water–cement ratio	0.36
HRWRA*	700 ml per 100 kg cement
WRA*	220 ml per 100 kg cement
AEA*	25 ml per 100 kg cement

* HRWRA = high range water-reducing admixture (superplasticizer). WRA = water-reducing admixture. AEA = Air-entraining Admixture.

	No. of tests	Average	Standard deviation	Range	
				Low	High
Slump (mm)	136	108	20	100	110
Air content (%)	136	4.8	0.7	2.8	6.2

28-day compressive strength data:

Number of tests	= 269
Average compressive strength	= 73.8 MPa
Overall standard deviation	= 5.17 MPa
Overall coefficient of variation	= 7.01%
Overall within-test coefficient of variation	= 2.86%

Table 7.19 Typical mechanical properties of high-volume fly ash concrete made with ASTM Type I and Type III cements (Malhotra [69])

	ASTM Type I cement	ASTM Type III cement
Compressive strength		
1 day	8 ± 2 MPa	14 ± 2 MPa
7 days	20 ± 4 MPa	27 ± 4 MPa
28 days	35 ± 5 MPa	40 ± 5 MPa
91 days	43 ± 5 MPa	50 ± 5 MPa
365 days	55 ± 5 MPa	—
Flexural strength		
14 days	4.5 ± 0.5 MPa	5.5 ± 0.5 MPa
91 days	6.0 ± 0.5 MPa	6.5 ± 0.5 MPa
Splitting-tensile strength		
28 days	3.5 ± 0.5 MPa	3.5 ± 0.5 MPa
Young's modulus of elasticity		
28 days	35 ± 2 GPa	37 ± 2 GPa
91 days	38 ± 2 GPa	—
Drying shrinkage strain		
at 448 days	$500 \pm 50 \times 10^{-6}$	—
Specific creep strain		
at 365 days (per MPa of stress)	$28 \pm 4 \times 10^{-6}$	—

compressive strength ranges from 14.1 to 44.7 MPa; the corresponding 7-day strength for the control Portland cement concrete containing 292–428 kg m^{-3} of cement ranges from 27.8 to 51.2 MPa. The 14-day flexural strength of the HP/HVSC is in general significantly higher than the control concrete at about similar water–cement ratio.

(c) Superplasticized fiber-reinforced shotcrete

Shotcrete is one of the most widely used methods for the repair of large-scale surface damage encountered in locks, embankments, highway bridges and overpasses. One of the serious drawbacks in the use of fibers (steel or polypropylene) in concrete has been the great reduction in workability that results when they are incorporated into the mix. The use of superplasticizers and silica fume in shotcrete mixes has significantly overcome the balling-up of the mix, improved sag resistance, reduced rebound and therefore led to the increased acceptance of superplasticized fiber-reinforced concrete. The fibers are added to improve the ductility, and impact resistance of the concrete.

Silica fume when used in conjunction with superplasticizers plays an important role in improving the cohesion of the shotcrete mix, allowing for the build-up of the sprayed concrete on vertical surfaces, within a short period, without sloughing off. In addition the combination of these admixtures has provided ease of pumping, better compaction and greatly reduced the rebound, thus minimizing waste and cleaning time. Typical mixture proportions and properties for wet-mix superplasticized shotcrete are presented in references 69 and 72.

(d) Self-leveling and self-compacting concrete and mortar

SELF-COMPACTING CONCRETE

High-flow self-compacting concrete (HFSCC) can be described as a concrete that has high deformability, good segregation resistance and excellent filling characteristics [73]. High flowability and adequate viscosity are achieved through the combined use of superplasticizers and non-adsorbent viscosity enhancing admixtures (NAVEA). Flowability is measured by the slump flow test (diameter of the concrete pat after removal of the slump cone) and is usually controlled to 60 ± 5 cm by adjusting the dosages of the admixtures. The properties of the concrete have been characterized by measurements of slump flow, air content, setting time and compressive strength at various ages [73].

The ability to consistently produce HFSCC has been made possible by the development of a new type of VEA. Most viscosity-enhancing admixtures (VEA) are readily adsorbed on cement grains and often form bridges between the cement particles. Consequently, they strongly affect the rheological properties of highly flowable mixes by reducing the effectiveness of the superplasticizer. This is due to the decreased adsorption sites available to the superplasticizer. The NAVEAs molecules on the other hand do not compete with the superplasticizer for the adsorption sites on the cement particles. Thus the admixture, while providing good cohesion to the mix, does not interfere with the deflocculation action of the superplasticizer. These statements apply only to the melamine formaldehyde sulfonate and acrylate polymer-based superplasticizers. Ultraviolet analysis of residual superplasticizer levels in centrifuged mortar indicates that naphthalene formaldehyde sulfonate-based superplasticizers are incompatible with NAVEA s because of the strong interaction between the admixtures [73].

Measured rheological parameters such as plastic viscosity and yield values show that while the admixture increases the apparent plastic viscosity, the flow value does not change. Therefore, it is easy to control viscosity while maintaining the flowability. The viscosity increase is proportional to the dosage of the NAVEA used and at a dosage exceeding 1.25% (by weight of cement) a mix with sufficient filling ability is obtained [73]. The good filling ability and cohesive concrete produced by the combined use of these admixtures finds wide application, particularly in repair and underwater construction, where access is limited and inspection of the quality of the poured concrete is not possible. Such concretes minimize the formation of stone pockets, cement washout, layering and also ensure that a good bond is formed with the existing concrete.

SELF-LEVELING MORTAR

Self-leveling mortars such as highly flowable grouts and underlayments are used to provide a smooth or level floor surface for machine base plates and floor coverings. Most products are applied over old or new floors constructed of cast-in-place concrete, precast concrete or wood and are also being increasingly used to repair floors that have deteriorated, sagged, scaled, or become worn [74]. The materials are easy to use, eliminate troweling and can take foot traffic in 4–6 h at temperatures ranging from 15° to 32°C (60° to 90°F) [73, 74]. Prepackaged proprietary materials (consisting of sand, cement and admixtures superplasticizers, polymer latex and gelling agents), when mixed with water, produce self-leveling slurry-like mixes that can be placed either by pumping or squeegee, or spread by trowel (Fig. 7.24).

Flow and self-leveling characteristics of these products are governed by the rheological behavior of the slurry-like materials. At the low water–cement ratios required to ensure proper suspension of the solids, most self-leveling compositions are characterized by a yield stress and thixotropic behavior [75]. To obtain self-leveling properties, the yield stress has to be reduced and this is achieved by the selection and combination of suitable mix ingredients at optimum levels. For example, the addition of a superplasticizing admixture results in the reduction of the yield stress and a simultaneous decrease in the thixotropy of the mix, while the amount and type of the finest fraction of the aggregate has a significant influence on the stability and flowability.

Fig. 7.24 Self-leveling underlayment placed by pumping (courtesy Monica Martyn).

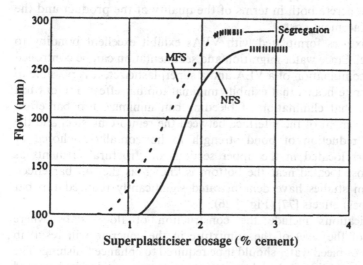

Fig. 7.25 Effect of superplasticizer type (MFS and NFS; melamine and naphthalene sulfonates) on flowing behavior.

The pivotal role that superplasticizers play in the formulation of self-leveling mortars is due to the dramatic effects they produce on flow behavior. Such effects are believed to be derived by the adsorption of the admixture on the surfaces of cement grains, thereby providing surfaces of a similar or zero charge which are mutually repulsive. They thus fully disperse cement particles, freeing more water for lubrication and reducing interparticle attraction. Both yield stress and plastic viscosity are decreased and the decrease is greater for yield stress; it may be completely eliminated if sufficient admixture is added so that Newtonian behavior is observed (Fig. 7.25) [75, 76].

7.7 Viscosity-enhancing admixtures

Rheology and fluid loss control form the basis of many applications for water-soluble polymers: in post-tensioning applications, the grouting of tunnel linings, in structural grouts used to provide leveled surfaces for machine baseplates, injection grouts, self-leveling underlays and under-water, self-compacting concretes. Concrete and mortar containing VEAs exhibit minimal bleeding tendencies despite having high slumps (> 180 mm) and flow table values. The concrete is cohesive enough to allow limited exposure to the water, yet has good mobility to move underwater with little loss of cement. Such non-dispersible concrete can be poured into a water-filled form without a tremie pipe to produce dense structural repairs. This type of material has particular advantages over conventional concrete both in terms of the quality of the product and the reduction in placement cost.

Concrete mixtures formulated with VEAs exhibit excellent bonding to reinforcing steel. Since water migration and sedimentation can be controlled by the proper combination of a VEA and a superplasticizer, it is possible to formulate concrete beams that exhibits minimal top bar effect. Pullout test results indicate that elimination of bleeding can minimize top bar effect through improvement of the interface between the reinforcing steel and the concrete. The reduction of bond strength to horizontally anchored or overlapped bars located in the upper sections of structural elements as opposed to those located near the bottom is known as the top bar effect. Full-scale beam studies have demonstrated significantly reduced top bar (U_{bot}/U_{top} values) effects [77] (Fig. 7.26).

Other applications include the construction of floor slabs where manipulation of the level of the admixture in the concrete will result in the production of bleed water should it be required to enhance finishing. The use of VEAs in grouting and underwater concrete applications is discussed below.

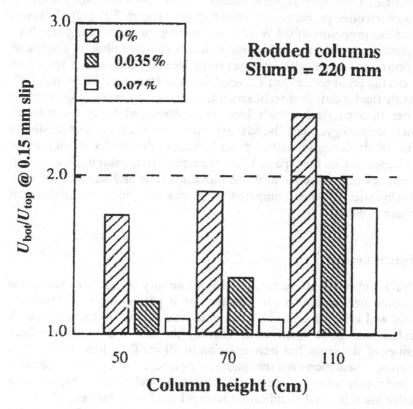

Fig. 7.26 Effect of VEA dosage and column height on top-bar effect (Khayat [77]).

7.7.1 Grouting applications

(a) Post-tensioning grouts

The bleed characteristics of neat cement grouts are well known [77]. With usual mixing and grout admixtures, it can be expected that water separation will occur due to sedimentation of the suspension of cement in water. This is particularly exaggerated in post-tensioning applications with strand-type tendons where the differential pressures cause the water to migrate from the grout into the voids between the strands. A substantial upward percolation of water can result in certain sections of the strand being entirely ungrouted. For these reasons, it is extremely important that these grout systems provide suspension, resist bleeding, and control fluid loss through steel strands without sacrificing workability,

Water-based polymer–superplasticizer blends can significantly increase the water-retentive properties of cement grouts. Figure 7.27 [78] illustrates the fluid loss properties of 0.4 W/C neat cement grouts containing blends of superplasticizer and a gum and a high-molecular-weight cellulose. The use of these polymers dramatically decreases static bleeding and reduces the water loss from the grout to 1.5% of the total water content used in the mix [79]. Extremely fluid grouts that resist separation and pressure-induced filtration of water through steel strands have been developed using water-based polymer technology. Such blends are currently used in post-tensioned grouting of structures to control grout blockages due to dewatering of the grout. Depending on the type of VEA–superplasticizer combination used in the grout, set time, air content in the hardened state and strength are only marginally affected when compared to grouts that do not contain such admixture blends.

(b) Injection grouts

An ideal injection grout must exhibit a low viscosity at injection shear rates to facilitate penetration, retain moisture as it passes through absorptive surfaces, and keep the cement particles suspended once injection ceases. A superplasticizer alone significantly reduces the low and moderate shear viscosities of the grout but only marginally affects fluid loss. It produces extensive sedimentation even though bleeding is reduced. Combinations of a VEA and superplasticizer generates a grout with excellent stability, an easily injectable viscosity, and significantly reduced fluid loss. An example of the impact combinations of admixtures can have on the performance of injection grouts is shown in Table 7.21 [78].

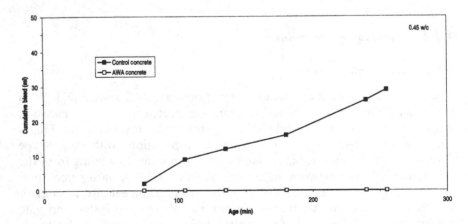

Fig. 7.27 Bleeding characteristics of washout-resistant concrete, total bleed = 0.0145 ml cm^{-2} (Rakitsky [78]).

Table 7.21 Water–cement injection grout formulations, Type I/II cement (Ralcitsky)

Formulation	Viscosity at 5 s⁻¹ (cP)	Viscosity at 170 s⁻¹ (cP)	Static bleeding* (ml)	Fluid loss† (ml)	Fluid loss‡ (%)
Control	270	21	28	77	85
0.60 wt% super	75	7	20§	73	81
0.10 wt% welan gum	670	61	30	75	83
0.60 wt% super					
0.05 wt% welan gum	75	25	9	21	23
0.60 wt% super					
0.10 wt% welan gum	255	50	0	12	13
EX 7112	255	58	0	7	8

* Bleed water present in 250 ml graduated cylinder after 2 h.
† Induced by applying a pressure of 10 lbf/in² (68.95 kPa) for 10 min to a cement slurry placed on a piece of filter paper.
‡ Percentage of total mix water lost.
§ Extensive sedimentation observed.

(c) Grouting of preplaced aggregate concrete

Preplaced aggregate concrete (PAC) is produced by first placing graded coarse aggregate in forms and then filling the voids in the aggregate mass with grout mixture. It differs from ordinary concrete in the manner of manufacture, the proportion of aggregates used in the mix, and the cement requirements. Normal compressive strength values range from 17 to 48 MPa (2500 to 7000 lbf/in^2) and densities generally higher than normal concrete, ranging from 2275 to 2483 kg m^3 (142 to 155 lb/ft^3) are produced [72].

PAC was originally developed for structural repairs, primarily because of its low setting shrinkage, reduction in drying shrinkage due to point-to-point contact of the preplaced aggregate particles, and good bond to existing concrete. It has been used in many different applications such as underwater concreting, mass concrete structures, repair of tunnel linings, underpinning foundations, resurfacing of dams, repair of piers, spillways and for intruding high-density aggregate for radiation shielding [79]. The method lends itself to use in vertical or overhead structures, can be produced as readily underwater as above ground, and in thick or thin sections. It is particularly useful for repair in locations where only a minimum of construction equipment can be used or where placing conditions are difficult, such as underwater concreting and also where a low volume change is required [23, 32]. Since this is a specialized type of construction, it is important that well-qualified personnel, experienced in this method of construction, carry out the work.

The primary function of the grout is to fill voids in the coarse aggregate, to bind them together upon hardening and to consolidate the entire mass. The quality of PAC concrete depends on the use of a grout mixture which is cohesive and workable and which develops sufficient strength in the hardened state. Grout mixtures consist of cement, sand (ratio of 1.6 by mass), supplementary cementing materials (fly ash silica fume and slag) water and VEAs [79, 80]. Fly ash and slag have been used to replace between 30–50% of the cement, and these materials contribute to reduced heat evolution, impermeability, higher ultimate strength and erosion resistance. Chemical admixtures provide air entrainment, delayed setting, increased flowability, homogeneity of the grout, lower water–cement ratios, reduction in setting shrinkage and expansion of the grout.

(d) Grouting of oil wells

Highly fluid cement slurries with good water retention characteristics are used to grout the space between the formation and the metal pipe in the construction of oil wells. The grout must be highly fluid to prevent excessive friction during placement, must not lose water to the formation, and must provide ample low-shear viscosity to prevent creating free water and sedimentation. Typical water–cement ratios range from 0.38 to 0.6; VEAs and superplasticizers are added to reduce friction pressure and eliminate free water and prevent sedimentation. Grouts containing these admixtures are easier to mix and pump. Once grouted the hydrostatic pressure of the cement column is the sole means of preventing fluids and gases from entering the well bore. Particle sedimentation reduces slurry density and causes a corresponding loss of hydrostatic pressure. Therefore it is critical to prevent sedimentation of the plastic cement slurry. VEAs improve fluid loss characteristics and provide good suspension of the aggregate in the plastic grout.

7.7.2 Underwater concrete

Underwater concreting can be done by three methods: the pumping of antiwashout concrete produced by the combined use of water reducing admixtures, superplasticizers and VEAs; the use of tremies also using antiwashout concrete; and the preplaced aggregate method discussed above.

(a) Antiwashout concrete

Concretes designed to resist cement washout when allowed to freefall through water incorporate a VEA admixture that prevents loss of cement fines and thereby preserves the quality of the concrete. The admixture increases the viscosity of water in the mix, resulting in an increased thixotropy of the concrete and an improved resistance to segregation. Dosage of the admixture ranges from 1 to 1.5% by weight of the water in the mix and it is frequently used in combination with a superplasticizer [78]. The magnitude of the effect produced is dependent on the admixture dosage and the molecular weight of the main component. It is usually discharged into the mixer at the same time as the other materials.

Often it is difficult to adjust the mixture proportions to achieve desired design parameters for all properties of concrete. Consequently the properties of colloidal underwater concrete are controlled by the addition of three chemical admixtures. Minimum water–cement ratios range from 0.36 to 0.40. Cement and fine-aggregate contents are usually higher than corresponding mixes placed on land, and silica fume may be used in conjunction with a superplasticizer or conventional water reducers to reduce segregation. The key to a non-dispersible concrete with self-leveling characteristics is the successful optimization of the VEA with the superplasticizer used to increase the slump.

Concretes containing VEAs have consistencies that are structured while static but become highly flowable, with self-leveling characteristics, when work is imparted to it. Data presented in Fig. 7.28 show that medium-strength concretes with equal slumps perform very differently in the washout test. This indicates that the slump test does not correlate to washout

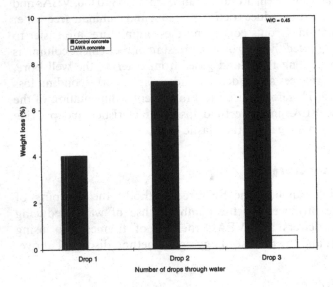

Fig. 7.28 Medium-strength concretes with equal slump perform differently in washout tests (Rakitsky [78]).

performance when antiwashout admixtures are formulated into the concrete. The washout weight loss test done in accordance with CRD C61, 'Test Method for Determining the Resistance of Freshly-Mixed Concrete to Washing Out in Water', showed that the use of a VEA at a level of 0.15% by weight of cement with a superplasticizer resulted in very fluid yet washout resistant concrete mixture. This contrasts with conventional flowing concrete which, depending on the viscosity and mixture proportioning, may have a washout weight loss ranging from 1 to 15% [79].

(b) Tremie concrete

VEAs have allowed placement of concrete underwater without the use of conventional tremies. Although this method is being replaced by pumping of non-dispersible concretes which are resistant to cement washout, it is still used in the casting of mass concrete in caissons, cofferdam seals and bridge piers. A tremie consists of a vertical pipeline, topped by a hopper which is long enough to reach the lowest point to be concreted from a working platform above the water (Fig. 7.29).

The quality and strength of tremie concrete is greatly dependent on proper mix design and placement. They should posses good cohesion and flowability. Concretes used for this purpose are usually cement rich

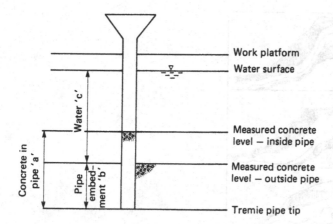

Fig. 7.29 Outline of a tremie pipe (Gerwick *et al.* [32]).

($>$ 400 kg m^{-3}) with slumps in the range of 150–225 mm. The fine-aggregate proportion is usually richer than for normal concrete (about 45% of total aggregate content) and the coarse-aggregate size is restricted to a maximum of 40 mm. Placing is continuous and no compaction by vibration is permitted. The lower end of the tremie pipe is always kept submerged in the fresh concrete to maintain a seal and to force the concrete to flow into position by pressure.

When concrete containing no admixtures is poured in this manner, usually poor edges result and much laitance is formed. Flow problems can be encountered, extending the placing time and causing gravel pockets due to the need to lift the pipe to facilitate concrete flow. Cohesiveness and flow properties are greatly improved by the use of VEAs, and silica fume, in combination with a SP. Previous work showed that a combination of these admixtures enhances uniformity of the placed concrete and realizes the following advantages [80, 81].

1. Increased flowability at a given water content allows greater horizontal movement (9–11 m) compared to concretes without admixtures. Flatter slopes are produced, enabling wider spacing of the tremie pipes (Fig. 7.30). The rate of placing is increased due to increased hydraulic pressure; concrete therefore moves further and faster. Increased pipe lengths are possible; consequently, substantial savings are achieved through a reduction in the number of tremies used.
2. Higher water reductions produce cohesive mixes which are less susceptible to segregation, laitance and washout of cement paste.

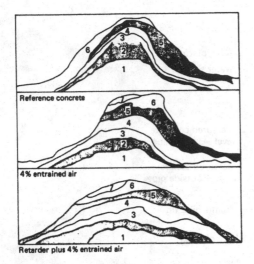

Reference concrete

4% entrained air

Retarder plus 4% entrained air

Fig. 7.30 Tremie flow patterns using concrete with and without admixtures [81]. Numbers indicate sequence of batches. Thick end sections are laitence and poor-quality concrete. Note reduction in thick end sections for pour containing retarding and air-entraining admixture.

Superplasticizers permit the use of wider pipes for faster placing and allow cement reductions so that less heat is generated.

3. Retardation of the rate of stiffening allows for the longer placing times and wider spacing of tremie pipes.

7.7.3 Formulation of construction products

The usefulness of VEAs has also been demonstrated in silica fume slurries and curing compounds. When formulated into a silica fume slurry, VEAs can keep the particles suspended so that constant agitation is unnecessary. Pigment dispersions, used in colored concrete and pigmented curing compounds, are other ideal applications for highly pseudoplastic polymeric suspending agents. Upon addition of the VEA to pigmented curing compounds, sufficient changes in the rheological properties occur so that:

● Pigment and latex particles remain well dispersed.
● Film formation on the slab is improved due to reduction of penetration; poorly coated areas become obvious.
● Curing compound can be applied to vertical surfaces due to enhanced sag resistance.

- Droplet particle size is increased while spraying to minimize loss of curing compound due to wind.

7.8. Dampproofing admixtures

Hydrostatic pressure, capillary action, wind-driven rain, a difference in a vapor pressure between the two sides of the concrete, or some combination of these can force water through concrete. The ingress and migration of moisture in liquid and vapor can be prevented or retarded to varying degrees. Treating of concrete to retard, not stop, the absorption of water or water vapor by concrete or to retard their transmission through concrete is considered to be dampproofing; treatment of a surface or a structure to prevent the passage of liquid water under hydrostatic pressure is called waterproofing [82, 83]. The positive prevention of the ingress and movement of water under pressure distinguishes waterproofing from dampproofing. Dampproofing of concrete through the use of an admixture can be achieved by the incorporation of an integral waterproofing admixture in the concrete mix.

Reducing the transmission of water vapor through concrete without stopping it entirely is a desirable feature in situations where the concrete is required to breathe, allowing the water to escape in the form of vapor. Although integral dampproofing admixtures are only marginally useful in watertight concrete, the significant role they play in maintaining the aesthetic qualities of concrete cannot be disputed. These admixtures will prevent the passage of rain and ground water by capillary action and will often reduce unsightly efflorescence. The enduring aesthetic effect produced by these admixtures due to their inhibition of disfiguring algal growth is shown in Table 4.8. This application is particularly useful in colored concrete products where it curbs the transport of lime to the surface by reducing the ingress and passage of moisture.

Dampproofing admixtures are water-repelling materials such as wax emulsions, soaps and fatty acids which react with cement hydrates [84, 85]. The most widely used water-repelling materials are the calcium or ammonium salts of fatty acids such as stearates. Proprietary products are available both as dry powders and liquids. Usually, a stearate soap is blended with talc or fine silica sand and used at the prescribed dosage per weight or bag of cement. In commercial liquid preparations, the fatty-acid salt (soap) content is usually 20% or less, the balance of the solid material is made up of lime or $CaCl_2$. Some proprietary admixtures combine two or more admixtures, e.g. stearates and non-chloride admixtures. The object of such composite mixtures is to effect a reduction in permeability and impart dampproofing without the strength reduction that occurs at early ages when soap-based materials are used.

A newer waterproofing admixture called hydrophobic blocking ingredient (HPI) based on two principal components – reactive aliphatic fatty acids, and an aqueous emulsion of polymers and aromatic globules – has been introduced to North America. The system has been widely used in Australia and Asia to combat severe marine and soil sulfate conditions. Both types of admixture increase resistance to water penetration, either by acting as pore fillers or by creating a hydrophobic coating within the pores or by combining both effects. A 20-year track record is claimed for the product [86].

Because of its limited effectiveness, dampproofing should be replaced by waterproofing under the following conditions [85]: if there is a likelihood that the treated concrete may later develop cracks; and if the concrete is subjected to a head of water at a later stage.

7.9 Recycling of cementitious wastes

The total volume of waste (including returned concrete plant mixer washout, washdown of truck, spillage of concrete and silt from yard washing) can exceed 140 t per year for a six-truck ready-mixed operation [87]. Environmental protection acts in Western Europe and North America now prohibit the disposal of concrete waste in landfill sites. Ready-mixed concrete operations have therefore altered procedures for disposal of alkaline waste and effluent. One such procedure where significant progress has been made is in the use of chemical admixture systems to eliminate washout from ready-mixed trucks.

Conventional procedures for washing out trucks use water which is either fresh or recirculated. Approximately 3000 liters of water are added by hose to each 6 m^3 truck at the end of each day. It is then discharged into wash troughs along with concrete waste from the truck mixer. Water and fines are allowed to drain into a slurry settlement pit and excess water drains off into the soil. At a typical ready-mix plant with six trucks, therefore, about 18 000 liters in total will be used each evening to wash out the trucks. Since this large volume of water is discharged in a short period of time, the excess highly alkaline wash water can run off to public sewers.

Chemical wash-water systems consisting of potent retarders (stabilizers) and accelerators (activators) developed in the last decade now enable the complete elimination of truck washout. The addition of the stabilizer to the truck at the end of the day allows for the incorporation of the residue into the concrete batched the following day. Chemicals used for the overnight retardation include phosphates, phosphonates, polyacrylate latexes and hydroxylated long-chain sugar acids. Safety glasses or goggles and rubber gloves must be worn when handling the stabilizer [88].

A fixed dosage of the chemical (stabilizer) together with a fixed amount of water is added to a nominally empty truck and mixed at high speed to contact all its internal surfaces for maximum fin cleaning and then stood overnight. The next day the residue and known volume of water are incorporated into batched concrete. In practice, most plants would probably use a combination of the chemical system and conventional wash-water system. Both laboratory and field data show that the quality of the concrete produced shows no adverse effect on the concrete batched using the chemical system. (See Table 6.15 and Section 6.7 for further details regarding effects on hydration.)

The use of the admixture system to reuse the previous day's wash water provides the following benefits:

1. Reduces water content to clean out each truck.
2. Eliminates concrete wash-water disposal.
3. Acts as a cleansing agent to reduce concrete build-up on fins thereby reducing maintaining costs
4. Reduces hauling charges.
5. Reduces/eliminates environmental concerns pertaining to the disposal of concrete wash water.

Other applications for the stabilizer and the recycling system include hot-weather concreting, the transportation of ready-mixed concrete over long distances (particularly in the Middle East countries) or through city traffic where unexpected delays can be encountered. Higher dosage of the admixture can be used to obtain the desired set extension and the appropriate dosage of the activator (carried on the truck) can then be added at the destination.

Comparing costs of the chemical system to the conventional water wash system is difficult because operating the latter shows large variations while the former costs are constant. However, a comparison that ignores all sources of water generation except truck mixer washout is given in Section 7.12.

7.10 Hot-weather concreting

Under moderate weather conditions, concrete remains workable long enough to allow its transporting, placing, compacting and finishing, without appreciable difficulties. In hot-weather conditions, however, the rate of stiffening, and the associated slump loss are both increased [89–91] and the initial and final setting times are decreased [92, 93]. Consequently, severe handling problems may be experienced on construction sites during the summer in temperate countries and in climates where hot weather prevails for most part of the year. Common problems encountered under hot weather conditions include the following:

1. Decreased strengths due to increased water demand, resulting from high mix temperatures.
2. The rapid decay of workability due to high slump loss, posing retempering (adding extra water on site) problems.
3. Difficulties in the handling and finishing of concrete as a result of accelerated setting due to increased rate of hydration.
4. The increased tendency for plastic shrinkage caused by rapid evaporation due to high mix temperature and low relative humidity.
5. Increased tendency for drying shrinkage and differential thermal cracking.

Accelerated slump loss and shorter setting times constitute the major problems of hot-weather concreting because they reduce the length of time during which fresh concrete remains workable and can be thoroughly compacted at the building site. Although stiffening of the fresh concrete, and slump loss, are brought about mainly by the hydration of the cement, some evaporation of the mixing water and absorption of water by dry aggregates may constitute additional causes. These factors reduce the amount of the free water in the mix decreasing the fluidity of the mix which produces stiffening. The accelerating effect of temperature on the hydration rate of Portland cement is very significant indeed. For example, it can be shown that with the rise of temperature from say 20°C to 40°C, the rate of the cement hydration increases by a factor of 2.41 [91] The effect of higher temperature on the rate of hydration and the resultant shorter setting times and a higher rate of slump loss is illustrated in Figs 7.31 to 7.33.

High temperatures during hot-weather concreting result in an increase of the amount of mixing water required to produce a given slump [91, 94]. This effect of temperature is attributable to its accelerating effect on the hydration rate of the cement. At temperatures in excess of 25°C, the mix

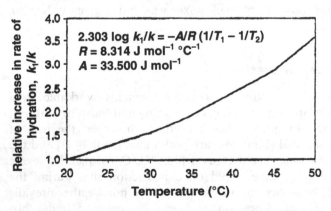

$$2.303 \log k_1/k = -A/R \left(1/T_1 - 1/T_2\right)$$
$$R = 8.314 \text{ J mol}^{-1} \text{ °C}^{-1}$$
$$A = 33.500 \text{ J mol}^{-1}$$

Fig. 7.31 Effect of high temperature on the rate of hydration.

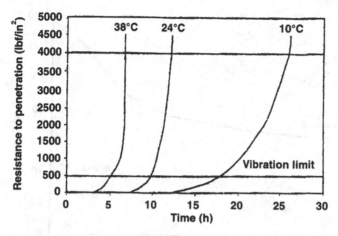

Fig. 7.32 Effect of high temperature on setting time.

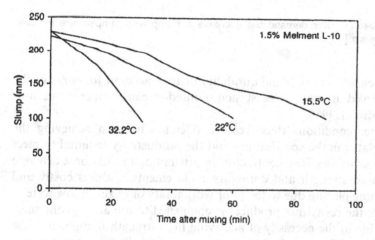

Fig. 7.33 Effect of temperature increase on slump loss (Mailvaganam [90]).

stiffens rapidly in the interval between the addition of water and the time slump is determined (Fig. 7.34). Hence, in order to overcome this accelerated stiffening, more water must be added to the mix, leading to the practice of retempering of the mix on site. Higher water contents increase the water–cement ratio and thereby adversely affects concrete properties. On the other hand, increasing the cement content, to obtain the required water–cement ratio, is also undesirable because of the increased cost involved, the increased drying shrinkage and attendant susceptibility to cracking. Previous studies have shown that traditional methods of increasing cement

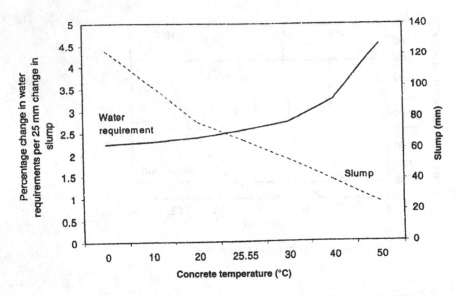

Fig. 7.34 Increase in water demand and decrease in slump with temperature increase (Soroka and Ravina [94]).

contents, to ensure strength and durability factors, account for some of the defects observed in both precast and poured-in-place concrete in hot-weather conditions [95].

Hot-weather conditions thus create a dilemma between achieving the quality stipulated in the specification and the productivity required to meet construction schedules. The contractor, in attempting to provide a concrete meeting desired strength and durability requirements, endures costly and time-consuming placing due to the poor workability of mixes of low water–cement ratio; the ready-mix producer is compelled to use an uneconomical mix design due to the necessity of achieving high strength margins and the precaster has to tolerate a high percentage of rejected units having severe blemishes and surface voids due to the poor compaction of dry mixes. It is self-evident that in order to avoid the poor compaction of low water–cement-ratio mixes, and the undesirable effects of the increased water demand, water-reducing admixtures (WRAs) and high-range water reducers (HRWR) should be considered. The use of such admixtures often provides cost savings both in raw materials and labor, enabling the production of quality concrete. Some of the commonly sought mix modifications under hot-weather conditions can be listed as follows:

- High water reduction and plasticizing effects for concretes where low water–cement ratios and minimum cement contents are specified.

- Increased cohesiveness in harsh mixes and mixes prone to bleeding.
- Improved watertight character of the concrete, to prevent the ingress of water containing sulfate and chloride ions.
- A retarded rate of initial set and loss of workability.
- A reduced rate of heat release, due to retardation of the rate of hydration.
- Reinstatement of the initial workability.

The current means employed to overcome field problems associated with accelerated slump loss and reduced stiffening time [94–96] is the use of chemical and mineral admixtures. Since retarding admixtures counteract the accelerating effect of temperature and thereby slow down the rate of slump loss, and as previously noted water-reducing admixtures are beneficial under hot-weather conditions, water-reducing and retarding (WRR) admixtures (i.e. ASTM C 494, type D) should be considered rather than admixtures that merely have a retarding effect and do not reduce water content (i.e. type B). The use of these admixtures, however, must also be considered with respect to the relevance of the ASTM C 494 compliance test data evaluated under laboratory conditions to use in situations where the temperature exceeds 30°C and their possible effect on plastic shrinkage cracking. These aspects are discussed below.

Although the conclusion that retarders reduce the rate of hydration is supported by experimental data from their effect on the setting times, penetration resistance in accordance with ASTM C 403) (Fig. 7.13) extrapolation of this delay in setting to reduced slump loss is not supported when the effect of retarders is evaluated under conditions which simulate ready-mixed concrete [94, 97, 98]. Relevant data of concrete tested accordingly are presented in Fig. 7.35. It is clearly evident from these data that the presence of WRR admixtures, depending on their specific type and dosage, actually increased rather than decreased, the rate of slump loss [98]. Similar data, of other researchers [99, 100], confirm this observation. It is therefore questionable whether type D admixtures, and perhaps also type G high-range WRR admixtures (ASTM C 494), are suitable to overcome slump loss in hot-weather conditions.

The seemingly contradictory behavior of type D admixtures in retarding setting and not slump loss may be attributed to the difference in the test procedures involved, i.e. while the increased slump loss was observed in concrete which was subjected to continuous agitation, the time of setting is determined on a concrete which remains undisturbed. The adsorption mechanism of retardation [91] suggests that the admixture adsorbs on the surfaces of the unhydrated cement grains, and thereby prevents the water from reacting with the cement, while the precipitation theory indicates that the retardation is caused by the formation of an insoluble layer of calcium salts of the retarder on the hydration products. Continuous agitation of the

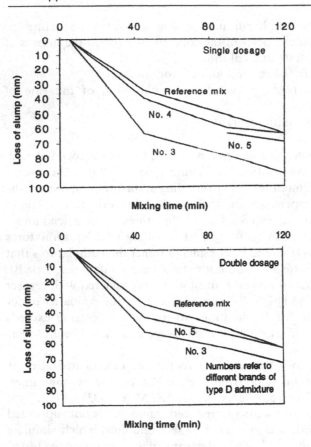

Fig. 7.35 Effect of type D admixture on slump loss; initial slump 95–115 mm, temperature 30°C (Soroka and Ravina [94]).

fresh concrete results in a 'grinding effect' which may cause 'peeling off' of the adsorbed layer of the retarder from the surface of the cement grains and hence, precludes the retarding mechanism.

While the above explanation may account for the anomalous behavior of retarders in continuously agitated concrete, it does not explain their accelerating effect under hot-weather conditions. The following mechanism has been suggested to explain this behavior [94]. Since the retarding mechanism is not operative in continuously agitated mixes, type D admixtures act merely as WRAs, lowering the mixing water content and concurrently reducing the spacing between the solid particles in the fresh mix. Consequently, in such a mix, due to the greater proximity of the particles, the forming hydration products are likely to bridge between the particles at an earlier stage than in a mix which contains a greater amount of water, and the spacing between the particles is greater. Hence, stiffening is brought about earlier and this, in turn, explains the accelerating effect of the admixtures on slump loss [94].

In view of the preceding discussion, it may be questioned whether it is prudent to use type D admixtures when prolonged mixing is involved – particularly in hot environments which further aggravate the problem. It should be noted, however, that despite the ineffectiveness of the admixture in minimizing slump loss it does play a significant role in preventing other problems that arise under hot-weather conditions. For example, once agitation is stopped, and the concrete is placed and compacted, retardation of the concrete occurs, and is advantageous in extending the vibrational limit and preventing the possibility of cold joints (Fig. 7.12). Furthermore, the accelerating effect of temperature can be counteracted by increasing the water content of mixes containing type D admixtures and utilizing more fluid mixes rather than mixes where the admixture is used to reduce water demand.

One of the primary concerns in the use of retarders under hot-weather conditions is their effect on plastic shrinkage cracking. Plastic shrinkage cracking occurs when the induced tensile stress, due to the loss of water from restrained concrete elements, and particularly concrete slabs, exceeds the tensile strength of the fresh concrete. The likelihood for cracking to occur increases with the increase in plastic shrinkage and the decrease in the rate of strength development. Since retarding admixtures delay the strength development, the susceptibility of plastic shrinkage cracking to occur increases. This has been confirmed in a study of retarded and unretarded cement mortars [101]; it was found that plastic shrinkage was greater in the retarded mortars (Fig. 7.36). Indeed, it was reported [102] that concretes and mortars whose set was delayed by admixtures nearly always showed serious plastic cracking in the laboratory as well as in the field. Hence, it may be concluded that the use of type D admixture aggravates the problem of plastic shrinkage cracking and that the likelihood of plastic shrinkage cracking occurring will be greater under hot–dry-weather conditions. Accordingly, extra care should be taken under such conditions to protect the fresh concrete from drying as soon as possible after placing and finishing.

7.10.1 Relevance of ASTM standards

When considering the use of retarding admixtures in situations where the temperature exceeds 30°C, the relevance of the ASTM C 494 compliance test data (done under laboratory conditions) to the field conditions the concrete will experience must be considered. The actual temperatures experienced in most hot-weather countries (> 40°C) go beyond the scope of many current recommended practices. For example the ACI guide to hot-weather concreting (ACI 305 R-89) does not meet the needs of, and is not practical

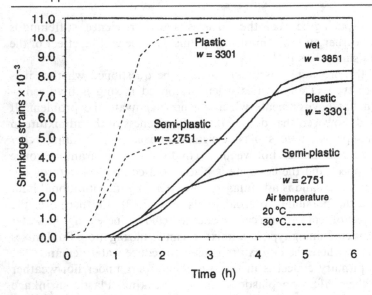

Fig. 7.36 Effect of temperature on plastic shrinkage (Soroka and Ravina [94]).

for, concrete operations in such climates. Furthermore, the ASTM C 494 specification for chemical admixtures for concrete does not provide adequate information for the selection of admixtures for hot-weather concreting, since the evaluation is based on only a minimum of mixing time at room temperature. The assumption that an admixture that retards at 21°C will do so at 33°C is not valid. Indeed, as mentioned previously, common retarders at 21°C often act as accelerators at 33°C [94]. Simulation of field conditions in the laboratory is usually not possible because of a multitude of varying ambient conditions. The best place for evaluation is, therefore, the actual job site, using the materials intended for the concrete to be used.

7.10.2 Use of fly ash

The replacement of Portland cement by fly ash class F (ASTM C 618) has been found to reduce the rate of slump loss in a prolonged mixed concrete, and the extent of the reduction is greater with increased cement replacement (Fig. 7.37). Fly ash also was found to be beneficial in reducing slump loss in concretes with conventional water-reducing and retarding admixtures [95]. The effect of fly ash on reducing slump loss can be attributed to chemical and physical factors. It was found that the surface of fly ash particles may be partly covered with a vapor-deposited alkali sulfate that is readily soluble [103, 104]. Thus the early hydration process of Portland cement is effected because sulfate ions have a retarding effect on the formation of the aluminates. Indeed, fly ash was found to be a more effective retarder than an

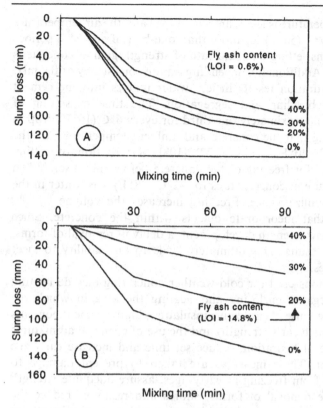

Fig. 7.37 Effect of replacing the cement with type F fly ash (ASTM C 618) on the rate of slump loss at 30°C. Loss of ignition of fly ash A and B is 0.6% and 14.8%, respectively (Soroka and Ravina [94]).

equivalent quantity of gypsum [105]. It seems that because the solubility of gypsum in water is low, it limits the amount of sulfate ions that go into solution, which allows the sulfate from the fly ash to be available for quite a long time to retard the hydration process of the C_3A. However, it must be realized that similar to the effect of retarders, the use of fly ash involves greater plastic shrinkage [94] and thereby increases the vulnerability of the concrete to plastic shrinkage cracking. Hence, when fly ash is used, and particularly when combined with WRR admixtures, extra care should be taken in order to prevent such cracking by protecting the fresh concrete from drying as soon as possible after being placed and finished.

7.11 Cold-weather concreting

Cold-weather conditions for construction is defined as a period when for more than 3 consecutive days, the following conditions exist [106, 107]: the average daily air temperature is less than 5°C (40°F) and the air temperature is not greater than 10°C (50°F) for more than one-half of any 24 h period. Under these conditions setting time and rate of strength gain of concrete is significantly delayed. Additionally, depending on the consistency of the mix, a reduced rate of hydration results in less water ingress into the cement particles, promoting bleeding and segregation. The time of setting of concrete increases by approximately one-third for every 6°C (10°F) drop in temperature (assuming that the concrete and ambient temperature are the same) down to 4°C (40°F) (Table 7.22) [108]. Retardation of setting increases the potential for freezing of the concrete before initial set. When the internal temperature of concrete falls to −2°C (28°F), free water in the pores begins to crystallize as ice. Freezing increases the volume by 9% generating stresses that incorporate defects within the concrete. Since retardation leads to low strength development, a delay in removal of forms, shores and reshores results. This ultimately leads to an inability to meet construction schedules.

Conventional methods used for cold-weather concreting include heating the water and aggregates, enclosing and heating the area in which the concrete is to be placed, use of protective insulation, higher cement contents or Type III cement (high early strength) and the use of chemical admixtures to accelerate the rate of hydration, reduce set time and increase early-age strength development. These measures are taken to prevent damage to concrete that results from freezing at early ages, assure adequate strength development for safe removal of forms, shores and reshores, reduce the potential for thermal cracking, and provide curing and protection to ensure the strength and durability of the concrete for the intended serviceability of the structure [106–109].

The choice of a particular option or options is determined by the desired set time and rate strength development. Accelerated time of setting is usually achieved by increasing the heat of hydration of the concrete mixture. Selection of appropriate chemical admixtures enables the production of cold-weather concrete mixtures with both accelerated setting and early-age strength development characteristics, similar to that obtained with plain

Table 7.22 Effect of temperature on the time of setting of concrete (Ref 108 with permission)

Temperature, °C (°F)	Approximate time of setting (h)
21 (70)	6
16 (60)	8
10 (50)	$10\frac{2}{3}$
4 (40)	$14\frac{1}{3}$
−1 (30)	19
−7 (20)	Set does not occur (concrete will freeze)

concrete mixtures at normal ambient and concrete temperatures (22°C). Concrete mixture proportions are usually the same as that used during other times of the year, with changes mainly in the types and dosages of chemical admixtures used. Depending on the type of cold-weather admixture used in the concrete mixtures, year-round construction with significant cost savings to owners and contractors can be achieved.

Although construction activity in Canada, America and Scandinavia is restricted to 8–9 months a year because of the severe cold-weather conditions that prevail during the winter, the lack of a track record in the use of antifreezing admixtures has precluded their use. Countries like Russia have three decades of experience in the use of these admixtures and the production of concrete all year. Non-chloride accelerating admixtures which provide protection against freezing in the plastic state at ambient temperatures as low as −7°C (20°F) are now an available and economical cold-weather concreting alternative to the conventional heating methods. In the following pages, the use of proprietary cold-weather admixture (CWA) over a range of above-freezing and sub-freezing temperatures is presented. The effectiveness of these admixtures as described in the projects cited provides support for the use of accelerating admixtures [110].

In order to withstand damage that results under cold-weather conditions, concrete should attain a compressive strength of 5 MPa before exposure to freezing conditions and 20.7 MPa before being subjected to freeze–thaw cycles [106]. Concrete that will be exposed to cyclic freezing and thawing in service should be air entrained. While this provides protection for hardened concrete, it is imperative that the plastic concrete be prevented from freezing so that the development of adequate strength and durability properties is ensured. In this regard, accelerating admixtures can be used to significantly increase the rate of hydration of Portland cement concrete, thereby reducing the need for some of the other cold-weather measures. Present-day cold-weather admixtures are often used below freezing temperatures and are expected to fulfill three roles: to depress the freezing point of the water in the mix; to increase the rate of hydration to offset the slowing down of the hydration reaction that occurs as curing temperature dips below 10°C; and to reduce the amount of freezable water present in the concrete by affording good water reduction, thereby minimizing frost damage. These roles are discussed below.

The main non-chloride, non-corrosive accelerating admixtures available on the market are of two types: (1) accelerating admixtures which accelerate hydration but do not depress the freezing point of water; and (2) accelerating admixtures for use in sub-freezing ambient temperatures which depress the freezing point of water. The former contain salts of formates, nitrates and nitrites and are effective for set acceleration and strength development. However, their effectiveness is dependent on the ambient temperature at the time of placement. The latter contain components that depress the freezing point of water and accelerating ingredients [110–117] and belong to the group of admixtures referred to as antifreeze admixtures. Chemicals used as antifreeze admixtures include sodium and calcium chloride, potash, sodium nitrite, calcium nitrate, urea, and binary systems such as calcium nitrite-nitrate and calcium chloride-nitrite-nitrate. However, only calcium nitrite-nitrate and calcium chloride-nitrite-nitrate are, reportedly, specially formulated for use as antifreeze admixtures in Russia [118]. They afford year-round construction and offer the most benefits for cold-weather concreting. Although their use in most industrialized western countries is more recent, they have been used for over 40 years in Russia in unheated concrete at ambient air or ground temperatures below 5°C (41°F), and minimum daily temperatures below 0°C (32°F) and down to −30°C (−22°F) [118, 119].

7.11.1 Acceleration of hydration and depression of the freezing point of the mix water

The primary benefit of antifreeze admixtures with regard to plastic concrete is the ability to prevent ice formation prior to setting, while providing accelerated setting and early-age strength development. The mode of action of antifreeze admixtures is twofold [110, 118]: first, by lowering the freezing point of water in the concrete; and second, to accelerate significantly the hydration of cement. The mechanism of ice formation in a normal salt solution can be used to illustrate how antifreeze admixtures lower the freezing point of the pore solution. During cooling of a normal salt solution, pure ice forms leading to an increase in the concentration of the solution. At a critical concentration – the eutectic point – the solution freezes. For a given antifreeze admixture, the freezing point is a function of the solution concentration; the higher the solution concentration, the lower the freezing point [117]. The eutectic point is also dependent on the type of antifreeze admixture; hence, the differences in effectiveness. Table 7.23 summarizes the eutectic point characteristics reported for some of antifreeze admixtures [118].

Similarly, in plastic concrete, the presence of dissolved calcium, sodium, potassium and sulfate ions lowers slightly the freezing point of pore water, which begins to crystallize as ice at a concrete temperature of about −2°C (28°F) [110]. The addition of the proper combination of salts such as those contained in antifreeze admixtures can further reduce the temperature at which ice formation begins.

One of the more widely used non-chloride accelerator in North America, the sodium thiocyanate-based multicomponent cold-weather admixture (CWA), is reported to be effective at sub-freezing temperatures, specifically, at ambient temperatures as low as −7°C (20°F) [110]. Data from laboratory studies [110] indicate that the temperature for ice formation in synthetic pore

Table 7.23 Eutectic point characteristics for some Russian antifreeze admixtures (Ratinov *et al.*)

Antifreeze admixture	Eutectic point characteristics	
	Solution concentration (%)	Freezing point °C (°F)
Sodium chloride	23	−21.2 (−6.2)
Calcium chloride	31	−55.0 (−67.0)
Sodium nitrate	28	−19.6 (−3.2)
Calcium nitrate	35	−18.5 (−1.3)
Urea	31	−8.4 (16.9)
Calcium nitrite-nitrate	35	−29.4 (−20.9)
Calcium chloride-nitrite-nitrate	30	−48.0 (−54.4)

water containing the CWA at a solids concentration of 47.4% was −19.1°C (−2.4°F). The investigators concluded from the data that for a 10% solids solution of the CWA, ice formation could be prevented down to −7°C (20°F).

The second mechanism by which antifreeze mixtures function, acceleration of cement hydration, was confirmed through calorimetry tests [110]. The data (Fig. 7.38) show that the CWA accelerated the time to reach peak temperatures by 8–10 h. It was also reported that, relative to plain cement,

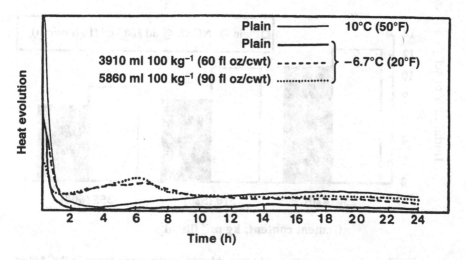

Fig. 7.38 Calorimetry test data for CWA-treated cement hydrated at −6.7°C and plain cement hydrated at −7 and 10°C (Brook *et al.* [110]).

the high and low doses of the CWA increases the total amount of heat generated by 625 and 569%, respectively, at −7°C (20°F). The increases in total heat generated comparing the high and low admixture treatments at −7°C (20°F) to the plain cement at 10°C (50°F) were 132 and 120%, respectively [110].

The effect of a non-chloride accelerating admixture (NCAA) on initial time of setting at an ambient and concrete temperature of 10°C (50°F) is illustrated in Fig. 7.39. A 250 kg m⁻³ (420 lb/yd³) cement factor concrete mixture that was treated with the NCAA produces an initial set time approximately equal to that for a plain concrete mixture containing 310 kg m⁻³ (520 lb/yd³) of cement. Slightly better performance was obtained with an admixture-treated concrete mixture containing 310 kg m⁻³ (520 lb/yd³) of cement compared to a plain concrete mixture with a cement content of 355 kg m⁻³ (600 lb/yd³). For the concrete mixtures evaluated, a 650 ml per 100 kg (10 fl oz/cwt) dosage of the NCAA was as effective as a 45 kg (100 lb) increase in cement content. More significant reductions in time of setting can be obtained by increasing the dosage of the accelerator.

The dosages used for the sodium thiocyanate-based CWA range from 325 to 5860 ml per 100 kg (5 to 90 fl oz/cwt). For acceleration of hydration at above freezing temperatures, dosages ranging from 325 to 3840 ml per 100 kg (5 to 59 fl oz/cwt) are recommended. The higher dosages of 3910–5860 ml per 100 kg (60–90 fl oz/cwt) are used for concrete placements in sub-freezing temperatures so that both protection against freezing, and acceleration of hydration can be realized. Previous work [110] has shown

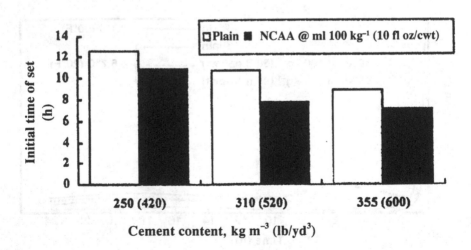

$Fig. 7.39$ Effect of cement content and a non-chloride accelerating admixture (NCAA) on initial time of set, at concrete and ambient temperatures of 10°C (Nmai [117]).

that the time of setting of concrete treated with the CWA, at an ambient temperature of $-7°C$ (20°F), was comparable to that for plain concrete at an ambient and concrete temperature of 21°C (70°F). Plain concretes placed at an ambient temperature of $-7°C$ (20°F) froze. Other antifreeze CWAs have been investigated [110, 111] and the US Army Cold Regions Research and Engineering Laboratory (CRREL) is working with two major admixture manufacturers in the United States to develop other antifreeze admixtures for commercial applications.

7.11.2 Reduction of freezable water

The duration of protection required to protect concrete from frost damage will depend on a number of factors. One of the critical factors is the degree of saturation of the freshly placed concrete and the duration of the saturated state. Under normal conditions the degree of saturation is reduced by both the consumption of water in the hydration process, and the evaporation of the free water. Therefore, any factor which reduces the degree of saturation below the level which would cause damage by freezing will benefit cold-weather concreting. Normal water-reducing and superplasticizing admixtures offer such benefit and this aspect is discussed below.

Approximately 65–70% of the water present in a concrete mix provides the workability required for placing operations. Such water is usually present as physically bound and free water filling the pores [118, 119]. If the required plasticity can be achieved with a much reduced water content, then the degree of saturation and vulnerability to formation of ice in the plastic state and frost attack is lowered. The 10–20% water reduction afforded by conventional water reducers and superplasticizers is significant in terms of pore water, and therefore offers a notable advantage in winter concreting. Normal water reducers, due to the wetting action, increase the rate at which water combines with the cement particles. Superplasticizers are particularly noted for the high deflocculation they produce. Consequently, a good amount of the water used in the mix is quickly bound in the physical form of films around cement particles, reducing the free water. It should be noted, however, that water-reducing admixtures will typically not accelerate the setting of concrete; therefore, for maximum benefit they should be used in combination with an accelerating admixture or an accelerating/water-reducing admixture (ASTM C 494 Type E) should be used.

When concrete is to be placed in cold weather, it is preferable that accelerators or antifreezers be used in combination with air-entraining agents and water-reducing admixtures. This, not only reduces the amount of freezable water in the mix but also generally reduces the quantity of antifreezers and accelerators needed to obtain desired effects compared to the amounts that have to be used when these are used separately. In addition these combination may be useful in increasing the resistance of concrete to frost action and to corrosive agents.

Cold-weather concrete mixtures incorporating non-chloride antifreeze accelerating admixtures have been used in a number of projects. Two of these projects are profiled. All the concrete mixtures described below were treated with the sodium-thiocyanate-based CWA mentioned earlier. These projects illustrate the impact of this admixture on normal concrete mixtures containing Class C fly ash, since fly ash typically delays time of setting and, hence, would not be a logical choice for cold-weather concrete.

7.11.3 Case studies

(a) Freezer floor

The project was a 9.2 m³ (12 yd³) concrete floor placement for a freezer at a dairy food manufacturing plant in Little Rock, Arkansas, USA. Although the ambient temperature was about 28°C (82°F), at the time of placement, the concrete temperature was 27°C (80°F) and the temperature in the freezer was −4°C (25°F). As shown in Table 7.24, the CWA was used at a dosage of 3910 ml per 100 kg (60 fl oz/cwt) and the design strength of 27.6 MPa (4 000 lbf/in²) at 28 days was achieved. The concrete contained approximately 23% Class C fly ash by mass of cement.

Table 7.24 Concrete performance data: freezer floor project (Nmai [117])

Mixture ingredient		
Cement:	440 lb/yd³	(200 kg m⁻³)
Class C Fly Ash	100 lb/yd³	(59 kg m⁻³)
Fine aggregate	1385 lb/yd³	(820 kg m⁻³)
Coarse aggregate	1900 lb/yd³	(1125 kg m⁻³)
Water	200 lb/yd³	(118 kg m⁻³)
CWA dosage	60 fl oz/cwt	(3910 ml 100 kg⁻¹)
Concrete temperature	80°F	(27°C)
Ambient temperature, Air	82°F	(28°C)
Freezer	25°F	(−4°C)
Slump	8.5 in	(215 mm)
Compressive strength		
3 days	3470 lb/in²	(24 MPa)
7 days	4030 lb/in²	(28 MPa)
28 days	4370 lb/in²	(30 MPa)

Table 7.25 Concrete performance data: parking garage project (Nmai [117]).

Mixture ingredient		
Cement	520 lb/yd³	(310 kg m⁻³)
Class C fly ash	100 lb/yd³	(59 kg m⁻³)
Fine aggregate	1300 lb/yd³	(790 kg m⁻³)
Coarse aggregate	1710 lb/yd³	(1015 kg m⁻³)
Water	240 lb/yd³	(142 kg m⁻³)
CWA dosage (see below)	5–20 fl oz/cwt	(325–1300 ml 100 kg⁻¹)
Third generation HRWR	18 fl oz/cwt	(1170 ml 100 kg⁻¹)
Air-entraining admixture	1 fl oz/cwt	(65 ml 100 kg⁻¹)
Air content	6.5%	(6.5%)
Slump	8.0 in	(200 mm)
Compressive strength		
3 days	4470 lbf/in²	(31 MPa)
7 days	5100 lbf/in²	(35 MPa)
28 days	6790 lbf/in²	(47 MPa)

Ambient temperature °F (°C)	Accelerator dosage fl oz/cwt (ml 100 kg⁻¹)
60–69 (16–21)	5 (325)
50–59 (10–15)	10 (650)
40–49 (4–9)	15 (975)
32–39* (0–4*)	20 (13005)

(b) Parking garage

This construction of a 62 710 m² (675 000 ft²), 30 580 m³ (40 000 yd³) parking garage in Detroit, Michigan, USA, required the production of high-workability pumpable concrete that would set in 4 hours throughout the year, regardless of the ambient conditions. In addition, the concrete had to develop 75% of the specified 28-day compressive strength of 34 MPa (5000 lb/in²) in 3 days. These requirements were achieved by using the CWA in combination with a third-generation high-range water reducer (HRWR) as shown in Table 7.25. To meet the 4 h time-of-set requirement, the dosage of the CWA was varied as shown.

Antifreeze CWAs are viable alternatives to conventional cold-weather practices. The reduction in labor and other costs related to conventional cold-concreting measures are discussed in the next section.

7.12 Economic aspects of admixture use

In estimating the economy that ensues by the use of an admixture, changes in mix composition of a unit volume of concrete and also the cost of handling, transporting, placing and finishing the concrete should be taken into account. The evaluation of any given admixture should be based on the results obtained with the particular concrete in question under conditions simulating those expected on the job. This is important, since the results obtained are significantly influenced by the characteristics of the cement and aggregate and their relative proportions, as well as by temperature, humidity and curing conditions.

If an admixture changes the yield of the concrete mix, the change in the properties of the concrete will be due to both the direct effects of the admixture, and to changes in the yield of the original ingredients. Therefore, its effect on the volume of a given batch should be taken into account. If there is an increase in the volume of the batch, the admixture must be regarded as effecting a displacement either of part of the original mixture or of one or another of the basic ingredients – cement, aggregate, or water. All such changes in the composition of a unit volume of concrete must be taken into account when testing the direct effect of the admixture and in estimating the benefits resulting from its use [119, 120].

Increased costs incurred from handling an additional ingredient, as well as the economic effect the use of the admixture may have on the cost of transporting, placing and finishing the concrete should be taken into account. Any effect on rate of strength gain and hence on the productivity of construction should be considered. An admixture may permit use of less expensive construction methods or structural designs to more than offset any added cost due to its use. In addition, placing economies are often realized. For example, water-reducing and set-retarding admixtures increase the ability to pump at greater heads and permit placement of large volumes of concrete over extended periods, thereby minimizing the need for forming, placing and joining separate units. Accelerating admixtures reduce finishing and forming costs, and required physical properties of lightweight concrete can be achieved at lower 'densities' (unit weight) by using air-entraining and water-reducing admixtures. The cost effectiveness of admixture use in some selected applications is discussed in the following pages.

7.12.1 Economies in mix proportioning

The concept of obtaining economic advantage from the use of admixtures is widely practiced in the concrete industry. Economies in mix design can be obtained through the use conventional, mid-range water-reducing and super-plasticizing admixtures. The areas of high-performance concrete (high strength, self-compacting) and concreting in hot, arid and cold-weather conditions offer the widest scope in this respect, since high cement contents, low water–cement ratios and high flow and self-leveling characteristics are desired objectives.

Water-reducing admixtures (WRA) can be used in three ways: (1) with a reduction in cement content but at the same workability as the control concrete to produce higher compressive strengths at all ages; (2) with a reduction of both cement and water contents and yet obtain a mix with similar workability and strength as the control concrete; (3) addition of a WRA with no alteration either to the cement or water content, to obtain similar strength development characteristics to the control concrete but with much higher workability. In all three ways, however, the admixture can be regarded as a cement saver.

The approximate range of effectiveness of the WRAs currently available is shown in Table. 7.4. The effectiveness will depend on the dosage used, temperature, cement chemistry, fineness and other mixture characteristics. While water reducer effectiveness is difficult to predict reliably, it can be determined readily from a few trial batches. Table 7.26 shows the water demand and cement contents for various levels of WRA effectiveness. It is seen that the cement content for water–cement ratio of 0.4 approaches 405 kg m^{-3} as the admixture effectiveness approaches 20%. This would require a superplasticizer. Using a mid-range WRA (15% effectiveness) the cement content can get down to approximately 430 kg m^{-3} for a water content of about 172 kg m^{-3}. Superplasticizers produce dramatic water reductions ($> 20\%$) at higher admixture dosages and will, therefore, afford greater cement reduction, or a switch from a more finely ground cement (Type III) to a coarser particle cement (Type I). This shown in Table 7.8 for mixes with cement contents exceeding 400 kg m^{-3} [54].

7.12.2 Economies from improved durability

The durability of many thousands of kilometers of concrete highways, airport runways and parking lots built in geographical areas where freeze–thaw conditions are experienced over a period of 30 years attests to the most dramatic cost benefits afforded by air entrainment. Reduced maintenance and replacement costs that derive from the use of both air-entraining and

Table 7.26 Cement content as a function of WRA effectiveness for a mixture incorporating 20 mm coarse aggregate, and a slump of 100 mm (Hover [119])

WRA effectiveness (%)	Water demand (kg m^{-3})	Cement content (kg m^{-3}) for various water–cement ratios				
		0.35	0.40	0.45	0.50	0.55
0	203	580	508	451	406	369
5	193	551	483	429	386	351
10	183	523	458	407	366	333
15	172	491	430	382	344	313
20	97	463	405	360	324	295

water-reducing admixtures has been a boon to transportation departments in Canada and the USA. What is remarkable is that the enhanced durability can be obtained at such low increase in cost. A consideration of the mix design aspects of air-entrained concrete is shown in Table 7.27 [120]. It can be seen that because of the water reduction afforded by the higher paste content and increase yield (due to air content) the total cost of the air-entrained concrete can be lower than the plain concrete, even though the cement content of the former is higher. This is not always the case however, as many factors affect the final mix design.

7.12.3 Economies from improved placing characteristics and construction methods

Frequently an admixture will allow the use of less expensive construction methods, a reduction of the numbers in the placing crew, or a reduction in the time required to carry out the operation, e.g. as previously mentioned, the placement of large volumes of concrete over extended periods minimizes the need for forming, placing and joining separate units. In this respect, the use of flowing concrete in slab-on-grade construction permits cost savings from a reduction in the number of operators in the placing crew or, alternatively, a reduction in the operational time when using the same number in the crew. One well-documented study [120] did demonstrate conclusively that the use of superplasticizers in standard 3000 lbf/in² (20.7 MPa) for residential slabs on grade could result in significant labor savings. The work involved the placing of two sets of 20 m slabs with normal concrete and superplasticized flowing concrete respectively. The properties of the resultant concrete and the labor requirements at each stage of slab production (chute, placing, screeding,

Table 7.27 Mix design/cost data of air-entrained mixes having similar compressive strengths (Rixom)

	Cost ($)	Plain mix, lb/yd³ (kg m⁻³)	Air-entrained mix (6.5%), lb/yd³ (kg m⁻³)
Cement	0.040/lb (0.088 kg⁻¹)	500 (298)	530 (315)
Aggregate	0.005/lb (0.011 kg⁻¹)	3090 (1839)	2810 (1672)
Water		275 (164)	291 (173)
Air entrainer	0.010/oz (0.35 kg⁻¹)	0 (0)	6.625 (oz) (0.25)
Cost ($)		35.45 (46.39)	35.32 (46.22)

jitterbug, floating) were noted and the percentage savings in labor time for these procedures are given below:

Activity	Saving in labor time (%)
Chute	52
Place	39
Screed	20
Jitterbug	22
Float	15
Combined	33

The report concluded that the standard placing crew could be reduced from 10 to 5 or 6 with some increase in hours worked, or the crew headcount could be maintained with an increase in the number of slabs produced per day. The overall labor savings were 33% and the quality of the concrete was improved in terms of compressive and flexural strength [120].

Smaller placing crews, faster placements, and concrete integrity are exceptionally good reasons to use a superplasticizer. Other accepted cost-effective uses revolve around lower shrinkage factors and decreased concrete permeability. Shrinkage factors can be reduced in many instances by 20–35% over conventional mixes. This reduction in shrinkage can allow extended joint spacing, which can be a major cost savings. Polyethylene is often used as a vapor barrier. Through the use of some superplasticizer (those that enable concrete to be produced with extremely low permeability), the vapor barrier can often be deleted from the specification because the water tightness of the finished slab is greatly increased.

Generally there are no major problems pumping concrete 10–30 stories vertically. Above these heights, though, problems begin to occur. The options that are usually explored for solving these problems are: (1) alternate mix proportions such as those containing higher cementitious contents; (2) larger pumps, or nine staging pumps – pumping up to another pump that lifts the concrete an additional 10–30 floors. These methods are frequently used successfully, but contractors who have used second- and third-generation superplasticizers have found significantly lower pumping pressures. This has allowed them to avoid the need for costly adjustments in the mix proportions, and has reduced the need for larger or staged pumps. For example, on the Texas Commerce Tower in Dallas, lightweight concrete was pumped 54 floors with no mix adjustment and with no increase in pump pressures. Extended working times and the ability to use lower than normal water–cement ratios enabled the contractor to proportion his mix around a second-generation superplasticizer [119].

7.12.4 Precast concrete

The process of precast concrete manufacture provides conditions which are more amenable to the monitoring of cost benefits afforded by the use of admixtures. The following benefits have been found:

1. High early strengths at reduced costs (use of lower cement contents, alternative cement) are attained so that forms are stripped earlier and rapid delivery time is ensured.
2. Suitable consistencies (e.g. self-compacting mixes) to place thin, highly reinforced structural elements by special concreting techniques may be used.
3. Lowering of curing costs in plants where accelerated curing is used.

Early strength is very important in precasting operations because, to a large extent, it controls the production cycle and the ability to accept and produce new work. This is particularly true of prestressed concrete where high early strengths are required prior to detensioning operations. Customary methods of assuring adequate strengths (about 30 MPa) have consisted of using excessively high cement contents, resulting in concretes having 28-day strengths far in excess of that required for structural or durability reasons. Consequently water-reducing admixtures are widely used to obtain requisite early strengths, but at reduced cement contents. This results in more economical mixes. Typical results that can be obtained by the use of hydroxycarboxylic-acid-based water-reducing admixtures are presented in Table 7.28.

More substantial increases in the rate of strength development are obtained by the use of superplasticizers. As mentioned previously, because of the dramatic water reduction that can be achieved by the use of higher dosage of the superplasticizers, significant cement reduction or a change in the type of cement (from a high early Type III to a normal Type I) is possible (Tables 7.28, 7.29 and 7.7).

Superplasticizers may be used to realize significant energy savings in accelerated curing or even eliminate the necessity for curing [120]. High early strengths can be obtained at considerably lower temperatures and curing times than are currently used. Figure 7.40 illustrates the advantage by presenting a comparison of two mixes, one containing a superplasticizer and the other a plain mix, both steam cured. Detensioning strengths for the superplasticized concrete can be achieved in six hours of curing as opposed to the 22 hours curing regime required for the plain concrete.

Superplasticizers have been used successfully to achieve savings in concrete consolidation in heavily reinforced precast concrete sections, e.g. the high-quality finish required for the precast tube elements (carrying the mechanical and electrical services) used in the construction of the Montreal Olympic Stadium was achieved through the use of a superplasticizer. Fig. 7.41 shows the high density of steel reinforcement used in these structures.

Table 7.28 The use of a hydroxycarboxylic-acid-based water-reducing agent to produce a more economical high-strength mix for prestressed, precast units

	Control mix	Modified mix
Mix composition		
RH Portland cement (kg)	285	255
Coarse aggregate 20 mm (kg)	830	860
Sand zone 2 (kg)	450	460
Added water (l)	80	73
Hydroxycarboxylic acid water-reducing agent (ml 50 kg^{-1} cement)	0	120
Concrete properties		
Slump (mm)	50	55
VeBe (s)	6.5	9.0
Compressive strength at: 1 day (steamed)	36.0	34.0
28 days	63.0	75.5

Table 7.29 Application of high range water-reducing admxitures to reduction of mix cement content (Hestor [62])

Slump (mm)	Accelerated cure compressive strengths (MPa) at age 16–17 h		28-day compressive strength (MPa)	
	Mix 1	Mix 2	Mix 1	Mix 2
216	23.8	33.2	34.5	47.2
	23.8	32.5	33.2	46.5
89	32.4	31.8	36.6	41.4
	31.8	32.2	35.6	41.9

Accelerated curing: 3 h precure 4.5 °C rise and 58 °C for 16–17 h, and subsequently moist cured.
Mix 1: Cement content 351.9 kg m^{-3} Type III cement, conventional carbohydrate type admixture at recommended dosage.
Mix 2: Cement content 286.6 kg m^{-3} Type III cement, recommended dosage of sulfonated naphthalene formaldehyde superplasticizer.
Note: Anomalous decrease in strength values for specimens from concrete mixes of lower slumps attributable to mix thixotropy. Identical minimal consolidation was used for both high and low-slump mixes.

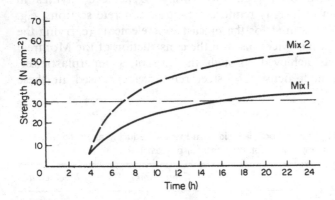

Fig. 7.40 Strength increases of cubes steamed under a normal heat profile showing that the destressing strength of about 30 N m^{-2} is obtained in 6–7 h as opposed to 22 h in the control mix.

Concrete products like bricks and blocks which are mass produced can also benefit from the use of air-entraining and water-reducing agents. These admixtures allow reductions in cement content up to 10% without altering the mechanical properties of the product. Unlike normal concrete, the strength of bricks and blocks is determined by the degree of interlocking the aggregates attain and is, therefore, more a function of compaction rather than cement content. The cement phase is a discontinuous phase that is present where the aggregates are in contact. In addition, on demolding the blocks from the machine, a considerable green strength is required which is also contributed by the cement. Reductions in cement contents up to 10% in the presence of a water-reducing admixture allows blocks to be produced and handled in the green state as well as having the required mechanical properties. The ideal requirement for the paste component of the block are that it should be as fluid as possible to allow maximum compaction and yet be stiff enough to maintain shape on demolding and give maximum strength. It is found that this compromise can be obtained by incorporating water-reducing admixtures into the mix. Typical mix designs and resultant properties of the blocks produced on an automatic machine are given in Table 7.30. Similar factors are applicable to extruded products where normal and retarding admixtures enhance lubricity and cohesion of the mix allowing easy extrusion and yet maintaining the product profile on emergence from the dye.

The economics of block production, particularly the 'egg-laying' process, is related to a large extent to the length of time before the blocks can be lifted out and stacked. Ideally this should be carried out the day after

Fig. 7.41 General view of the high-density steel reinforcement used in the tube element of the Olympic Stadium in Montreal, Canada (courtesy SKW, Trostberg, Germany).

production, but in winter conditions the strength attained is often inadequate to permit this. The use of accelerating admixtures to achieve the required strength is commonplace and indeed, in certain cases, accelerating and water-reducing are used simultaneously to reduce the cement content.

7.12.5 Economic benefits of cold-weather admixtures

In the winter months, significant benefits are derived from the use of concretes that contain cold-weather accelerating (CWA) admixtures with antifreeze properties. The economic benefits are derived from their ability to

Table 7.30 A lignosulfonate-based water-reducing agent allows lower cement contents to be used in concrete blocks whilst maintaining the compressive strength

Mix composition					
RH Portland cement (kg)	90	83		80	
Zone 2 sand (kg)	550	550		550	
Limestone aggregate (kg)	650	650		650	
A/c ratio	13.3 : 1	14.5 : 1		15.0 : 1	
Admixture (lignosulfonate at 140 ml 5 kg^{-1} cement)	No	No	Yes	No	Yes
Compressive strength of blocks (N mm^{-2}) at: 33 h	2.5	2.4	2.7	2.3	2.5
28 day	6.0	5.4	6.5	4.8	6.5

eliminate heating costs, afford the use of less costly mixes and enable the extension of the construction period. This is illustrated in the example cited in Fig. 7.39 in the section on cold-weather concreting. The specific benefits to the ready-mixed concrete producer, contractor and owner, include the following [117].

(a) For the ready-mixed concrete producer

- Ability to provide concrete for use in sub-freezing ambient temperatures.
- A viable, economic alternative to other cold-weather concreting options (increased cement content (Fig. 7.39) and hot water).
- Increased productivity and profitability.

(b) For the contractor

- Ability to place concrete in sub-freezing ambient temperatures.
- Accelerated setting and increased early-age strength development.
- Earlier finishing of slabs.
- Reduced heating and protection costs (insulating blankets, heaters, etc.).
- Earlier stripping and reuse of forms.
- Faster construction, increased productivity and greater flexibility in scheduling work.
- Overall reduction in construction costs (especially for labor).
- Reduced construction loan finance charges and increased profitability.

(c) For the owner

- Earlier use of structure.
- Reduced loan interest payments.

7.12.6 Economic benefits from the recycling of plastic concrete and wash water

The high tipping and sewage charges for the disposal of concrete waste and the ban on dumping of plastic concrete in landfill sites by the environmental protection agencies of many countries makes the chemical system for recycling concrete waste a cost-viable if not cost-saving system. A study done by a US admixture company [87] indicates the following cost savings based on the waste generated by one ready-mix truck. Total washes for one year for each truck (1 062 000 liters) and water and sewage charges equate to about $3200.00 assuming a worst case scenario. Using the chemical washout system, although 3000 liters are used per wash, the cost in relation to washout is zero. This is because the water is used in subsequent batches of concrete. The total cost of the washout chemical (assuming at a 1.5 liters overnight dosage and an admixture cost of $3.55 per liter for the same number of washouts as the conventional method) is $2326 per truck per annum – a cost saving of about $874 per annum. Similar cost savings are achieved in recycling of the volume of solid waste.

The admixture industry has made a major impact on the economics of the construction industry. Admixtures of various chemical classes provide a means to control and improve the quality of concrete, permit concreting to continue under adverse climatic conditions and facilitate design and construction techniques not readily practicable with concrete not containing such materials. The topical high-performance concretes often contain a mixture of admixtures that has enabled the development of specific applications that not only enable placing in areas of limited access but also ensure proper consolidation of the concrete and provide labor savings. However, it must reiterated that when mixtures of admixtures are used, compatibility of the admixtures becomes a critical parameter and the concrete producer must determine suitability prior to field use. The concrete engineer must ensure that the materials chosen for a specific application have the necessary background data to support the choice.

7.13 Guidelines for the use of admixtures

The effectiveness of each admixture may vary, depending on its concentration in the concrete, the time of addition in the mixing cycle and various constituents of the concrete. Although each class of admixture is defined by its main effect (i.e. water reduction, set acceleration), it may have one or more secondary effects (retardation of set, increased bleeding, air entrainment) and its use may result in side effects. Side effects are those modifications of properties produced in the concrete that, even though unsought, are both inevitable and independent of an admixture's main function. Prior to selecting an admixture for an intended application, these auxiliary effects must be quantified through laboratory and field testing with the materials that will be used on the job.

In this section, information on the operational aspects pertaining to the use of admixtures in the field is presented. In addition to details concerned directly with the physical aspects of batching and dispensing equipment and their operation, considerable attention is directed to the material-related problems that arise in the use of admixtures both in mainstream and special applications. It also offers guidance in the selection, use and control of uniformity of the product. Most of the issues discussed here apply equally to the three major fields of concrete construction, namely, site-batched and -placed concrete, ready-mixed concrete, and precast concrete.

7.13.1 Evaluation and selection

Admixtures should be purchased under specifications that stipulate the desired properties, exclude adverse effects and provide evaluation of uniformity of the admixture from batch to batch. The user should ensure that the admixtures being considered conform to ASTM or other applicable specifications and the manufacturer of the admixture should be required to certify that individual lots meet the requirements of applicable standards or specifications. The following information should be provided by the admixture manufacturer or supplier: (1) composition of the admixture in terms of the generic type of its main constituents; (2) incompatibility with other admixtures or special cements; (3) typical dosage; (4) detrimental effects of overdosage or underdosage; (5) chloride content; and (6) whether air is entrained at the recommended dosage.

Data should be provided by the admixture supplier on the results of laboratory performance tests, detailed in the relevant standard specifications. The information should include mix proportions of the concrete materials, dosage of the admixture, slump or compacting factor value obtained, set times, air content, compressive strength, and the effects of admixtures on other concrete properties. However, it should be noted that standard tests deal primarily with the influence of admixtures on standard properties of hardened concrete under specific laboratory conditions and only afford a valuable screening procedure for selection of admixtures. The concrete supplier, contractor, and owner of the construction project may be interested in other properties of concrete, primarily workability, pumping qualities, placing and finishing qualities, early strength development, and features of concrete construction such as reuse of forms or molds, appearance of formed surfaces, etc. These secondary functions and additional features may often be of great importance when the selection of an admixture is being considered and also in determining dosage of the admixture.

Suitable admixtures being considered for a specific application should, therefore, be evaluated with job site materials through mix trials conducted under concrete plant operating conditions, so that compliance with specifications and uniformity of the product is determined and a proper measurement of the desired engineering properties is obtained. In evaluating the admixture, the following items should be taken in to account:

- A change in type or source of cement or amount of cement, or a modification of aggregate grading or mixture proportions, may be needed.
- Many admixtures affect more than one property of concrete, sometimes adversely affecting desirable properties.
- Temperature conditions affect time of setting and early strength development.
- The effects of some admixtures are significantly modified by such factors as water content and cement content of the mixture, by aggregate type and grading, and by type and length of mixing.
- When more than one admixture is to be used, the dosages of the individual admixtures may require adjustment to avoid adverse effects on the concrete.

Trial mixtures can be made with a range of cement contents or water–cement ratios or other properties to bracket the job requirements. The evaluation will permit adjustment of the manufacturer's recommended dosage required to meet the difference in air content and setting time produced in bulk mixes and also the dosage that may be necessary to enhance a particular secondary effect. It will also enable determination of the cost of using the admixture for the specific application.

Such an evaluation is particularly important where: (1) the admixture has not been used previously with the particular combination of materials; (2) special types of cement are specified; (3) more than one admixture is to be used; (4) mixing and placing is done at temperatures well outside generally recommended concreting temperature ranges; (5) there is a limit on the amount of chloride ion that is permitted in concrete as manufactured.

Many manufacturers describe their admixtures as being 'chloride free'. However, no truly chloride-free admixture exists, since admixtures often are made with water that contains small but measurable amounts of chloride ion. Thus, to evaluate the likelihood that using a given admixture will jeopardize conformance of concrete with a specification containing such a limit, one needs to know the chloride-ion content of the admixture being considered, expressed in terms relevant to those in which the specification limit is given. Such limits are expressed as maximum percentage of chloride ion by weight (mass) of cement or chloride ion per unit weight (mass) of concrete, or 'water-soluble' chloride ion per unit weight (mass) of cement or concrete. If, in using the available information on the admixture and the proposed dosage rate, it is calculated that the specification requirement will be exceeded, alternative admixtures should be considered.

7.13.2 Admixture uniformity

Admixture uniformity plays a pivotal role in minimizing the variation of the concrete produced. It is therefore important to insure product compliance with uniformity provisions, standards (ASTM, BS, CSA) specifications and with the producer's own finished-product specifications. The degree of control of the uniformity of the concrete will depend on the critical nature of the structure being built, but any substantial use of admixtures in operations which produce concrete continuously should be accompanied by routine sampling and testing of the admixture to determine within-lot and batch-to-batch uniformity of the product. Samples for testing and inspection should be obtained by random sampling from plant production, from previously unopened packages or containers, or from fresh bulk shipments.

Quality control tests, for the most part, will be based on physical and chemical index tests, e.g. solids content, specific gravity of solutions and infra-red spectroscopy of active constituents. The trace obtained from infra-red spectroscopy is particularly useful in this regard, since it serves as a fingerprint of the material. Uniformity of the admixture batches supplied during construction can be readily determined by matching the trace of the sample supplied during purchasing negotiations to those obtained from the job site. Such tests, although not definitive of the quality of the product, are usually determined mainly to assure uniformity of the product being supplied.

7.13.3 Precautions in the use of admixtures

The full potential of admixtures in both economy and desired engineering objectives can often be realized only by changes in the proportioning of the concrete mix otherwise employed. Not all admixtures are compatible and the manufacturer's advice should be sought before two or more admixtures are used together or separately. Therefore it is important that laboratory and field trials be carried out to select the appropriate type and dosage to be used. The effects obtained with the admixture using job materials and job-site conditions should be determined. As previously mentioned, this is particularly important when special type of cements are specified, when more than one admixture is to be used, or when mixing and placing is carried out at ambient temperature conditions above or below generally recommended temperatures. For such trials, dosages up to three times normal should be included so that the potential problems resulting from over dosage can be appreciated. In evaluating the results obtained for each significant property of the admixture or the concrete, it is prudent to remember that uniformity of results is as important as the average result.

The manufacturer's instructions should be carefully followed when using admixtures. Specific effects produced in concrete by admixtures are dependent on the instructions provided by the manufacturer of the admixture and a number of factors such as cement composition, aggregate characteristics, the presence of other admixtures and ambient conditions. The interaction of the selected admixture with these factors and the extent of the side effects should be verified prior to field use. The desired effect should not be too sensitive to small variations in the amount of admixture used, or in the amounts of other concrete constituents used. Since the effects produced may vary with the point of addition in the mixing cycle, a standard mixing sequence should be established and admixtures must not be added during transportation, placing or compaction. Admixtures of all classes may be available in either powder or liquid form and since relatively small quantities are used, it is important that suitable and accurately adjusted dispensing equipment be employed. All liquids, particularly emulsion types, should be protected from exposure to drastic temperatures.

Although the continuing development of new and improved admixtures warrant re-evaluation concerning the benefits of admixture use, currently used admixtures that modify the properties of fresh concrete may also cause problems through early stiffening or undesirable retardation. Early stiffening often is caused by incompatibility between the cement and admixture used (Section 7.13.6(b)). Retardation can be caused by an overdose of admixture or by a lowering of ambient temperature, both of which delay the hydration of the calcium silicates. Air content and time of setting of job concrete can differ considerably from laboratory concrete with the same materials and mixture proportions. The cause of abnormal setting behavior and variation in air content should be determined through studies of how such admixtures affect the cement to be used. All parties should then be alerted to this possibility at the start of a job and be ready to make adjustments in the addition rates of materials (particularly air-entraining admixtures) to achieve the specified properties of the concrete at the project site.

Specific guidance for use of accelerating admixtures, air-entraining admixtures, water-reducing and set-controlling admixtures, admixtures for flowing concrete, and admixtures for special purposes are given in the relevant chapters of this book. It should be emphasized that admixtures provide additional means of controlling the quality of the concrete, but they cannot correct the adverse effects caused by the use of poor-quality materials, unsatisfactory proportioning of the concrete mix and inappropriate procedures.

7.13.4 Mix proportioning using computers

Modern concretes are flowable, requiring little compaction; they can be placed in the depths of winter, under Arctic conditions and remain workable in hot arid conditions. The number of properties that have to be matched is greater and meeting such demanding performance dictates the use of a variety of supplementary cementing materials and often a blend of admixtures. Designing such concretes would require a understanding of the precise relationships between mix composition and properties at all ages. The usual method of ascertaining such information is through trial mixes. Although such trials are necessary and will remain so for a long time, computer simulation now offers a quick and safe way of designing a mix to give desired results at the lowest price. It serves as a guide to provide orders of magnitude and general trends[121, 122].

Two main stages are involved in computer simulation: (1) establishing mathematical models to determine precise relationships between mix composition and properties of concrete; (2) incorporation of these models into computer software, so that numerical optimization becomes possible. With a consistent set of such mathematical models, it is possible not only to design mixes but also simulate the effects of changes in the mix on the material engineering properties. The user has to define the components used in the simulations and may also fix the composition of the cement (normal Portland or blended cement), strength, the maximum size and nature of the aggregate (rounded or crushed), and the unit costs of the constituents. The computer then predicts the properties of any mix defined by the binder contents, admixture contents, coarse- and fine-aggregate ratio and free water content. Hundreds of batches may be simulated in a few minutes, so that an appreciation of the behavior of the concrete may be quickly gained. The list of rules of mix proportioning given below is an example of information derived from computer simulations[121].

- The change in slump due to the addition of water is non-linear, i.e. the slump is more sensitive to water for highly workable mixes.
- There is an optimal cement content with regard to slump, where the volume of cement exactly fills the voids created due to aggregate packing.
- If the concrete is too lean, adding a filler will allow the water content at constant slump to be decreased, so reducing the water–cement ratio and increasing the strength. As the porosity of a well-proportioned aggregate mix depends primarily on its grading curve, the critical cement content will be controlled by the aggregate. It will be higher for crushed than for round aggregate, and will decrease as the maximum size increases.
- Partial replacement of cement with fly ash improves the workability but reduces the strength up to 28 days. At 90 days the strengths attained are very similar to 100% Portland cement mixes.

- High dosages of a superplasticizer permit the replacement of a large amount of the cement by fly ash, while maintaining workability and strength. This is typical high-volume fly ash concrete.
- Slag in partial replacement essentially retards early-age strength development.
- Without a water-reducing admixture, condensed silica fume dramatically decreases the workability.
- Silica fume offers a greater increase in strength with superplasticizer.
- For normal-strength concrete, a larger maximum-size aggregate (MSA) reduces cement consumption.
- For high-performance concrete, the effect of MSA is not significant. Some researchers claim that a smaller aggregate size is preferable.
- Air entrainment is good for workability, but not for strength.
- Air entrainment reduces maximum achievable strength, although high performance is feasible.

7.13.5 Safety and hygienic aspects in the handling of admixtures

The national organizations for health and safety of most countries mandate the supply of information necessary to ensure the health and safety of personnel handling admixtures. Such information is usually contained in the hazards data sheet required to accompany shipments of admixtures. The information required includes the following[123]: nature of the hazard, e.g. caustic, inflammable, etc.; stipulated protective clothing, e.g. gloves, goggles, boots, etc.; toxicity and required first aid in the event of ingestion, prolonged exposure, skin or eye contact; action to be taken in the event of spillage, e.g. type of fire extinguisher if flammable, washing of slippery floors, etc.[124].

7.13.6 Admixture problems – limitations and incompatibility

Modern concretes often incorporate a mixture of chemical and mineral admixtures, each of which may interact with the various constituents of cements and influence cement hydration reactions. The admixture–cement interactions may in fact be viewed as the reaction between two complex chemical systems – the multicomponent, multiphasic inorganic materials in the cement and the organic compounds of multicomponent admixture systems. For example, lignosulfonate water-reducers are intrinsically complex mixtures of chemical compounds derived from the chemical degradation of lignin, while synthetic admixtures such as superplasticizers contain species with a broad distribution of molecular weights, reaction products, or other chemicals added for a specific purpose [125]. The performance of an admixture in concrete is highly dependent on many factors such as the nature and amount of admixture, the composition, specific surface and history of the cement, the nature and proportions of the aggregate, the sequence of adding water and admixture to the mix, the compatibility of the admixtures present in the concrete, the water–cement ratio and the temperature and conditions of curing.

It is important to understand admixture–cement and also admixture–admixture interactions so that optimum use of these materials can be made, admixture–cement incompatibility can be prevented, better troubleshooting of field problems is enabled, and the prediction of concrete properties is made possible. In the following pages some examples of problems that arise from admixture–cement and admixture–admixture interactions are cited, and an outline of the physicochemical concepts involved in the interference with cement hydration and interparticle interactions that limit admixture performance, and cause incompatibility and field problems is presented.

(a) Admixture–cement interactions

The overall process of cement hydration and setting results from a combination of solution processes, interfacial phenomena and solid-state reactions which lead to the formation of complex products. Some of the hydration products formed from the different mineral components of cement are shown in Table 7.31 [125]. Admixture–cement interactions are essentially interactions between admixtures and the initially formed cement hydrates; the influence of admixtures on cement hydration is best considered by reference to the evolution of the reaction with time. Five stages can be identified [125, 126]:

I.   Initial hydration processes (0–15 min).
II.   Induction period or lag phase (15 min–4 h).
III. Acceleration and setting (4–8 h).
IV.   Deceleration and hardening (8–24 h).
V.   Curing (1–28 days).

When water comes into contact with the cement (stage I), wetting of the highly hygroscopic cement particles and the solubilization of a variety of ionic species, e.g. Na^+, K^+, Ca^{++}, SO_4^{--}, OH^-, by complete or selective solubilization (surface hydrolysis) of the various phases present in cement occurs. Surface hydrolysis quickly leads to the formation of a thin layer of both amorphous and gel products. Beyond the initial solubilization phenomena, the formation of any of the solid hydration products (Table 7.31) will be governed by nucleation processes which may occur 'homogeneously' from the solution phase or 'heterogeneously' at a solid–solution interphase [64, 125].

Subsequent to nucleation, hydration products will grow at a rate determined by the bulk concentration of the solution species, the availability

Table 7.31 Hydration products for mineral components of Portland cement (Jolicoeur et al. [125])

Component	Typical content in OPC (wt%)	Hydration product
C_3S	55	C–S–H, CH
C_2S	20	C–S–H, CH
C_3A	6	Monosulfate ettringite
C_4AF	9	C_6AFH_{12}
$CaSO_4$ (soluble anhydrite)	—	$CaSO_4 \cdot 2H_2O$
$CaSO_4$ (insoluble anhydrite)	5	—
$CaSO_4 \cdot \frac{1}{2}H_2O$	—	$CaSO_4 \cdot 2H_2O$
Na,KSO_4	1	—
CaO	—	CH
Typical OPC	—	C–S–H, CH Monosulfate ettringite

of water and ionic species at the reaction sites (i.e. diffusion through the solution or reaction boundary), the activation energy for the molecular processes and the orientational requirements involved in the crystallization. As might be expected, the presence of admixtures which can interfere with nucleation and/or the growth processes will influence the hydration reaction rate, the reaction products, or both.

In the latter part of stage I, the cement particles in the paste become fully coated with a layer of hydrate products. This 'protective' layer hinders the diffusion of the reacting species in and out of the reaction interphase, thus sharply reducing the rate of the various reactions. The system enters into a period of 'latency' referred to as the induction or dormant period; processes initiated during stage I, however, will continue throughout the induction period.

In the early part of stage II, reactions of the aluminate phase will predominate. It is at this stage that the SO_4^{--} concentration plays a predominant role in the growth of ettringite crystals and the setting of the paste. If the SO_4^{--} concentration is too low, excessive nucleation and growth of C-A-H products may occur, producing flash set; if the SO_4^{--} concentration is too high (due to the presence of hemihydrate and alkali sulfates) prodigious nucleation and growth of gypsum crystals (Ca-$SO_4.2H_2O$) occurs, resulting in false set. In the presence of adequate SO_4^{--} content and availability, several physicochemical processes occur: continued growth of ettringite crystals; increased production of C-S-H gel (which coats the initial aluminate-rich gel layer); increased concentrations of Ca^{++}, OH^- in solution; and the development of osmotic and mechanical pressures as the hydration front moves inwards in the cement particle [127].

The above mentioned processes will determine the rheological and setting characteristics of the system and the interaction of a chemical admixture with any of the reactive species, or its interference with diffusion, nucleation and growth processes can significantly influence the behavior of concretes during the induction period.

Near the end of the dormant period, the rate of cement hydration increases sharply. The onset of the acceleration period has been attributed to the following effects [125]:

- Disruption of the hydrate protective layer by physicochemical transformations of the hydrates.
- Breakdown of the protective layer by osmotic pressure effects.
- Nucleation and growth of C-S-H products.
- Nucleation and growth of calcium hydroxide $Ca(OH)_2$.

These processes may be affected by chemical admixtures, particularly the formation and properties of the protective layer. Also, admixtures remaining in the pore solution may further influence nucleation and growth of the hydration products, causing volume expansion, outward mechanical pressure on the protective gel layer and its subsequent disruption.

The admixture may remain in a free state as a solid, in solution, interact at the surface, or chemically combine with the hydrates. Physicochemical and mechanical properties of the concrete may be influenced by the type and extent of the interaction. Thus, the early hydration reactions of cement may be affected in diverse ways and it is possible that more than one effect occurs at the same time. These are summarized below [125].

- Chemical interference with hydration reactions and or physical interaction with the hydration products may occur, resulting in the alteration of the rate of hydration of cement constituents, or in the composition and morphology of the hydrated products formed.
- The admixture may react with cement constituents to precipitate insoluble products, and these may form slightly permeable films or coatings on the cement grains, acting as protective barriers with respect to further hydration.
- The admixture itself may also form a similar protective coating by adsorption on the surface of the cement particles.
- The hydration reactions may also be influenced by changes in the pH value of the solution in contact with hydrated cement, which may alter the solubility or stability of some hydrated cement compounds or inhibit the formation of protective coatings.
- Since solution phenomena are dominant in the very early stages, the influence of various admixtures on early hydration reactions may be reflected by changes in the composition of the liquid phases in contact

with hydrated cement, e.g. only extremely small amounts of Al_2O_3, SiO_2 and other oxides have been found to be present in the aqueous phase, and it appears that these oxides can be disregarded in normal cement paste liquid. However, when an admixture is also present in solution there is the possibility of appreciable concentrations of these oxides being produced, e.g. organic hydroxy compounds are known to form soluble complexes with aluminum and iron oxides (Al_2O_3, Fe_2O_3).

(b) Admixture–cement incompatibility

Admixtures that modify the properties of fresh concrete may cause problems through early stiffening or undesirable retardation of the time of setting. Early stiffening often is caused by changes in the rate of reaction between tricalcium aluminate and sulfate. Retardation can also be caused by an overdose of admixture or by a lowering of ambient temperature, both of which delay the hydration of the calcium silicates. As more chemicals are added to the concrete mix, compatibility becomes the central parameter governing selection. Side effects or reactions between chemicals due to the sequence of addition, cement type, temperature change and batching equipment can all affect performance.

In the following sections, physicochemical concepts which apply to admixture–cement interactions and the manner in which chemical admixtures can interfere in the cement hydration process and in particle–particle interactions are described.

EFFECT OF TYPE OF GYPSUM AND GYPSUM HEMIHYDRATE OR SOLUBLE ANHYDRITE RATIO (ABNORMAL SETTING AND LOSS OF WORKABILITY)

The relative reactivity of the different mineral phases of cement with water is usually given as $C_3A > C_3S > C_2S > C_4AF$. Aluminate phases and their hydration products therefore play an important role in the early hydration process. Because of the high reactivity of calcium aluminate, the aluminate hydration reaction is carried out in the presence of sulfate ions. The latter provide control of the reaction rate through the formation of mixed aluminum sulfate products (ettringite and monosulfoaluminate) Calcium sulfate which is added to the cement clinker hence controls the properties of the aluminate hydration products. Sulfates thus play a crucial role in cement hydration and the influence of chemical admixtures on any process where sulfates are involved may be expected to be significant [127].

In order for normal set to occur in Portland cement paste, mortar, or concrete, calcium sulfate must be present in the cement–water system. In today's cements, most of the calcium sulfate introduced into the system as a component of the cement, can be present in one or more forms: gypsum ($CaSO_4.2H_2O$), hemihydrate ($CaSO_4.\frac{1}{2}H_2O$), soluble anhydrite (or natural anhydrite ($CaSO_4$)). It is very important that the calcium sulfate be sufficiently soluble in the cement–water aqueous phase to provide calcium and sulfate ions for the formation of calcium aluminosulfate (ettringite – $3CaO.Al_2O_3.3CaSO_4.32H_2O$) and prevent, the tricalcium aluminate phase of the cement from reacting directly with water, which results in flash set[125].

Flash set is distinguished from false set in that (1) it evolves considerable heat and (2) the rigidity of the mix cannot be dispelled by further mixing without addition of extra water. False set is also a form of abnormal setting of cement, caused by the presence of hemihydrate or anhydrite. Dehydration of gypsum when it is interground with too hot a clinker can produce these two forms of gypsum in the cement [128, 129]. When such a cement is mixed with water the hemihydrate and anhydrite hydrate to gypsum. A plaster set takes place with a resulting stiffening of the paste.

Reports of incompatibility between Portland cement and chemical admixtures have increased over the past 10 years. Rapid set, accelerated stiffening, increase in time of set, lack of water-reduction are some of the reported problems. Such effects produced by the interaction of the composition of the cement and that of the chemical admixtures has been often referred to as 'cement–admixture incompatibility'. This is discussed below.

The use of natural anhydrite ($CaSO_4$) as a substitute or partial replacement for gypsum ($CaSO_4.2H_2O$) in the manufacture of Portland cement is gaining in popularity. Although Portland cement containing large amounts of natural anhydrite will perform normally in the absence of admixtures, in many instances, however, the use of chemical admixtures with such cements produces problems. Work done by Dodson and Hayden [130] showed that all of those cements used in concrete which exhibited incompatibility had one common denominator, the presence of natural anhydrite. The natural anhydrite, whether interground purposely or as an impurity in gypsum, with Portland cement clinker can produce rapid set in paste, mortar or concrete in the presence of calcium lignosulfonate (CLS). The results of this study also indicate that when the cement contains a ratio of gypsum to natural anhydrite less than 2, rapid stiffening will occur when chemical admixtures are added to concrete. A similar effect is noted in the cement containing the mixture of gypsum and natural anhydrite.

The soluble sulfate versus time curve for the cement containing the natural anhydrite is radically changed when CLS is present [130, 131]. The rate of solution of natural anhydrite, which is much slower than that of gypsum or calcium sulfate hemihydrate, is further retarded in the presence of chemical admixtures, which leads to a 'sulfate-starved' system in the concrete, often producing rapid set and an increase in rate of concrete slump loss (Fig. 7.42). Apparently the adsorption of the lignosulfonate by the natural anhydride plus the rapid reaction between the soluble SO_3 and the tricalcium aluminate keeps the SO_3 level near zero concentration for the first 10–20 min causing rapid stiffening and quick set (approximately 15 min). It is also reported that chemical admixtures based on salts of hydroxylated carboxylic acids as well as carbohydrates (sugars, corn syrups) act very much like CLS when added to concrete containing natural anhydrite[128].

The uncontrolled variation in the gypsum–hemihydrate or soluble anhydrite ratio (G-H or A) that occurs during production, storage and transportation of cement can cause compatibility problems with certain superplasticizers, particularly a reduction in workability. The extent of the reduction on workability properties is dependent on the G-H ratio (ranging from 80: 20% to 20: 80%), level of C_3A and alkalis present in the cement and the fineness of the cement. For the most reactive type of cement with a high content of both C_3A and alkalis, a reduced G-H or A ratio affects the yield stress, while plastic viscosity is not much affected. These effects are much less pronounced for a less reactive cement, but with the lowest G-H or A ratio, false set can occur [130–132]. An increased fineness of cement increases the effect of the G-H or A ratio. The effect has been found to be most pronounced with melamine-based superplasticizers [130, 133].

INFLUENCE OF SOLUBLE ALKALIS ON THE STABILITY OF THE AIR-VOID STRUCTURE

Water-soluble alkalis in Portland cement or fly ash can have a harmful effect on the characteristics of the air-void structure (AVS), particularly the stability of the spacing factor. Because of the more restrictive environmental

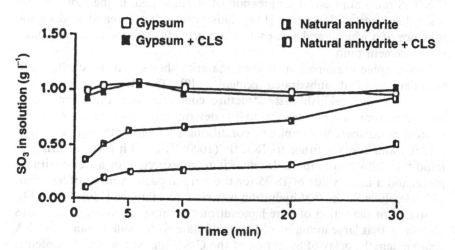

Fig. 7.42 SO_3 solubilities of gypsum and natural anhydrite in saturated lime water with and without CLS admixture (Dodson [127]).

regulations affecting the manufacture of cement, alkali contents of cements have risen in some regions. As with cement, increasing use of fly ashes that may contain significant amount of soluble alkalies increases the likelihood of related AVS problems. This alkali-induced breakdown of the AVS makes concrete more vulnerable to freeze–thaw damage.

Although previous investigations [132, 134, 135] have shown that increased soluble alkalis could have an important influence on the AVS, more recent work [136, 137], however, indicates that soluble alkalis *per se* do not influence the production of the AVS. It is only when the sulfate level (inherent or by extraneous addition) is high that the AVS production is difficult. An increase in the soluble-alkali content of cement has been found to improve significantly the stability of AVS, particularly in concrete mixtures to which a superplasticizer (SP) is added [138].

SULFATE-RESISTANT CONCRETE (SUPPRESSED STRENGTH DEVELOPMENT)

Low dosages of CLS-based water-reducing agents (WRAs) and SP are generally used in concretes made with Type V cement because of the low C_3A contents in the cement. Problems arise in the field when workers trying to cope with placing and pumping difficulties add further WRA and SP dosages without knowledge of the synergistic effect produced on the concrete's plastic properties by admixture combinations, particularly with Type V cement. In some instances the addition of higher than normal WRA dosages or a combination of WRA and SP in air-entrained concrete containing Type V cement with unusually low C_3A (<1%), low C_3S and C_3S/C_2S ratio can lead to suppression of strength and, in the more drastic cases, the arrest of hydration [139]. This latter effect is manifested by the presence of a whitish and soft paste structure reflecting the effect of a high water–cement ratio.

Petrographic examination of such material shows it to be chalky with abundant residual unhydrated cement [140]. The degree of hydration achieved in such whitish paste structure compared with the normal gray paste structure can be ascertained by determining non-evaporable water content on comparable samples by conditioning to constant weight at 230°F (110°C) and furnace firing to 1832°F (1000°C) for 4 h [140]. Bearing in mind that 20% non-evaporable water is to be expected for a well-hydrated paste, and a mean value of 18.25 for the normal paste, a mean of less than 13% clearly indicates that hydration was severely inhibited [139] (Fig. 7.43).

Studies of the action of pure lignosulfonate acting on pure C_3A and C_3S also show that large amounts of lignosulfonate with small amounts of C_3A result in lengthy delay of hydration of the C_3S [130]. At lower C_3A contents, smaller amounts of the WRA are adsorbed, leaving larger amounts of the admixture to affect the C_3S component. Alkalis may affect dissolution and

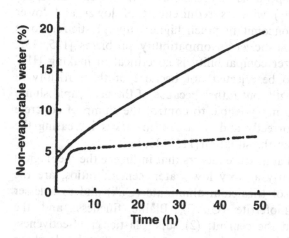

Fig. 7.43 Non-evaporable water content of concretes with (— · — ·) and without (— —) admixtures (Johnston [140]).

interaction reactions. Thus, there is some evidence to suggest that, in a cement with very low C_3A combined with low (but not unusually low) alkalis, a high dose of lignosulfonate may create the conditions for permanently suppressed strength development caused by inhibited hydration of the C_3S. This appears to be a case of the 'cement killer effect' of too much lignosulfonate being left free to act on relatively little C_3S as a consequence of an unusually low C_3A content.

The above example demonstrates the inadequacies inherent in present cement and admixture specifications when cement compositions chemically on the fringe of normal requirements are combined with high doses of chemical admixtures. Minimum as well as the present maximum limits on cement compound composition in ASTM C 150 might best address this problem, but a requirement in ASTM C 494 that admixture compliance be established for the job cement when it is not a Type I or Type II cement might also be useful[138].

HIGH-PERFORMANCE CONCRETE (CEMENT–SUPERPLASTICIZER COMPATIBILITY)

With some cement–superplasticizer combinations, various problems have been reported such as low fluidification effect, rapid slump loss, severe segregation, extended set retardation and loss of entrained air. These are briefly reviewed below.

Loss of workability in high-performance concrete The advent of high-performance concrete (HPC) with its requirement of lower and lower water–cement ratios and concomitant much higher superplasticizer doses accentuated cement–superplasticizer incompatibility problems [125, 132, 135]. Cement–superplasticizer compatibility is so critical in making HPC that some cements have to be rejected not because of the difficulty in achieving the required strength, but rather because of the very rapid slump loss. In some instances it is not possible to control the slump of concrete long enough to place it correctly and in a few instances increasing the admixture dosage did worsen the situation [125, 135, 137].

The principal cement and admixture factors that influence the rheological behavior in HPC, particularly at very low water–cement ratios, are the following: (1) the quantity of tricalcium aluminate (C_3A) and to a lesser degree tetracalcium aluminoferrite (C_4AF), Blaine fineness, and the solubility of the sulfates in the cement; (2) superplasticizer effectiveness relating to the position of the of the sulfonate groups in the molecule of the main resin (naphthalene, melamine, lignosulfonate), the content of monomers and lower-molecular-weight fractions, the length of the polymer chains, the extent of cross-linking, the amount of residual sulfate present and the nature of the counter ion that has been used in the neutralization [125, 133, 139]. In addition to their physical role of cement particle dispersion, these admixture properties also interfere with cement hydration kinetics and the solubility of gypsum. Since the availability of sulfate ions in the interstitial solution directly influences hydration, gypsum solubility is of critical importance to concrete rheology. As mentioned previously, excess sulfate can result in false set, while insufficient quantity can cause flash set.

The calcium sulfate content of Portland cement has been regulated to comply with national standards, primarily on the basis of rheological and strength behavior of standard pastes and mortar made at water–cement ratios (W/C) of approximately 0.5. This practice has provided satisfactory results for the most commonly used concretes at W/C >0.4 and having compressive strength in the range of 20 to 40 MPa. Superplasticized HPCs, however, are characterized by much lower W/C (0.2–0.4) with compressive strengths ranging from 50 to 100 MPa. In such concretes, the rate of solubility of calcium sulfate present in the cement becomes a critical factor in controlling rheological behavior, since they contain less water to accept sulfate (SO_4^{--}) ions and yet more C_3A that must be controlled [125, 141].

Segregation of the mix This mix can occur when larger doses of the admixture are added to offset the lack of initial fluidizing of the concrete or to reinstate workability. The segregation phenomenon is always related to a superplasticizer overdose, beyond the saturation point. In some instances the surface of the hardened concrete is covered with a white layer composed of lime, calcium sulfate and some calcium carbonate. The saturation point varies for different cement types and cement blends, e.g. Type V and blends of fly ash and Portland cement, require much less admixture to produce the fluidizing effect than Type I. This is due to the smaller amount of C_3A contained in Type V cement compared to Type I and the C_3A dilution effect in blended mixtures. A mix showing signs of segregation can be recovered by the addition of additional quantities of cement to the mix to consume the excess of superplasticizer [125, 142].

Destabilization of the air system in superplasticized concrete   The addition of a superplasticizer (SP) modifies the relationship between the air content and the spacing factor and sometimes destabilizes the air-void system significantly (Fig. 7.44). Low dosages of SPs appear to be less problematic than higher doses [136]. It is reported [130, 133, 135] that in field concretes containing certain naphthalene-based SPs combined with AEAs, the total volume of air decreases after initial mixing (75 seconds) and the air content of the hardened concrete is reduced to less than half of that measured in the fresh state; low specific surface and high spacing factor are also produced. The possible cause postulated for this decrease is the escape of air from the fluidized concrete and the coalescence of the air bubbles. However, a study comparing the performance of a combination of two types of SPs with an AEA shows that such adverse effects were not produced when a melamine-based SP was used [135]. A much better and more stable AVS is produced using the latter SP.

From the ready-mix producer's point of view, considerations about instability of the air-void system must be linked to a practical measurement of the air content. This raises the question as to how unacceptable spacing factor values can be detected from air content measurement. Since each combination of cement and AEA will produce its own values and the addition of SP will modify the relationship between the air content and air-void spacing, the only way to assess the stability of the AVS in superplasticized concrete is to carry out preliminary test for each cement–AEA–SP combination. It is probable that with several tests of a specific combination of cement–AEA–SP, a relationship between the air content and spacing factor could be estimated with a reasonable degree of precision. But any change of AEA type or brand of cement will make necessary a new evaluation of this relationship.

Abnormal retardation of superplasticized mixes   The factors responsible for extended retardation of superplasticized mixes are closely related to the Blaine surface area and to the chemical composition (mainly the C_3A content) of the cement used to produce it; it is also dependent upon the SP dosage, the prevailing ambient and concrete temperature. Low-C_3A cements like Type V are susceptible to extension of their set even at dosages that are considered normal for other cements with higher C_3A contents [125, 137, 139]

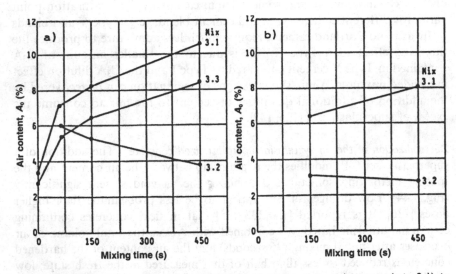

Fig. 7.44 Effect of mixing time after addition of air-entraining admixture (mix 3.1) in combination with superplasticizer (mix 3.2) on the total air content in (a) fresh concrete and (b) hardened concrete (Mark and Gjoerv [131]).

When confronted with cement–SP compatibility problems, the following action could be taken to identify whether the problem is primarily due to the reactivity of the cement or the poor performance of the admixture [125, 140]:

- Check the reactivity of the cement–C_3A content, type of gypsum (e.g. hemihydrate, anhydrite, the dihydrate to hemihydrate ratio) present and Blaine fineness.
- Determine performance of the cement–SP combinations by cross-testing with other SP and cements.
- Try small doses of a hydroxycarboxylic-based retarder.
- Try blending the cement with fly ash or slag.
- Finally, if none of these measures proves fruitful, the W/C should be progressively increased while varying the dosage of the SP to achieve the desired workability and required strength.

(c) Admixture–admixture incompatibility

UNDERWATER CONCRETE – COMPATIBILITY PROBLEMS WITH VISCOSITY ENHANCING ADMIXTURES (VEAs)

Combinations of VEAs, WRAs and SPs are used in controlling sedimentation in highly flowable slurry like concrete and in underwater concrete used in repair. Although many of the polymers shown in Table 6.1 can be used to increase the viscosity of concrete, many of them are not capable of maintaining the pseudoplastic flow behavior for extended periods. The key difference between the polymers that provide pseudoplastic flow behavior for extended periods and those that do not is the compatibility with Portland cement. The former admixtures are tolerant of high salt concentration and a wide pH range [73]. These admixtures can be further differentiated by the stability of the paste viscosities to shear conditions, variations in temperature and compatibility with SPs.

Many VEAs cause workability and set retardation problems because of the mode of action by which they provide increased viscosity of the concrete. Most tend to increase the viscosity by forming gel cement slurries, rather than increasing the viscosity of the water in the mix [73]. The VEA molecules of these materials compete with the SP for active sites on the cement particle (Fig. 7.45) and some have surfactant properties with a potential to entrain air. Consequently, compatibility problems such as drastic set retardation, marginal resistance to sedimentation, increased air contents, reduced strengths and the production of concrete consistencies that cannot be pumped or mixed easily (because its viscosity increases with shear) can occur. Increased cohesion which impairs significantly workability and strength reduction occurs when the dosages of these admixtures exceed 0.035%, particularly with high-alkali cements [78, 143]. Compatible materials, on the other hand, are not affected by the presence of an SP and are able to build sufficient viscosity at low shear rates to limit separation without making mixing, placing and pumping operations difficult.

The effect of the interaction between two types of SPs (naphthalene formaldehyde sulfonates–NFS and melamine formaldehyde sulfonate–MFS) and the VEA on the fluidity of mortar is seen in Fig. 7.46. In the case of NFS, the mortar flow decreases with increasing addition of the VEA. This result indicates an appreciable interaction between NFS and the VEA. On the other hand, the fluidity of the mortar containing the MFS does not change. Measured amounts of the SPs showed that there was good correlation between the amount of SP adsorbed and observed fluidity.

DAMP-PROOF CONCRETE (COMPATIBILITY PROBLEMS WITH WRA/DAMP-PROOFING ADMIXTURE COMBINATIONS)

Damp-proofing admixtures include soaps and fatty acids which react with the cement hydrates to modify workability, bleeding and settlement, air content, compressive strength and durability characteristics. Mix proportions, mix consistency, admixture dosage and poor mixing influence the effects produced by the admixture. In cement-rich mixes void content is often increased, resulting in increased permeability. Since the admixture

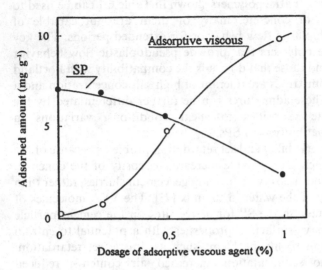

Fig. 7.45 Adsorption of the superplasticizer and viscosity-enhancing admixture molecules on the cement particles (Yammuro et al. [73]).

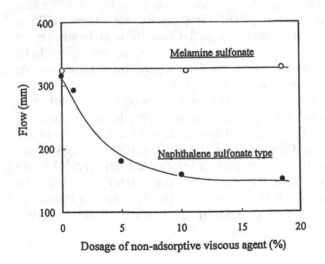

Fig. 7.46 Effect of interaction between superplasticizer and viscosity-enhancing admixture on flow value (Yammuro et al. [73]).

produces air, mixes with high fines content may promote air entrainment. Higher workabilities produce a frothing action with soaps, waxes and bituminous emulsions, particularly if large admixture doses are used. The use of higher than recommended dosages results in reduced density, strength and watertightness [82, 83].

When fatty-acid or wax emulsion type damp-proofers are used in conjunction with lignosulfonate or hydroxycarboxylic-based admixtures, heavy air entrainment results with attendant strength reduction. Both these effects decrease waterproofing characteristics. At higher workabilities, lignosulfonates counteract the reduced bleeding effect obtained with waxes and increase bleeding rates[83].

SHRINKAGE-COMPENSATING CONCRETES

Superplasticizers and accelerators when combined with expansion producing admixtures have been shown to reduce expansive potential of calcium sulfoaluminate (CSA)-based admixtures significantly [144–146]. The magnitude of the effect is dependent on whether the admixture retards or accelerates set. Retarders tend to increase ultimate expansion at normal temperatures [145]. Under hot-weather conditions, however, retarders offset the accelerating effects of high temperatures and allow the normal level of expansion to occur. Air-entraining agents do not influence the expansive reaction, although higher air contents may result when a CSA is used in air-entrained concrete. The inclusion of silica fume in grout compositions made with Type K cement or CSA type expansive agents may reduce expansion.

The silica fume is said to decrease the formation of ettringite by reducing the concentration of calcium (Ca^{++}) and hydroxyl (OH^-) ions involved in the formation of ettringite [146]. The expansion of lime-based admixtures is not affected by water-reducing admixtures to the same extent as observed for CSA-based admixtures. However, set retardation may result due to the increased amounts of calcium hydroxide produced in the presence of the WRA.

CORROSION MITIGATION IN REPAIR CONCRETE MIXES (ACCELERATED SETTING IN CONCRETES CONTAINING CALCIUM NITRITE AND SUPERPLASTICIZERS)

Calcium nitrite is mainly used as a corrosion inhibitor, but its ability to accelerate set and strength is often overlooked. Consequently, the incidence of rapid setting of very low-water–cement-ratio HPC repair mixes is not infrequent. Calcium nitrite is both an accelerator and a corrosion inhibitor which accelerates set time significantly particularly in superplasticized low-water–cement-ratio concrete mixes at dosages exceeding 5% by weight of cement. When such mixes are used at ambient temperature in the range 20–25°C, it is prudent to incorporate a retarding admixture in the mix.

SILICA FUME CONCRETE (DECREASED AIR CONTENTS IN SILICA FUME CONCRETE)

The replacement of cement with silica fume results in reduced air contents in the fresh concrete compared to control concrete containing no silica fume. Achieving the required air content necessitates an increase in the AEA dosage, but the AEA demand decreases in the presence of a WRA or SP. There is also an evident increase in SP and WRA amounts required to produce desired modifications in the presence of silica fume. The mechanisms responsible for such effects are described as follows.

The air-bubble generating and stabilizing process requires a minimum paste consistency. Silica fume particles are smaller than those of Portland cement and addition of silica fume therefore increases the fine fraction of the particles. The higher fraction of smaller particles then increases the surface area causing a greater binding of the water in the mix. This removes the water required for the bubble-generating process.

The basic mechanism of plasticizing effect on fresh cement mixes is explained by forming a temporarily stable double layer on cement particles. Since the formation of the double layer is also connected with the surface of particles, the increased demand for AEA, WRA and SP, in silica fume concrete and the decreased demand for AEA in the presence of a WRA or SP can be directly correlated to the specific surface increase of silica fume–cement blends and the dispersing action of WRAs and SPs in achieving a mortar consistency that enables the air-bubble-generating and stabilizing process [147, 149]. Concrete producers are now cognizant of the effect of these factors. Field silica fume concrete with a satisfactory, stable air-void system can therefore be produced consistently.

(d) Problems caused by admixture side effects and poor practice

ABRASION-RESISTANT CONCRETE (SURFACE FINISHING PROBLEMS ASSOCIATED WITH AIR-ENTRAINED AND RETARDED CONCRETE)

For floor construction requiring dry-shake surface hardeners, the air entrained by lignosulfonate water-reducers may cause finishing problems, such as blistering of the finished surface. The slow bleeding, resulting from entrained air in the concrete, increases the chances of blistering when a hard-trowelled surface is required. Even small amounts of entrained air can cause blistering when a dry-shake surface hardener is used. Because the freshly placed concrete bleeds slowly, the finishers are likely to start finishing operations prematurely and seal the surface before bleeding has stopped. Consequently, the use of a water-reducer that entrains more than 2% air should be avoided for heavy-duty industrial floors.

The customary approach to the timing of finishing operations (of waiting until the bleed water has evaporated from the surface) can cause problems if applied to air-entrained concrete slabwork because, as previously stated, air-entrained concrete undergoes little bleeding. Often workers wait for bleeding to appear until the concrete has set up to an extent that makes finishing difficult or impossible. In general, finishing should always commence sooner with air-entrained concrete. Early finishing is even more important when the air temperature is high, humidity is low or a wind is blowing.

Retarding water-reducers particularly hydroxylated carboxylic acids and salts delay initial setting by $1\frac{1}{2}$ h. Delayed setting helps keep concrete workable during hot weather, but under dry conditions surface crusting may occur in flatwork. Bleed water disappears and the surface appears to be ready for finishing. Because the underlying concrete has not set, however, the surface is rubbery and feels spongy. When this happens, power tools may cause waviness and tearing during floating and finishing. Prevention of surface drying or switching to a water-reducer that does not retard as much will help control this problem [148, 149].

HOT-WEATHER CONCRETING (PLASTIC SHRINKAGE IN FLAT-SLABS DUE TO THE USE OF RETARDING ADMIXTURES)

The contraction that occurs in the prehardening stage as fresh concrete dries out is known as plastic shrinkage. Plastic shrinkage may cause cracking during the first few hours after the concrete has been placed, usually at a stage when its surface becomes dry. Such cracks are characterized by a random map pattern but sometimes they develop as diagonal cracks at the edges of slabs or along the reinforcement, particularly when the reinforcement is close to the surface [142]. Plastic shrinkage depends not only on the intensity of the drying but also on the stiffness of the mix and the length of time it takes the mix to set, i.e. the stiffer the mix, and shorter the setting time, the lower the expected shrinkage under the same conditions. Thus, plastic shrinkage is determined by the net effect of the environmental factors on both the rate of drying and rate of setting [141].

In view of the above discussion, it may be expected that the use of set-retarding admixtures will increase plastic shrinkage and, indeed, this is confirmed by data in Fig. 7.47, which compares the shrinkage of retarded and non-retarded cement mortars. An increased plastic shrinkage is associated with an increased risk of plastic cracking. Hence, the use of retarders should preferably be avoided under environmental conditions such as hot dry weather, which favors high plastic shrinkage. The use of a water-reducing admixture with some retarding characteristics has been found to be more effective in providing the necessary set extension and maintenance of workability under hot-weather conditions but without the attendant cracking observed with retarding admixtures [94]. This fact is of practical

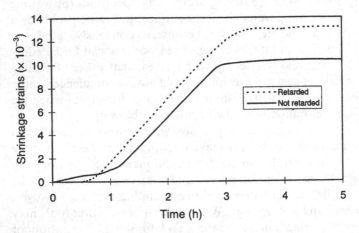

Fig. 7.47 Plastic shrinkage of retarded and unretarded cement mortars of plastic consistency and OPC content of 550 kg m^{-3}. Air temperature 30°C, wind velocity of 20 km h^{-1} and IR irradiation (Soroka and Ravina [94]).

importance because the use of retarders (particularly hydroxycarboxylic-acid-based materials) under hot dry conditions in order to counteract the accelerated effect of such conditions on slump loss in fresh concrete is widely practised.

It has been suggested that the mechanism by which plastic shrinkage occurs in concrete involves tensile stresses which develop in the capillary water when menisci are formed in the capillaries of the concrete surface on drying. This suggested mechanism is supported by experimental data which demonstrate the relation between shrinkage and capillary pressure [150]. Recently developed shrinkage-reducing admixtures (SRA) which appear to be effective in reducing the tensile stresses in the capillaries and also provide good water-reduction with some retardation show promise and their use under hot-weather concreting may prove fruitful.

EFFECT OF ADDITION PROCEDURE (COMPATIBILITY OF CONVENTIONAL ADMIXTURES ON THE AIR-VOID SYSTEM)

In spite of the decades of use of air-entraining agents (AEA) for the improvement of frost resistance of concrete, there has been a significant increase in reported cases of poor performance of concrete exposed to freezing and thawing, particularly where de-icing salts are used. Examination of specimens from a large number of concrete structures showed that only about half of the structures had an air-void system (AVS) that met the requirements for adequate frost resistance [147, 148]. Experience indicates that the AVS in fresh concrete is unstable during mixing, transportation, retempering, compaction and finishing. Furthermore, the combination of two or more mineral and chemical admixtures in the same concrete mix makes compatibility of the different admixtures and the procedure of addition key parameters that need to be controlled. The effects of these parameters on the AVS of field concrete are reported below and they clearly show the importance of using a compatible combination of materials.

When a lignosulfonate water-reducer (WRA) is also added to the concrete containing an AEA, a substantial increase in the air content occurs compared to when the AEA is added alone. This happens whether the AEA and WRA are added separately with the mixing water or the WRA is delayed relative to the AEA addition. Despite the doubling of the air content, the specific surface of the bubbles is found to be substantially reduced. Thus, the presence of the lignosulfonate WRA appears to give a poorer AVS. The presence of the WRA also produces a more unstable AVS and the time of addition of the WRA also affects the stability. When the addition is delayed, the stability of the AVS is further reduced. These adverse effects are mainly due to the sugars and other contaminants present in commercial lignosulfonates and superplasticizers [125].

RETEMPERING OF CONCRETE (EFFECTS ON AIR-VOID SYSTEM)

Concrete often arrives on site more than half an hour after mixing. The period for completion of placement operations can range from 15 to 60 min depending on the size of the load and field conditions. When the slump decreases to an unacceptable level during operations, water is added to the mix and very often field inspectors will tolerate 'reasonable' retempering, i.e. enough extraneous water to increase the slump by 50–60 mm. If the air content of the fresh concrete is below the required minimum percentage when it arrives on site, it is also common to add an amount of AEA – just enough to reach the minimum acceptable value.

The positive influence of an AVS with an adequate spacing factor on freeze–thaw durability of concrete is well established and specified in North American and Scandinavian standards. Since the spacing factor is measured on hardened concrete, ready-mix producers and contractors need to know what materials and procedures affect the stability of the AVS. Does retempering 30–60 min after initial mixing have an effect on the AVS (particularly the spacing factor) and how much AEA must be added during retempering to increase the air content and decrease the spacing factor below the 200 μm limit? Some of the data from both lab and field tests are as follows [136, 147, 148, 151]:

- Using 2–5% of the original dosage of AEA is sufficient to increase the air content to the original level and provide a suitable spacing factor [136].
- Adding water to increase the slump by an average of 25 mm had little influence on the spacing factor.
- Air-content variations do not necessarily correspond to spacing-factor variations and air-content variations of the order of 1–2% should not be used to predict spacing-factor variations.
- Air content of field mixes varies very little for mixing periods between 10 and 25 min and retempering with water to increase the slump from 50 to 100 mm has no significance on the air-void spacing factor, although it results in a small increase in the air content.
- It is possible to correct an unsatisfactory AVS by adding more AEA after 45 min, but to lower significantly the value of the spacing factor, the quantity of the admixture that is added must be significant (>30–50% of the normal dosage) to cause a marked increase in air content.
- Neutralized vinsol resin, sulfonated hydrocarbon, and fatty-acid (detergent)-based AEAs give satisfactory AVSs, but at low dosages, only synthetic detergent-based materials can produce an adequate and stable spacing factor.

7.14 Batching and dispensing of admixtures

In concrete construction operations, one of the most important factors affecting the quality of the finished product is the use of appropriate methods of preparation and batching of admixtures. Achieving a consistent admixture dosage in subsequent batches plays a pivotal role in minimizing the variation of the concrete produced and neglect in these areas may affect properties, performance and uniformity of the concrete significantly.

Most of the currently marketed admixtures furnished as ready-to-use liquids are added with the gauging water after the ingredients have been mixed together. First- and second-generation superplasticizers (or high-range water-reducers), which are added just prior to discharging of the mix containing the full volume of water, are exceptions to this general rule. Certain admixtures, such as pigments, expansive agents and pumping aids, are used in extremely small dosages and most often are batched by hand from premeasured containers. Other hand-added admixtures may include accelerators, permeability reducers, and bonding aids, which often are packaged in amounts sufficient for proper dosage per unit volume of concrete. Powdered admixtures are usually spread on the dry batch immediately prior to mixing. Calcium chloride, however, must be dissolved in water before addition to the mix.

In North America and certain Scandinavian countries, the severe winter conditions require the use of both water-reducing and air-entraining admixtures in the same concrete mix. Modern-day high-performance concrete mixes also incorporate two or more admixtures. These admixtures are introduced into the concrete mixture at the concrete plant or into a truck-mounted admixture tank for introduction into the concrete mixture at the job site. Although measurement and addition of the admixture to the concrete batch or into the truck-mounted tank often is by means of a sophisticated mechanical or electromechanical dispensing system, a calibrated holding tank should be part of the system so that the plant operator can verify that the proper amount of admixture has been batched into the concrete mixer or into the truck-mounted tank.

7.14.1 Manufacture

The quantity of an admixture used is small relative to other constituents of the mix but its effect depending on the amount used and the time it is added during the mixing cycle may be quite potent. Therefore, admixtures should be prepared to assure the user of a consistent concentration of the involved chemicals. Proprietary admixtures are generally furnished as ready to use liquids which require no further preparation at the location of use. Indeed, site preparation with these products is strongly discouraged by their manufacturers. Certain admixtures, however, are exceptions to this rule and these are described below:

1. Admixtures supplied in concentrated form to minimize high freight costs that would otherwise be incurred in the transport of significant amounts of water.
2. Flake calcium chloride that is dissolved in water at the ready-mix plant by some concrete producers.
3. Water-soluble powder admixtures which require mixing at the job site or point of use. Such job mixing may require that low-concentration solutions be made due to difficulty in mixing.

Manufacture may involve making standard solutions (as in the case of calcium chloride and powder admixtures) or dilution to provide the manufacturer's recommended dosage to facilitate accurate dispersion [126, 152]. Problems such as difficulty in mixing and sedimentation of insoluble or active ingredients may arise in the large-scale manufacture of admixtures due to incorrect charging of the ingredients or a mismatch of mixer type with the viscosity of the product. The recommendations of the manufacturer should be followed if there is any doubt about procedures being used.

All mixing of powdered admixture must be done in a mixing tank which is separate from the admixture storage tank from which the admixture is fed into the dispenser. The density of the admixture mixed in this manner should be checked daily or when new material is mixed, using hydrometers. The storage tanks should be equipped with agitators to keep the admixtures in suspension [151, 152]. Since many low-concentration solutions contain significant amounts of finely divided insoluble materials or active ingredients, which may or may not be readily soluble, it is important that precautions be taken to insure that these be kept in uniform suspension before actual batching.

7.14.2 Packaging and delivery

(a) Liquid materials

Most proprietary admixtures are neutral or slightly alkaline solutions that are supplied in various forms of packaging depending on the amount contained and the ease of use required (Fig. 7.48). These are as follows [126]:

Fig. 7.48 The various forms of packages commonly used for the delivery of admixtures: (a) 5 and 25 l containers; (b) 210 l lined drums; (c) 1000 l portable tanks; (d) 25 kg multi-ply sacks.

- Plastic containers, 5 or 25 liter capacity: these are made from low- or high-density polyethylene or polypropylene for small jobs.
- Lacquer-lined mild-steel drums 210 liter capacity: such drums should have a small opening available for fitting a tap in addition to the normal filing bung. It is useful to have these drums in a cradle so that handling is made easy. Often second-hand relined drums are quite commonly used for this purpose.
- Portable tanks, 1000 liter capacity: these are made of mild- or heavy-duty polypropylene with a steel frame. The tanks can be delivered filled in a flat bed truck and are useful for medium-sized precast and cast on site projects. A crane or a fork lift truck is usually use to handle these tanks on site.
- Cylindrical or rectangular bulk storage tanks, 2500–15000 liter capacity: these are commonly fitted with delivery and ventilation pipes and an outlet pipe complete with a tap a few inches from the bottom. Although plastic tanks are sometimes used, those made with mild steel are more widespread. Clear marking of the tank is essential, and this is usually done with a label that cannot be altered and that is placed in a conspicuous area such as near the access opening. Safety and handling aspects in a printed form should always be readily available, preferably in a plastic holder, secured to the tank.

Deliveries to fixed tanks are made from road tankers or occasionally 20 t rail tankers. More commonly, admixtures are delivered by specialized vehicles containing several compartments with individually calibrated meters with print-out facilities (Fig. 7.49). This method of delivery

Fig. 7.49 Special admixture delivery vehicle.

removes the necessity for inaccurate tank dipping or vehicle weighing before and after deliveries. It is required by law that such tankers should carry advice on hazards presented by accidental spillage of the materials being carried.

(b) Powdered products

Typical powdered materials are waterproofing admixtures based on stearates or stearic acid, accelerators like calcium formate, some air-entraining agents, latex materials, the spray-dried solid forms of super-plasticizer, corrosion inhibitors like sodium nitrite, water-retaining floccu-lating agents, expansion-producing admixtures and some caustic materials used as shotcrete admixtures. Most of these materials are often deliquescent or hygroscopic and require storage conditions which exclude moisture and carbon dioxide. They are, therefore, usually packed in either multi-ply paper sacks or small fibrite kegs.

7.14.3 Labels

Since admixtures are potent modifiers of the characteristics of concrete, it is imperative to minimize potential mistakes. Thus it is important that all containers in which admixtures are delivered or stored should be clearly labeled and marked in such a way that the information is readily understood and cannot be altered. An example of a typical label is shown in Fig. 7.50. The label should feature the standard information given below [126, 152]:

- Manufacturer's and distributor's name, address and telephone number.
- Brand name of the product, reference number and/or letter.
- The relevant standards (ASTM, BS, CSA) number the admixture complies with.
- Designated type as given by principal and secondary functions (or the relevant standards), e.g. accelerating, water-reducing or water-reducing/retarding admixture (e.g., ASTM, Type D).
- Basic information for use and precautions regarding the handling of the admixture.
- The chloride content of the admixture. This should be expressed as a percentage (of anhydrous chloride ion) of cement at the recommended dosage.
- Data relating to shelf life or special storage requirements.
- Manufacturer's recommended dosage or dosages and the maximum amount not to be exceeded.

Fig. 7.50 A typical label for admixtures

7.14.4 Storage

Three items are important in the storage of admixtures, namely, the ease of identification, humidity and temperature at which they are stored. To avoid confusion between different admixtures, drums should be delivered clearly labeled as to the contents; identification can be made easier by a colorant which has no influence on the concrete. Powdered admixtures are more sensitive to moisture and carbon dioxide than are Portland cements. The materials should, therefore, be packed in waterproofed bags and always stored in areas free of high relative humidity and temperature extremes to avoid condensation. Storage tanks should be vented properly, and fill nozzles and any other tank openings should be capped when not in use to avoid contamination.

Most admixtures with the exception of some damp-proofers and non-chloride accelerators and air-entraining agents are aqueous solutions which freeze at $-3\,°C$. They therefore require protection against freezing. Some, like emulsions, require greater care in storage to prevent freezing and the manufacturer's instructions should be carefully followed. Typical data for minimum storage temperatures for the various types of admixtures currently marketed are presented in Table 7.32. In the UK, winter conditions are mild and the majority of the materials can be stored in drums in unheated enclosed buildings without freezing. North American winters, however, are much more severe, and proper provision should be made to prevent freezing of liquid admixtures. Although bulk storage outdoors in lagged steel or plastic tanks have been found satisfactory in the milder climate of the UK

Table 7.32 Approximate minimum storage temperatures for various admixtures before separation or solidification occurs

Category	Type	Approximate minimum storage temperature (°C)
Water-reducing	Lignosulfonate	−3 to 0
	Hydroxycarboxylic acid	−5 to 0
	Hydroxylated polymer	−5 to 0
	Superplasticizers	−3 to 0
Air-entraining	Neutralized wood resins	−3 to 0
	Fatty acid soaps	
	Alky-aryl sulfonates	
Waterproofers	Fatty-acid emulsions	1
	Wax emulsions	
Accelerators	Calcium chloride solution	−10 to −5

and some coastal and southern parts of North America, in Canada and the Northern States they should be inside an enclosed, heated building. Protection must be afforded to the whole system, including storage areas, lines to the dispenser and lines to the mixer [126, 153].

In some plants where bulk storage outdoors in lagged steel or plastic tanks is still practiced, freezing of the admixture is prevented by heating the storage tank and its contents. This method, however, is less preferred than storage in a heated building for the following reasons. Some heating probes can overheat the admixture and reduce the effectiveness of certain admixture formulations or overheat locally and pyrolize certain constituents, producing explosive gases; electrical connections to heating probes, bands or tapes can be disconnected, allowing the admixture to freeze and damage equipment; the cost of operating electric probes, bands, tapes, etc., is normally higher than the cost of maintaining above-freezing temperatures in a heated storage room; a heated admixture storage room protects not only storage tanks, but pumps, meters, valves and admixture hoses from freezing; if plastic storage tanks or hoses are used, care must be taken to avoid heating these materials to the point of softening and rupture. Since the storage temperature of heated buildings is subject to less widespread variation throughout the year, admixture viscosity is more constant and dispensers require less frequent calibration.

Although most of the frozen admixtures mentioned in Table. 7.32 may be used in concrete after thawing and thorough mixing, some emulsions cannot be reconstituted on thawing [151, 152]. Thawing of frozen admixtures is usually done by gradually raising the material's temperature to at least 10°C above their freezing temperature and maintaining this temperature for a period of time. The contents of the drum or tank should be agitated, either manually or using a low-pressure air spurge, to produce a homogeneous solution or emulsion. It should be emphasized that frozen admixtures require both thawing and remixing prior to use. The use of admixtures which have only been thawed with no remixing may result in a number of problems including severe retardation of concrete, bleeding and segregation and heavy air entrainment. Dispenser lines may also be clogged because of the separation of the admixture into a supernatant water layer and thick lower liquid that occurs on thawing.

Solid admixtures are not normally affected by winter conditions, but should be stored in areas free from dampness, condensation or rain, preferably in an elevated location. Materials in ripped bags should be used as soon as possible after the bags are damaged or should be discarded.

Although elevated temperatures, even those prevailing in hot-weather countries, do not markedly change the shelf life or effectiveness of the majority of the admixtures, some because of their chemical make-up are more prone to bacterial and fungal attack than others. The following basic admixture raw materials which can support fungal and bacterial growth are vulnerable to such attack; impure lignosulfonates, hydroxycarboxylic acids, hydroxylated polymers and fatty-acid-based materials. Since bacterial and fungal activity is usually accelerated at elevated temperatures, the use of bactericidal and fungicidal agents is important to prevent deterioration of admixtures under hot-weather conditions. Materials such as formaldehyde or sodium-o-phenyl phenol tetrahydrate are normally added to prevent this form of attack.

Other admixtures prone to damage by the high temperatures that prevail in hot-weather countries include emulsified products, such as waxes, fatty acids and waterproofers. The emulsion breaks down at elevated temperatures due to coagulation of the suspended particles and most often it is difficult to reconstitute the material on lowering the temperature and remixing. The use of such materials is generally discouraged in hot climates. However, when their use is necessary, proper care should be exercised in the storage and use of the products and advice should be sought from individual manufacturers [127, 154–156]. Admixture manufacturers ordinarily can furnish either complete storage and dispensing systems or at least information regarding the degree of agitation or recirculation required with their admixtures. Timing devices are commonly used to control recirculation of the contents of storage tanks to avoid settlement or, with some products, polymerization.

7.14.5 Dispensing of admixtures

Batching and mixing plants encountered in construction work vary over a wide range. The plants may be large-capacity, fully automatic plants with several stationary mixers; they may be small portable, manually operated plants dry-batching materials for a single end use such as a paving mixer or they may be a combination between these two extremes. It is important that batching and dispensing equipment meet and maintain tolerance standards to minimize variations in concrete properties and, consequently, better performance of the concrete. Tolerances of admixture batching equipment should be checked carefully. ASTM C 94 requires that volumetric measurement of admixtures shall be accurate to approximately 3% of the total volume required. ASTM C 94 also requires that powdered admixtures be measured by weight (mass), but permits liquid admixtures to be measured by weight (mass) or volume. Accuracy of weighed admixtures is required to be within 3% of the required weight (mass) [156].

(a) Time and sequence and point of addition

The incorporation of admixtures into a concrete batch involves control of the rate of discharge, timing in the batching sequence and the amount of material used. To insure uniform distribution throughout the concrete mixture during the charging cycle, the rate of admixture discharge should be adjustable. In North America and other cold-weather countries, where the use of two admixtures in a single batch is the norm, the point of addition in the mixing cycle is paramount because of the dire consequences that can result from an incorrect addition sequence, e.g. the difference between adding a retarder in the initial gauging water and adding it after a premixing period can be two or three times the normal retardation time. For any given condition or project, a procedure for controlling the time and rate of the admixture addition to the concrete batch should be established and adhered to closely.

The compatibility of the admixtures intended to be used in the same concrete batch when discharged in the same water phase should be determined and a procedure for controlling the time and rate of addition of the different admixtures to the concrete batch should be established. It should be emphasized that the mixing of admixtures prior to introduction into the mix is strongly discouraged because of the possible reaction between the mixed admixtures, which very often results in nullifying of the desired effects these admixtures were expected to produce individually. This applies particularly in situations where rapid repetitive mixing occurs and it is often not possible to adhere closely to these recommendations. Thus, when two or more admixtures are to be used in the mix, the dispenser system must be designed so that an appropriate delay is built into the system to avoid intermixing of admixtures prior introduction into the concrete and they are incorporated at different times during the mixing cycle. Likewise, in a manual system, the operator must be instructed in methods to prevent such comingling of admixtures. Table 7.33 summarizes the preferred times of admixture addition in the mixing cycle to obtain the best results [126, 153, 157].

Table 7.33 Preferred point of addition in the mixing cycle for different admixtures types

Admixture type	Point of addition	Notes
(a) All water-reducing admixtures (except superplasticizers used for flowing concrete production) particularly when a retarding effect is required	After initial mixing period of up to 30s of aggregates, cement and part of the gauging water. Should be dissolved in a proportion of the remainder of the gauging water	Premixing with moist aggregate is often sufficient to partially hydrate the cement
(b) Air-training agents (c) Accelerators (except powders) (d) Emulsified waterproofers	Dissolved in gauging water which can be added directly to the premixed cement and aggregates	These materials are not particularly sensitive to the point of addition and the main ambition is to obtain the maximum dispersion through the mix
(e) Powdered waterproofers (f) Powdered accelerators	Premixed with dry mix ingredients before addition of the gauging water	In order to aid the dispersion of this type of mix it is advisable to sprinkle the admixture over the mix during the dry cycle
(g) Superplasticizer for flowing concrete	After the mixing cycle just prior to placing	
(h) Third-generation superplasticizers	With the gauging water	

Discharge of the admixture from the calibration tube to the concrete batch should be to the point where the admixture achieves the greatest dispersion throughout the concrete. The discharge end of the water line leading to the mixer is a preferred location, as is the fine-aggregate weigh hopper or the belt conveyor carrying fine aggregate. Often, the calibration tube is emptied either by gravity or by air pressure and the admixture may have considerable distance to flow through a discharge hose or pipe before it reaches its ultimate destination. Therefore, the dispenser control panel should be equipped with a timer-relay device to insure that all admixture has been discharged from the conveying hoses or pipes. If the admixture dispenser system is operated manually, the plant operator should be furnished a valve with a détente discharge side to prolong the discharge cycle until it is ascertained that all admixture is in the concrete batch.

Due to the rapid slump loss associated with first- and second-generation superplasticizers (high-range water-reducers), job site addition of such admixtures is the norm. Such addition may be from truck-mounted admixture tanks or job site tanks or drums with the associated components of the dispensing system (e.g. pumps, meters, pulse transmitters, and counters) If truck-mounted tanks are used, the proper dosage of admixture for the concrete in the truck is measured at the batch plant and discharged to the truck-mounted tank at a special filling station. At such a station, a series of lights or other signals tells the driver when the admixture batching is complete and when his tank contains the proper amount. At the job site, the driver sets the mixer at mixing speed and discharges the entire amount of admixture from the truck-mounted tank into the concrete. To insure that all the admixture is introduced and thoroughly distributed throughout the concrete, air pressure should be used to force the admixture into all parts of the mixer drum and the truck mixer should operate at maximum speed, preferably over 19 rpm [151, 155].

7.14.6 Dispensing equipment

Due to the potent effects of chemical admixtures and the small amounts involved, accurate and reliable dispensing of these materials is essential for their proper use in concrete mixes. A means must be provided for reliable and accurate batching of the admixtures into the concrete batch. Uniform dispersion of the admixture is essential and dosage should be reproducible so that the quantities are controlled with very little variation. The type of dispenser must be relevant to size of the project and determined by the volume of concrete, batch size and type of admixture itself. The greatest demand is for the medium-dosage-range water reducers but the use of low-dosage products such as air-entraining agents and high-dosage products such as superplasticizers will require equipment that is suited for measuring and controlling small and large dosages respectively. Small amounts should be diluted at the point of use. Suitable admixture batching systems are usually supplied by the admixture manufacturer and are generally of two types [155].

- Liquid batching systems for materials introduced in the liquid form.
- Dry batching systems for materials which are powders.

Weight batching of liquid admixtures ordinarily is not used because the weight batching devices are more expensive than volumetric dispensers. In some cases, it is necessary to dilute admixture solutions to obtain a sufficient quantity for accurate weighing (determination of mass).

(a) Liquid dispensing systems

Currently used dispensing systems include simple pumps, timers, visual volumetric displacement systems and weight batching. Liquid admixture dispensing systems are available for manual, semiautomatic and automatic batching plants. Essentially a liquid dispensing system consists of equipment for moving the liquid from the feed tank to volumetric containers (usually by pneumatic or electric pumps), a batching or dispensing device, controls and appropriate interlocking devices. It is necessary to interlock the discharge valve so that it will not open during the filling operation or when the fill valve is not closed fully. Usually the fill valve is interlocked with the discharge valve so that it will not open unless the discharge valve is fully closed. A secondary circuit is usually provided to come into play in the event of a failure of the primary control. Visual volumetric containers in the form of a calibrated cylindrical sight-glass for visual checking of the required quantity should also be provided for all liquid systems. With certain water-reducing and air-entraining admixtures, it may be necessary to incorporate an antifoaming device to ensure accuracy.

Measuring devices are either positive displacement devices or timer controlled systems and a very limited number use weight batching systems. Positive displacement devices consist of flowmeters and measuring containers equipped with floats or probes. They are well suited for use with automatic and semiautomatic batchers because they may be operated easily by remote control with appropriate interlocking in the batching sequence. Flowmeters may be fitted with pulse-emitting equipment which operates at a preset electrical counter on the control unit located near the batching console. Often they are set by inputting the dosage per unit of cementitious material. The amount of cementitious material input to the panel combined with dosage rate sets the dispensing system to batch the proper amount of admixture. Such flowmeters are usually geared in a manner which facilitates variable admixture dosage and are calibrated for liquids of a given viscosity. Errors caused by viscosity change due to variations in temperature can be avoided by recalibration and adjustment made by observation of the volumetric container or calibration tube. Figure 7.51 shows a typical displacement flowmeter operated by pneumatic power, while Fig. 7.52 shows a control unit on which the required dosage can be preset.

A measuring container operates by the principle whereby the linear movement of a float in a visual volumetric container of given cross-section meters out a volume of solution required for the batch. Usually the floats are connected to a potentiometer or pulsating switches which operate electrical preset counters [157–159]. A typical unit is shown in Fig. 7.53 where it can be seen that the measuring device consists of a sight glass containing a movable electrode, which in this case is connected to a scale calibrated in terms of cement content. The admixture is pumped into the sight glass until it touches the upper electrode when the pump is automatically switched off and the material is allowed to discharge into the water pipe. Figure 7.54 shows a series of visual volumetric measuring containers with displacement flowmeters installed in a ready-mix plant.

Fig. 7.51 Positive displacement flowmeter (courtesy Euclid Chemicals, USA).

Timer-controlled systems are simple, low-cost systems which involve timing of flow through an orifice. Most include a sight glass or other means of checking the amount batched. A number of variables, such as changes in power supply, constriction of the measuring orifice and increased viscosity of the admixture can introduce considerable error. Timer-controlled units must be recalibrated frequently, and the plant operator must be alert to verify the proper admixture dose by observation of the calibration tube. They have been successfully used with high-volume dilution admixtures [151, 153, 156]. However, due to these inherent disadvantages, their use is generally discouraged [157]. The systems are permitted only for calcium chloride in the NRMCA Check List.

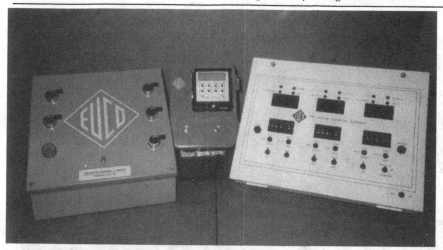

Fig. 7.52 Manual, semi-automatic and automatic control units for dispensing admixtures (courtesy Euclid Chemicals, USA).

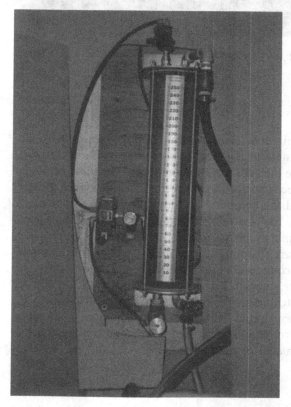

Fig. 7.53 Measuring container with calibrated scale to show volume (courtesy Euclid Chemicals, USA).

Fig. 7.54 A series of dispensing units installed in ready-mix plant (courtesy Master Builders Technologies, Canada).

Weight batching of admixtures using beam and dial scales with an indicator in the batching system to register the discharge of the material is occasionally used to measure the addition of admixtures. The disadvantages of the system include the necessity for conversion of the admixture dosage from volume to weight, and the necessity to dilute admixture solutions to obtain a sufficient quantity for accurate batching. The method has been found suitable for air-entraining admixtures.

Most dispensing systems are custom-made to meet the method of control or degree of automation required for the particular operation. Prior to installation of the dispenser, the system should be analyzed carefully to determine what possible batching errors could occur and, with the help of the admixture supplier, they should be eliminated. Commonly available dispensing systems and controls can be discussed under three main groups – fully automated, semi-automated or manual systems.

FULLY AUTOMATED

A fully automatic or pneumatic dispenser based on either electric or pneumatic systems has good accuracy and can cope with more than one type of admixture with facilities for flushing. It is therefore the ideal method of adding liquid admixtures to a concrete. One form of dispenser feeds the admixture by an electronically driven pump controlled by a timing device, and can be fitted with a sensing unit for automatic cut-off. This type of dispenser is shown installed in a ready-mixed concrete plant in Fig. 7.55. The control release can be operated automatically so that when the batching console is used in the normal way a signal is picked up by the dispenser relay and the material is automatically fed into the mix. Precasting operations usually employ an electrical dispenser while ready-mixed batching plants where a high rate of production and fast accurate measurement is essential may use the pneumatic type. These systems include the required combination of automated individual volumetric devices, but they should also have the following features [158, 159]:

Fig. 7.55 Electrically operated fully automatic dispenser (courtesy Master Builders Technologies, Canada).

1.   A single starting mechanism or separate starting mechanisms for admixtures not batched at the same time as other ingredients.
2.   A mechanism for signaling empty and resetting to start by volumetric devices.
3.   A device that prevents the discharge of any ingredient until all individual batches have been cleared of the previous batch and returned to zero within tolerance.

SEMI-AUTOMATED

These systems (Fig. 7.56) can operate either electrically or pneumatically and they contain the necessary volumetric devices, the controls of which are either semi-automatic, interlocked or automatic, or a combination of these. With this type of equipment the level of admixture is allowed to rise in the sight-glass by the operation of a switch or valves until a correct level is

Fig. 7.56 A portable semi-automatic dispenser (courtesy Eucid Chemicals, USA).

reached. The valve is then opened to allow dispensing of the admixture into the water pipe or mixer. Although this type of dispenser is widely used, some precautionary action is required in the installation and operation of the system:

- The arrangement of the piping system in the dispensing equipment requires careful attention to ensure that variable amounts of material are not trapped in the lines.
- It must be checked that the valves will not leak and that the amount of material measured is actually discharged into each batch [158, 159].

Modern high-performance concretes (such as non-dispersible underwater concrete, self-compacting concrete, and concretes used at below freezing temperatures) have necessitated the use of special admixtures, often requiring on-site addition. Figure 7.57 shows a special dispensing unit with storage tank used for this purpose.

MANUAL

Simple manual dispensing systems designed for low-volume concrete plants depend solely on the care of the concrete plant operator in batching the proper amount of admixture into a calibration tube and discharging it into the batch. These types of dispensing equipment are used in small-scale operations where admixtures are only occasionally used. For small and occasional additions, hand-dispensing equipment may be used to measure reasonably accurate quantities. A hand-operated plunger can be used to

Fig. 7.57 On-site dispensing unit with tank and dispenser for speciality admixture (courtesy Euclid Chemicals, USA).

dispense water-reducers and air-entraining agents for normal work but for intermittent use, a mechanical dispenser is the best.

- Hand pumps are normally constructed to fit a 210 l drum bung hole. Two types are available: one is simple which will require hand dispensers of the types described below to measure the actual quantity; the other is a type in which one stroke of the pump delivers a set volume. It can be used for adding the admixture directly to the concrete mix [159].
- Hand dispensers are containers of standard and correct volume that may be used in small applications or special mixes. It is important that a dispenser of the correct size for one mix be used rather than a measuring cylinder or other partly filled vessel.

The complexity and cost of the dispenser system used is largely dictated by the size of the operation. The basic requirements of all methods, however, are that they function at the temperature ranges at which normal concreting is permitted, have an insensitivity to changes in viscosity for the admixture and an accuracy of $\pm 3\%$ [154–159]. All components should be constructed with materials which will not be corroded by the admixture being used and of proven low failure rate to ensure reliability in the long term. A clear indication should be given on the dispenser as to the range throughout which it will effectively operate within the specified limits. The equipment should also comply with Plant Manufacturer's Bureau (CPMB) standards and provide a device for routine diversion of a measured dosage into a small container for verification of the batch quantity. Specific requirements pertaining to dispensing systems are detailed in BS 5328 [154], CPMB Standards [155], ASTM Publication C94 [154], the NBS Handbook [157] and the checklist [161].

(b) Dispensing systems for solid admixtures

Solid admixtures such as calcium formate or some waterproofers are measured by weight. Since relatively small quantities of such admixtures are used, a greater degree of measurement control is required. Therefore, it is wise to blend them with finely divided inert or pozzolanic material such as fine sand, pulverized fly ash or ground blast furnace slag. This not only facilitates more accurate weighing but also prevents flotation of the powder on the mix water thus ensuring better dispersion in the mix. In the use of powdered admixtures, care must be taken to prevent the adherence of the powder to the sides of the mixer or the formation of clumps. These problems can be avoided if the powder admixture is not dispensed with the mixing water but is added after the fine aggregate in the mixing sequence [161].

A suitable measuring device for powders is the automatic pigment scale. Most of the units commercially available consist of a part that operates volumetrically (e.g. screw, shaking gutter) and a suitable metering scale. Space is essential in positioning the metering unit and powder admixture containers as closely as possible to the mixer or raw-material supply lines. Care must be taken to ensure that the powder that enters the concrete mixer is free flowing [161]. These automatic scales are quite expensive.

7.14.7 Calibration and maintenance of batching systems

Batching and dispensing equipment should meet tolerance standards to minimize variations in concrete properties; such tolerances should also be checked carefully and maintained by regular calibration, thus ensuring better and consistent performance of the concrete. ASTM C 94 requires that volumetric measurement of admixtures shall be accurate to approximately 3% of the total volume required. ASTM C 94 also requires that powdered admixtures be measured by weight (mass), but permits liquid admixtures to be measured by weight (mass) or volume. Accuracy of weighed admixtures is required to be within 3% of the required weight (mass).

Dispensing systems require routine periodic maintenance to prevent inaccuracies developing from such causes as sticky valves, worn parts, build-up of foreign matter in meters or in storage and mixing tanks, or worn pumps. To facilitate maintenance of the equipment and prevent the liquid admixture thickening within the dispensing system when the batch plant is not in operation, means should be afforded for washing out the complete system. Tanks, conveying lines, and ancillary equipment should be drained and flushed on a regular basis, and calibration tubes should have a water fitting installed to allow the plant operator to water flush the tube so that divisions or markings may be clearly seen at all times.

It is important to protect components from dust and temperature extremes, and they should be readily accessible for visual observation and maintenance. All measuring devices must be kept in serviceable condition, zeroed daily and calibrated monthly. Although admixture batching systems are installed and maintained by the admixture producer, it is desirable that the supplier makes available detailed advice on regular maintenance so that plant operators can thoroughly understand the system and be able to adjust it, perform simple maintenance and recalibrate the system on a regular basis. Specific requirements pertaining to dispensing systems are detailed in BS 5328 [154], CPMB Standards [155], ASTM Publication C 94 [156], the National Bureau of Standards (NBS) Handbook [157] and the NRMCA checklist [160].

A checklist for monitoring dispensing equipment and procedures should include:

1. Type of admixture used, how it is furnished (powder or liquid), brand name, manufacturer and date of manufacture,

2. Was manufacturer's dosage used?
3. Description of dilution procedures used to obtain recommended dosage. Was the admixture tested and approved?
4. Is the admixture prevented from freezing?
5. What type of dispenser is used? Are the dispenser controls interlocked to discharge automatically and how is interlocking achieved? How often is the dispenser calibrated and when was it last done? Are the dispensing operations suitably controlled to ensure proper sequence of addition of the different admixtures?
6 Are the supply tanks of diluted admixture mechanically agitated?

7.14.8 Computer batching

Concrete is becoming increasingly complicated; the number of ingredients has increased and the properties to be attained are both varied and many. Consequently, the potential for mistakes using conventional manual methods of batching is heightened. Specifications for most projects that require performance concrete therefore insist on computerized batching (Fig. 7.58). The merits of computerized batching reside in three aspects of concrete manufacture: (1) production, (2) quality control and (3) recording. The ultimate purpose, however, is to gain a competitive advantage through cost effectiveness. Most of the savings are volume related but the trend in the industry is to computerize plants, even those with smaller annual volumes. Today, almost all plants with annual production in excess of 25 000 m^3 are computerized and savings in five main categories, (1) materials, (2) load time, (3) labor, (4) reduction in rejected loads and (5) product consistency are afforded [122].

Batching cement manually and poor management of cement stocks can result in large losses. Modern batch plant computers have provision for entering the amounts of all received materials and producing a report of all

Fig. 7.58 Computerized batching in a ready-mix plant (courtesy W.R. Grace, USA).

entries. All material amounts batched automatically and manually are used to calculate an accurate material balance on-hand. Alarms are triggered automatically when material stocks fall below a prescribed limit.

Savings in truck loading time can be significant (as much as 50%) for plants supporting over 15 trucks. Controlling the rate and sequence of introduction of materials in to the truck can substantially reduce truck mixing time, and shorter mixing time means the load can be checked sooner for proper slump and air content.

Factors such as the number of mix ingredients (fly ash, slag silica fume and a mixture of admixtures in the same mix), variability in material properties and temperature make it difficult to achieve the required slump and air content. Furthermore, the use of two or more admixtures poses compatibility problems, particularly if the sequence of addition is not properly controlled. Thus, there are too many ingredients used in a single batch to handle manually without risking mistakes. Computerization not only controls batching sequence and rate, but also automatically adjusts mix ingredient contents (such as aggregate and water) to ensure stipulated fresh property requirements are achieved. Product consistency means less variation from load to load, resulting in less variation in product strength [121, 122]. This creates an opportunity to reduce the over design of mixes by decreasing the amount of cementitious material to achieve the required strength.

The reduction in rejected loads is a major savings in material and transport costs. More significant savings are achieved by avoiding the placing of concrete that does not meet specification requirements. The cost of concrete removal and replacement can exceed the cost of computerization many times over. Thus, specifications for high-performance concrete dictate the use of computerized batching when such concrete is stipulated for a project. The product consistency attained in controlled introduction of materials, reduced mixing time, the automatic adjustment of aggregate and mixing water to maintain a proper slump and air content all ensure compliance with specifications.

Reduction of labor due to the use of computers is becoming a trend. Many concrete producers take advantage of the features offered by computers – direct entry of customer orders, automatic delivery docket printing, and remote workstations – to reduce personnel. Customer orders can be taken and entered at the head office using a remote terminal.

Plant computerization, in addition to providing improvement to the batching process, affords more time to the plant operator for the managing of other yard operations. Taken together, the significant savings mentioned above, the product consistency and labor-saving advantages are compelling reasons for ready-mix plants to computerize.

References

1 Gordon, W.A. (1966). *Freezing and Thawing of Concrete, Mechanisms and Control*, ACI Monograph No. 3.
2 *Design and Control of Concrete Mixtures* (1978). Engineering Bulletin of the Portland Cement Association, Skokie, Illinois, USA.
3 Madderom, W.F. (1980). *Concrete International*, February, 110–14.
4 Malhotra, V.M. (1981). *Report MRPIMSL 81-53 (J)*, Draft CANMET, Ottawa, Canada.
5 Ramachandran, V.S. (1994). *Concrete Admixture Handbook*, Noyes Publications, 150.
6 Taylor, T.G. (1948). *ACI Journal*, Proceedings, **45**, 469–87.
7 Blanks, R.F. and Gordon, W.A. (1949). *ACI Journal*, Proceedings, **45**, 489–97.
8 Kobayashi, M. (1967). *Proceedings of the International Symposium on Admixtures for Mortar and Concrete*, Brussels, 87.
9 Malhotra, V.M. and Malanka, D. (1977). CANMET Report, Department of Energy, Mines and Resources, Ottawa, Canada, 65–77.
10 Hester, T.W. (1979). *Superplasticizers in Ready-Mixed Concrete. (A Practical Treatment for Everyday Operations)*, Publication No. 158, NRMCA, Silver Springs, Maryland, USA.
11 Mielenz, R.C. and Sprouse, J.H. (1979). *Superplasticizer in Concrete*, ACI Publication SP-62, 171–92.
12 Curtis, R.J. (1975). *Contract Journal*, **10**, 24–5.
13 Newman, K. (1975). *Proceedings of the Conference on Ready-Mixed Concrete*, Dundee, Scotland.
14 Keeley, C. and Holdsworth, R. (1978). *Middle East Construction*, October, 104–7.
15 Dodson, V. (1990). *Concrete Admixtures*, Van Nostrand Reinhold, 45.
16 Dodson, V. (1990). *Concrete Admixtures*, Van Nostrand Reinhold, 49.
17 Freedman, S. (1970). *Modern Concrete*, November, 170–6.
18 Howard, E.L., Griffiths. K.K. and Moulton, W.E. (1959). *Journal of Materials*, **14**, 220–30.
19 Aitcin, P.C. (1980). *Concrete Construction*, **6**, 220–30.
20 Dodson, V. (1990). *Concrete Admixtures*, Van Nostrand Reinhold, 199.
21 Valore, J.R.R.C. (1978). *Significance of Tests and Properties of Concrete and Concrete Making Materials*, STP 169 B, ASTM, 860–72.
22 Mailvaganam, N.P. (1994). *Concrete Admixtures Handbook*, Noyes Publications, New York, 994–1000.
23 Anon. (1962). Army Engineering Waterways Experiment Station, Mississippi, 45-030, Cement and Concrete Association, AD-756299.
24 Browne, R.D. (1973). *Proceedings of the Conference on Large Pours for RC Structures*, University of Birmingham, 44–8.
25 Fitzgibbon, M.E. (1975). *Proceedings of the CS Symposium 'Tomorrow's Concrete'*, Doncaster, UK.
26 Forbrich, L.R. (1940). *ACI Journal*, Proceedings, **37**, 161–83.
27 Morgan, H.D. (1958). International Commission on Large Dams, *Question No. 23*, No. R-12, 10–24.
28 Shutz, R.J. (1959). *ACI Journal*, **11**, 769–81. *Concrete Technology*, Monte.

29 Fisher, C.H. (1972). *ACI Journal*, **69**, 556–61.
30 O'Brien, J. (1973). Cement and Concrete Association of Australia, Technical Report No. 33.
31 Mather, B. (1978). *Superplasticizer in Concrete*, ACI Publication, SP-62, 15 866.
32 Gerwick, J.R.C.B. (1980). *Research Requirements for Concrete in Marine Environments*, ACI SP-65, Symposium on Performance of Concrete in Marine Environment, St. Andrews by the Sea, Canada, 577–88.
33 Mailvaganam, N.P., Bhagrath, R.S. and Shaw, K.L. (1983). *Effects of Chloride and Non-chloride Admixtures on Superplasticized Silica Fume Concrete*, Annual Meeting of the Transportation Research Board, Washington, D.C.
34 Anon (1968). *Concrete Construction*, December, 860–72.
35 *Tilt-Up Construction*, (1980). ACI Compilation No. 4.
36 ACI Committee 201. (1977). *Guide to Durable Concrete* (ACI 207 2R 77), ACI Publication SP-62, 559-608, Detroit.
37 Daugherty, E.K. and Kowalewsky, Jr. M. (1967). *Journal of Materials*, **11**, 161–6.
38 Rosskopf, P.A., Linton, F.J. and Peppler, P.B. (1975). *Journal of Testing and Evaluation*, **11**, 330–2.
39 Benstead, J. (1982). *Silcate Industry*, No. 45, 67–9.
40 Rosenburg, A.M., Gaidis, J.M., Kossivas, T.G. and Previte, R.W. (1977). *A Corrosion Inhibitor Formulated with Calcium Nitrite for Use in Reinforced Concrete*, STP-629, ASTM, 228-40, 58.
41 *Superplasticizing Admixtures in Concrete*, (1976). Joint Working Party Report, 45-030, Cement and Concrete Association and Cement Admixtures Association.
42 Bonzel, J. (1974). Directives for the production and manufacture of flowing concrete. Translation from *Beton Herstellung Verwendung* **24**, 4.9.S. 342–4.
43 Hester, W.T. (1978). *First International Symposium on Superplasticizers in Concrete*, Ottawa, Canada, ACI Publication SP-62, 533–58.
44 Collepardi, M. (1976). *Cement and Concrete Research*, **6**, 401–7.
45 *High Range Water-Reducer Committee Recommended Practice*. (1980). Draft of the Prestressed Concrete Institute.
46 Previte, R.W. (1977). *ACI Journal*, Proceedings, **124**, 361–6.
47 Ravina, D. (1975). *ACI Journal*, Proceedings, **62**, 291–5.
48 Collepardi, M. (1998). *Journal of Cement, Concrete and Composites*, **20**(2/3), 103–12.
49 Rixom, M.R. and Wadicor, J. (1981). *Developments in the Use of Superplasticizers*, ACI Publication SP-68, 359–80.
50 Mailvaganam, N.P. (1978). ACI Publication SP-62, 389.
51 Khalil, S.M. and Ward, M.A. (1979). *Fourth International Symposium on Concrete Technology*. Monterrey, Mexico, 35.
52 McCurrich, L.H. and Lamiman, S.A. (1979). *Symposium on Concreting in Hot Weather Conditions*, Cement and Concrete Association.
53 Admixtures for concrete. (1981). ACI Report 212-IR-81. *Concrete International*, 24–52.
54 La Fraugh, W.R. (1978). *Proceedings of the First International Symposium on Superplasticizers in Concrete*, Ottawa, Canada, 161–82.

55 Mielenz, R.C. (1960). *Symposium on the Effects of Water-Reducing and Set Retarding Admixtures on the Properties of Concrete*, ASTM SP-266, 161–82.

56 Dodson, V.H. and Hayden, T.D. (1989) *Cement, Concrete, and Aggregates*, CCAGDP, **11**(1), 52–56.

57 Khalil, S.M. and Ward, M.A. (1989). *Ceramic Bulletin*, **57**(12), 1111–22.

58 Mielenz, R.C. and Sprouse, J.H. (1979). *Proceedings 1st International Symposium on Superplasticizers in Concrete*, ACI SP-62, 137–55.

59 Langley, W.S., Gilmour, P. and Tromposch, E. (1995). The Northumberland Strait Bridge Project, ACI SP-154, 543–64.

60 Naik, T., Ramme, B.W. and Tews, J.H. (1992). Pavement construction with high-volume fly. *Fourth International Conference on The Use of Fly Ash Silica Fume, Slag and Natural Pozzolans in Concrete*, Istanbul, Turkey, May 3–8, 65–80.

61 Hester, W.T. (1979) *Field Applications of High Range Water Reducing Admixtures*, ACI SP-62, 533–57.

62 Hester, W.T. (1978) *J. Prestressed Concrete Institute*, **23**(4) 68–85.

63 Hansen, J.A. (1963). *ACI Journal*, Proceedings, **60**(1), 75.

64 Aitcin, P.C. (1997). *Mario Collepardi Symposium on Advances in Concrete Science and Technology*, Rome, Oct., 108–26.

65 Davis, H.E. (1940). *Proceedings ASTM*, **40**, 1103–10.

66 Tazawa, E., Miyazawa, S. and Shigekawa K. (1991). *Proceedings Cement and Concrete*, Cement Association of Japan, No. 45, 122–7.

67 McGrath, P. (1990). *Ceramic Transaction, Advances in Cementitious Materials, American Ceramic Society*, **16**, 489–500.

68 Tazawa, E. and Miyazawa, S. (1997). *Magazine of Concrete Research*, **49**, No. 178, 15–22.

69 Malhotra, V.M. (1997). *Mario Collepardi Symposium on Advances in Concrete Science and Technology*, Rome, Oct., 271–314.

70 Hoff, G.C., Walum, R., Elimov, R. and Hisada, M. (1994). *Production of High-Strength for Hibernia Offshore Concrete Platform*, ACI SP-149, 37–62.

71 Malhotra, V.M. (1995). *Concrete Admixtures Handbook*, 2nd Edn, Noyes Publications, 800–36.

72 Mailvaganam, N.P. (1992). *Repair and Protection Of Concrete Structures*, CRC Press, 231–4.

73 Yammuro, H., Izumi, T. and Mizunuma, T. (1997). *Fifth CANMET/ACI International Conference on Superplasticizers and other Chemical Admixtures*, Rome, Ed. V.M. Malhotra, 425–44.

74 Wallace, M. (1987). *Concrete Construction*, March, 268–77.

75 Sasse, H.R. and Steppler, K. (1983). *Second International Conference on Superplasticizers in Concrete*, Montebello, Canada, 73–81.

76 Alexsanderson, J. (1990). *Concrete International: Design and Construction*, **13**, 43–7.

77 Khayat, K. (1995). *Use of VEAs in Cement Based systems – An Overview*, University of Sherbrooke, Quebec, Canada, 1–36.

78 Rakitsky, W.G. (1993). *Proceeding Conchem International Exhibition and Conference*, Karlshrue, Germany, 155–82.

79 Gerwick, Jr. B.C., Holland, T.C. and Komendant, G.J. (1981). *Tremie Concrete for Bridge Piers and Other Massive Underwater Placements*, Report No.

FHWA/RD-81/153, Federal Highway Administration, US Department of Transportation, Washington, D.C.

80 Tynes, W. (1967). *Evaluation of Admixtures for Use in Concrete to be Placed Under Water*, Technical Report C-67-3, US Army Engineer Waterways Experiment Station, Vicksburg, Mississippi, USA.

81 Anon. (1968). *Concrete Construction*, July, 815–23.

82 Mailvaganam, N.P. (1994). *Concrete Admixtures Handbook*, Noyes Publications, New York, 978–80.

83 Hewlett, P.C., Edmeades, R.W. and Holdworth, R.L. (1977). *Concrete Admixtures, Use and Applications*, Construction Press, 55–66.

84 Permeability Reducing Admixtures (1977) *Miscellaneous Publication* MP-20, Part I, Standards Association of Australia.

85 Bruere, G.M., and McGowan, J.K. (1958) *Australian Applied Science*, **9**, 127–40.

86 Aldred, J.M. (1988) *Concrete International*, **10**, 52–57.

87 Kinney, F.D. (1989). *Proceedings Third International Conference on Super-plasticizers and Other Admixtures in Concrete*, ACI SP 119, 19–40.

88 Anon. (1988) *Concrete Construction*, **33**, 316–19.

89 Previte, R.W. (1977). *ACI Journal, Proceedings*, **74**, 361–7.

90 Mailvaganam, N.P. (1979) *Superplasticizers in Concrete*, ACI SP 62, 389–403.

91 Soroka, I. (1993). *Concrete in Hot Environments*, E. & F.N. Spon, London, 35.

92 Ravina, D. and Soroka, I. (1994). *Cement and Concrete Research*, **24**, 1455–62.

93 Ramachandran, V.S., Feldman, R.F. and Beaudoin, J.J. (1981). *Concrete Science*, Heyden & Sons Ltd., Philadelphia, 137–8.

94 Soroka, I. and Ravina, D. (1998). *Journal of Cement, Concrete and Composites*, **20**(2/3), 129–36.

95 El-Rayyes, M.S. (1990). *Proceedings RILEM Symposium, Admixtures for Concrete*, Chapman and Hall, London, 120–34.

96 Fattuhi, N.I. (1988). *Construction and Building Materials*, **2**, 27–30.

97 Tuthill, L.H. and Cordon, W.A. (1955). *ACI Journal*, **52**, 273–86.

98 Tuthill, L.H., Adams, R.F. and Hemme, J.M. Jr. (1960). *Water Reducing Admixtures and Retarding Admixtures*, ASTM Special Technical Publication No. 266, Philadelphia, 101–8.

99 Hampton, J.S. (1981). *Development in the Use of Superplasticizers*, ACI SP-68, 409–22.

100 Keely, C. and Holdsworth, R. (1978). *Middle East Construction*, October, 104–7.

101 Ravina, D. and Shalon, R. (1961). *Proceedings RILEM/CEMBUREAU Colloquium Shrinkage of Hydraiilic Coticretes*, Vol. 11, Edigrafis, Madrid, 61–9.

102 Ravina, D. and Shalon, R. (1968). *Proceedings JACI* 65, 889–92.

103 Ravina, D. (1995). *Concrete International*, **17**, 25–9.

104 Allen, G.C., Lee, B.J. and Wild, R.D. (1982). *Power Industry Research*, No. 2, 35–42.

105 Cabrera, J.G. and Plowman, C. (1982). *International Symposium on the Use of Pulverised Fly Ash in Concrete*, University of Leeds, Dept. of Civil Engineering 111–20.

106 ACI Committee 306 (1988). *Cold Weather Concreting*, ACI 306.R-88, American Concrete Institute, Detroit, Michigan.

107 Rixom, M.R. and Mailvaganam, N.P. (1986). *Chemical Admixtures for Concrete*, Second Edn, E. & F.N. Spon, London.
108 ACI Committee 318 (1992). *Building Code Requirements for Reinforced Concrete*, ACI 318 (Revised 1992), American Concrete Institute, Detroit, Michigan.
109 Haddad, J.G. (1975). *Conditions for the Production, Placing, and Curing of Winter Concrete*, ACI Chapter Meeting, Ottawa.
110 Brook, J.W., Factor, D.F., Kinney, F.D. and Sarkar, A.K. (1988). *Concrete International – Design and Construction*, 10(10) 44–9.
111 Smith, P. (1987). *Corrosion, Concrete, and Chlorides, Steel Corrosion in Concrete: Causes and Restraints*, SP-102, American Concrete Institute, 25–34.
112 Chin, D. (1987). *Corrosion, Concrete, and Chlorides, Steel Corrosion in Concrete: Causes and Restraints*, SP-102, American Concrete Institute, 49–77.
113 Popovics, S. (1987). *Corrosion, Concrete, and Chlorides, Steel Corrosion in Concrete: Causes and Restraints*, SP-102, American Concrete Institute, 79–106.
114 Ramachandra, V.S. (1995). *Concrete Admixtures Handbook – Properties, Science, and Technology*, Second Edition, Noyes Publications, New Jersey, Chs 5, 11.
115 Korhonen, C.J., Cortez, E.R. and Smith C.E. (1991). *Cold Regions Engineering, 6th International Specialty Conference/TCCR-ASCE/*, West Lebanon, NH, Feb. 26–28, 200–9.
116 Korhonen, C.J., Cortez, E.R., Charest, B.A. and Smith C.E. (1994) *Proceedings of the 7th International Cold Regions Engineering Specialty Conference*, Edmonton, Alberta, Canada, Mar. 7–9, 87–96.
117 Nmai, C.K. (1998). *Journal of Cement, Concrete and Composites*, 20(2/3), 123–8.
118 Ratinov, V.B., Rozenburg, T.I. and Kucheryaeva, G.D. (1981). *Beton i zhelezobeton*, 9, 9–11.
119 Hover, C.K. (1998). Concrete mixture proportioning with water-reducing admixtures to enhance durability: a quantitative model, *CANMET/ACI International Workshop on Supplementary Cementing Materials, Superplasticizers and Other Chemical Admixtures*, Toronto, Canada, April 6–7.
120 Rixom, M.R. (1998). Economic aspects of admixture use, *CANMET/ACI International Workshop on Supplementary Cementing Materials, Superplasticizers and Other Chemical Admixtures*, Toronto, Canada, April 6–7.
121 Furlong, J. (1997). *Concrete*, 9(3), 32.
122 Larrard, de F., (1997). *Concrete*, 9(3), 35–8.
123 Russell, P. (1983). *Concrete Admixtures*, a Viewpoint Publication, Eyre & Spottiwoode Publications, 95–6.
124 *Admixtures for Concrete*. (1992). ACI Report 212-IR-92, Concrete International.
125 Jolicoeur, C., Nkinamubanzi, M.A., Simard, M.A. and Piotte, M. (1994). *4th International CANMET/ACI Conference on Superplasticizers and Other Chemical Admixtures*, Montreal, Canada, October, 63–88.
126 Rixom, M.R. and Mailvaganam, N.P. (1986). *Chemical Admixtures for Concrete*, Second Edition, E. & F.N. Spon, 259–63.
127 Dodson, V. (1990). *Concrete Admixtures*, Van Nostrand Rheinhold, 36–42.
128 Hansen, W.C. and Hunt, J.O. (1949). *ASTM Bulletin*, No. 161, 50–8.

129 Manabe, T. and Kawada, N. (1960). *Semento Konkurito*, No. 162, 24–7.
130 Dodson, V.H. and Hayden, T.D. (1989). *Cement, Concrete, and Aggregates*, CCAGDP, **11**(1) 59–6.
131 Mork, J.H. and Gjoerv, O.E. (1997). *ACI Materials Journal*, March–April, 142–6.
132 Khalil, S.M. and Ward, M.A. (1989). *Ceramic Bulletin*, **57**(12), 1111–22.
133 Hansen, W.C. (1956). *The Role of Calcium Sulfate in the Manufacture of Portland Cement*, NRMCA Publication No. 63, National Ready Mix Concrete Association, Silver Spring, MD, 9.
134 Mielenz, R.C. and Sprouse, J.H. (1979). *Proceedings 1st International Symposium on Superplasticizers in Concrete*, ACI SP-62, 137–55.
135 Blank, B., Rossington, D.R. and Weiniand, L.A. (1983). *Journal of the American Ceramic Society*, **46**(8) 395–9.
136 Baalbaki, M. and Aitcin, P.C. (1994). *4th International CANMET/ACI Conference on Superplasticizers and Other Chemical Admixtures*, Montreal, Canada, October, 47–62.
137 Plante, P. and Pigeon, M. (1983). *Proceedings 3rd International Symposium on Superplasticizers and other Chemical Admixtures in Concrete*, ACI SP-119, 118–43.
138 Hansen, W.C. and Pressier, E.E. (1947). *Industrial and Engineering Chemistry*, **39**(10), 1280–2.
139 Ramachandran, V.S., Beaudoin, J.J. and Shihva, Z. (1989). *Materiaux et Constructions*, No. 22, 107–11.
140 Johnston, C.D. (1987). *Concrete International*, **9**(9), 43–51.
141 Ravina, D. and Shalon, R. (1961). *Proceedings RILEM/CEMBUREAU Colloquium on Shrinkage of Hydraulic Concretes*, Madrid, Vol. II, Edigrafis, Madrid, 46–51.
142 Ravina, D. and Shalon, R. (1968). *Journal of ACI*, **65**(4), 282–92.
143 Mailvaganam, N.P. (1994). *Concrete Admixtures Handbook*, Noyes Publications, New York, 981–4.
144 Standards Association of Australia. (1977). *Expanding Admixtures for use in Concrete, Mortar and Grout*, SAA MP. 20-3.
145 Hoff, G. (1972). *Practical Applications of Expanding Cements*, CTIAC Report 8 US Army Engineer Waterways Experimental Station, Miss., USA.
146 Xie, P. and Beaudoin, J.J. (1992). *Cement and Concrete Research*, **22**, 845–54.
147 Pigeon, M., Saucier, F. and Plante, P. (1990). *ACI Materials Journal*, **87**(83), 252–9.
148 Burg, G.R.U. (1983). *ACI Journal*, Proceedings, **80**(4), 332–9.
149 Langan, B.W. and Ward, M.A. (1976). *Canadian Journal of Civil Engineering* (Ottawa), **3**(4) 570–7.
150 Shoya, M., Sugita, S. and Sugawara, T. (1990). *Proceedings of the International RILEM Symposium*, 484–95.
151 *Admixtures for Concrete*. (1992). ACI Report 212-IR-92, Concrete International.
152 Russell, P. (1983). *Concrete Admixtures*, a Viewpoint Publication, Eyre & Spottiwoode Publications, 95–6.
153 Gaynor, R.D. (1978). *Ready-Mixed Concrete*, ASTM STP 169 B, 471–502.
154 British Standards (1976) BS 5328, 21.

155 Concrete Plant and Manufacturer's Bureau (1977). Sixth Revision, Silver Springs, Maryland, USA, 34.
156 ASTM (1980). Standard Specification C 94.
157 *NBS Handbook 44*, National Bureau of Standards, USA.
158 Anon. (1976). *Certification of Ready-Mix Production Facilities*, National Ready-Mixed Concrete Association, Silver Springs, Maryland, USA, 18.
159 Parsons, J.S. (1973). *Report on Ready-Mixed Concrete Equipment Ministry of Transport and Communications*, 36–112.
160 Bray, L.S. and Keifer Jr, O. (1964). *Journal of the ACI*, May, Title No. 61-36, 625–42.
161 Anon (1983). *Metering of Inorganic Pigments in the Manufacture of Coloured Concrete Products*. Technical Information Leaflet, Mobay Chemical Corporation, Pittsburg, PA, USA, 6.

Index

Abietic acid 106, 112
Abrams rule 46
Accelerating water reducing admixtures,
    definition 3
Accelerators
    chemistry 163–4
    definition 162
    effect of temperature 179
    effect on
        compressive strength 181, 196
        corrosion of reinforcement
            187–192
        degree of hydration 171
        durability 182–196
        heat evolution 178
        porosity 177
        setting time 170, 178–9, 180–1
        shrinkage and creep 196–7
        workability/rheology 164
    mechanism of action 176
    purpose of use 162
Admixtures
    selection 378
Adsorption isotherm 16, 18–19
Aggregates
    coarse, effect on air entrainment 126
    fine, effect on air entrainment 126
    grading deficiencies 127
    lubrication of 32
Aggressive liquids, attack by 51, 138
Air-detraining agents 9
Air-entraining agents
    effect on
        aqueous phase 115

bleeding 133
carbonation 144
cement hydration 116
cohesion 105
compressive strength 135
creep 145
durability 104, 137
flexural strength 136
modulus of elasticity 137
paste viscosity 110
reinforcement protection 143
resistance to aggressive liquids 138
shrinkage 144
standard deviation 291
sulfate resistance 139
tensile strength 136
water reduction 132
workability 125
Air entraining water-reducing admix-
    tures
    definition 3
    effect on
        bleeding 42
        compressive strength 46
        freeze–thaw resistance 138–43
Air entrainment
    by dampproofers 157
    effect of
        air entraining agent dosage 120–1
        cement characteristics 113, 122, 125
        coarse aggregates 126
        factors effecting 282
        fine aggregates 126
        fine fillers 127

Air entrainment (*continued*)
   mixer capacity 123
   mixing technique 122
   pozzalons 129
   temperature 126
   water–cement ratio 132
   water-reducing admixtures 128
   water-soluble alkalis 389
   workability 132
  mechanism of action 119
  mix design requirements 133
Alkali aggregate expansion reducing
   admixtures 200
Alkali aggregate reaction 200
Alkyl aryl sulfonates 107
  use in air-entraining agents 107
  use with lignosulfonates 4
Alkyl sulfates 106, 107
Aluminium stearate 151
Antifreeze admixtures 208
Anti-washout admixtures 212
  chemical types 212
  washout resistance 216
Applications of admixtures 199
Aqueous phase
  effect of
   air-entraining agents 116
   water-reducing admixtures 24
Arabinose 8

Beef tallow 152
Bentonite 153
Bitumen 151
Bleeding
  effect of
   air-entraining agents 133
   water-reducing agents 42
Bond
  between admixture and cement 16–20
Borate esters 9
Bubble size 130
Bubble spacing 142–3
Butyl stearate 151

C_3A hydrates 19
  adsorption isotherms 18
  effect of calcium lignosulfonate,
   morphology of 27

C_3A phases, adsorption of calcium
  lignosulfonate 17
Calcium chloride
  effect on
   cement hydration 165, 167–9
   compressive strength 181
   creep 197
   flexural strength 181
   freeze–thaw resistance 187
   microstructure 174
   morphology 172
   permeability 175, 182
   pore structure 175
   reinforcement bond 190
   reinforcement protection 187
   shrinkage 196
   sulfate resistance 183
  manufacture 163
  use with
   hydroxycarboxylic acids 5
   hydroxylated polymers 5
   lignosulfonates 5
Calcium formate
  effect on
   cement hydration 166
   morphology 173
   reinforcement protection 192
   shrinkage 196
  manufacture 163
  use with lignosulfonates 5
Calcium ion concentration 173
Calcium lignosulfonate
  adsorption by cement 17–20
  adsorption isotherm 18
  bonding with cement 20
  effect on
   morphology 26–7
Carbonation, effect of air-entraining
  agents 144
Cement characteristics, effect on air
  entrainment 122
Cement content, effect on air
  entrainment 123–4
Cement economies concept of 2
Cement, effect on water-reducing agents
  29–39
Cement hydration, effect of
  air-entraining agents 167

calcium chloride 167
calcium formate 169
calcium nitrate 169
triethanolamine 168
dampproofers 151
Citric acid 10
Cohesion,
    effect of
        air-entraining agents 133
        water-reducing agents 13, 41
    rheological considerations 12
Cold weather concreting 359–67
    reduction of freezable water 361
Compatibility of admixtures 394–7
Compressive strength effect of
    air entrainment 135
    calcium chloride 181
    dampproofers 157
    water-reducing admixtures 46–7, 66
Contact angles 149, 154
Corresponding mixes, definition 2
Corrosion inhibitors 219
    nitrites 222
    organic based 226
    phosphates 225
Creep
    effect of
        air-entraining agents 146
        calcium chloride 197
        hydroxycarboxylic acids 68
        lignosulfonates 67–9
        lignosulfonates with accelerators 68
C_3S phases, adsorption of calcium
        lignosulfonate 17

Dampproofers 149, 345
    chemistry of 150–3
    definition 149
    effect on
        compressive strength 157
        freeze–thaw resistance 159
        plastic concrete 156
        shrinkage 159
Degree of hydration, effect of
        accelerators 172
Desorption isotherms 21
Dibutyl phosphate 9
Dibutyl phthallate 9

Differential thermal analysis 21
Dispensers 402, 409–23
Dispersion of cement 22
Dosage
    effect on
        air entrainment 120
        water-reduction 38
        workability 33
Drying shrinkage, see shrinkage
Durability
    effect of
        air-entraining agents 137–43
        dampproofers 158

Economic aspects 367–72
Efflorescence 149
Entrained air, stability of 128
Expanding admixtures 227

Fatty acids 349
    effect on air entrainment 114
Fatty acid salts, solubility 114
Fatty acid soaps, use with
        lignosulfonates 9
Flexural strength
    effect of
        accelerators 181
        air entrainment 136
Flowing concrete 315, 324
Fluidity, of cement paste 13
Forces, interparticle 22
Freeze thaw resistance,
    effect of
        air-entraining agents 138–43
        calcium chloride 187
        dampproofers 159
        water-reducing agents 55–61
    mechanism of 141
Fructose 8
Fungal attack 409

Galactose 8
Gluconic acid 10
Glucose 8
Grouting 341–5

Heat evolution
    effect of

434   Index

Heat evolution (*continued*)
    accelerators 169, 170
    calcium lignosulfonate 27, 29
    sodium gluconate 27
Heptonic acid 10
High alkali cement, effect on workability
    loss 36
High performance concrete 292–7,
    330–7
High workability concrete, *see* flowing
    concrete
Hot weather concreting 351–9, 399
Hydrocarbon resins 151–2
Hydroxycarboxylic acids
    effect on
        bleeding 42
        compressive strength 46
        creep 68–72
        freeze–thaw resistance 58–9
        modulus of elasticity 48
        setting time 40
        shrinkage 68
        sulfate resistance 54
        tensile strength 46
        workability 31–3
    water reduction by 37
    workability loss 35
Hydroxylated polymers 11
    formula 11
    use in formulations 11

Impact pressure of raindrops 156
Inisitol 11
Initial slump for flowing
    concrete 91
Isothermal calorimetry 27

Labels 406
Large pours 301
Lignosulfonates
    adsorption isotherrns 16
    air entrained by 9, 31
    commercial types 6
    degree of ionization 6
    effect on
        aqueous phase 24
        bleeding 42
        compressive strength 46

creep 68–72
    freeze–thaw resistance 55–8
    modulus of elasticity 47
    reinforcement bond 63
    reinforcement protection 62–3
    shrinkage 67
    sulfate resistance 52
    temperature rise 27
    tensile strength 46
formula 6
molecular shape 7
molecular weight 7
rheology of cement paste 13
use in formulations 7
water reduction by 38
Lime–silica ratio, effect of calcium
    chloride 173
Lithium salts 201

Malic acid 10
Mannose 8
Manufacture 403
Marine structures 309
Melamine formaldehyde sulfonate
    adsorption on cement 85
    effect on
        compressive strength 94
        flow table spread 92
        initial slump 92
    molecular weight 79
    preparation 80
    rheology of cement paste 83
    typical characteristics 80
Mix design
    economies in 45
    for air-entrained concrete 133
Mixer capacity, effect on air
    entrainment 122
Mixing technique, effect on air
    entrainment. 122
Modulus of elasticity
    effect of air-entraining agents 137
    effect of dampproofers 157
    water-reducing admixtures 47–8
Molecular layers of water-reducing
    admixtures 19
Morphological forms of cement
    hydrates 27, 29

Morphology
  effect of
    calcium chloride 172
    calcium formate 173
    triethanolamine 172
Mucic acid 10

Naphthalene formaldehyde sulfonates
  adsorption isotherms 85
  effect on
    compressive strength 94
    paste viscosity 81
  molecular weight of 79
  solids content of 79
  water-reduction by 91
Neutralized wood resins
  effect on
    air void characteristics 113
    freeze–thaw resistance 142–3
    shrinkage 145
Non-chloride accelerators 163, 312

Oleic acid 107, 151

Packaging 404
Permeability
  effect of
    air-entraining agents 144
    water-reducing admixtures 48
Piling 300
Polyacrylates 80
Polymer-based admixtures 235
Pore structure effect of calcium
    chloride 175
Pozzalons
  effect of water-reducing agents 45
  air entrainment 129
Precast concrete 372–5
Prestressed concrete 329
Pumping air-entrained concrete 283

Reinforcement bond
  effect of
    calcium chloride 190
    water-reducing agents 63
Reinforcement corrosion, effect of
    calcium chloride 187
  thiocyanate 192

Reinforcement protection,
  effect of
    air-entraining agents 143
    calcium chloride 197
    calcium formate 192
    water-reducing agents 62–3
Retardation
  effect on aqueous phase 22
  relationship to adsorption
    isotherms 18
Retarding water-reducing admixtures
  definition 3
Retempering 401
Rhamnose 8
Rheology of pastes
  effect of
    air-entraining agents 109
    water-reducing agents 14–15

Salicylic acid
  complex 21
  formula 10
Segregation
  effect of air-entraining agents 133
Setting time
  effect of
    retarding admixtures 40
    water-reducing admixtures 40
Shrinkage
  effect of
    air-entraining agents 144
    calcium chloride 196
    calcium formate 196
    calcium lignosulfonate 67
    dampproofers 159
    hydroxycarboxylic acids 68
    lignosulfonates 67
    triethanolamine 197
Shrinkage reducing admixtures 265
Shotcrete admixtures 252
  aluminates 257
  carbonates 256
  hydroxides 256
  non-alkaline 257–8
  silica fume 259
  silicates 256
Slump relationship to flow table
    spread 92

Sodium abietate, effect on aqueous
    phase 116
Sodium dodecyl benzene sulfonate
    effect on air void characteristics 113
Sodium dodecyl sulfate
    effect on
        aqueous phase 114
        air-void characteristics 113
Sodium gluconate
    adsorption isotherm 18
    effect on
        aqueous phase 22
        monolayer formation 19
        paste viscosity 20
        rheology of cement paste
            containing 20
        water reduction by 37–8
        viscosity reduction index 20
Sodium lauryl sulfate effect on
    freeze–thaw resistance 142
Sodium lignosulfonate
    effect on
        paste viscosity 14–16
        setting time 40
        solubility 8
        typical analysis 6
Stearates, durability 158
Stearic acid
    effect on compressive strength 157
Stiffening time, effect of
    accelerators 179
Storage of admixtures 407
Sulfate resistance
    effect of
        air-entraining agents 138
        calcium chloride 183–7
        water-reducing agents 52–4
Sulfite lye, sugars in 8
Superplasticizers
    acrylate based 319
    adsorption on cement 85
    air entrainment by 90
    chemical types 78
    definition 77
    effect on
        bleeding 319
        cement paste viscosity 81–2

compressive strength 94
creep 99
freeze–thaw resistance 95, 101
setting time 93
shrinkage 99
sulfate resistance 101
water reduction 91
workability 91–2
workability loss 91, 321
zeta potential 82
    mode of action of 88–9
Surface area of cement constituents 19
Surface area of cements, effect of water
    reducing agents 29

Tartaric acid, formula 10
Temperature
    effect on
        accelerators 179
        air entrainment 126
Tensile strength
    effect of
        accelerators 181
        air-entraining agents 136
        plain concrete 47–8
        water-reducing agents 46–7
Tributyl phosphate 9
Triethanolamine
    effect on
        cement hydration 166
        morphology 172
        shrinkage 197
    manufacture 164
    use in formulations 9, 166
    with hydroxylated polymers 11

Underlayments 339
Underwater concrete 345–8

Van der Waal forces 16
Vegetable oils and fats 151–2
Vibration
    effect on air void characteristics 130
    recommendations 132
Viscosity reducing index 20
Volume deformations, see creep and
    shrinkage

Water-reducing agents
    air entrainment by 3, 4, 9, 30
    effect on
        air entrainment 30
        cohesion 13, 41, 43
        compressive strength 46
        mix-design requirements 45
        mode of action 29
        standard deviation 291
        workability 31
        workability loss 34
Water reduction
    effect of
        air-entraining agents 132
Water repellancy 149
Waterproofed concrete, *see*
    Dampproofers
Watertight concrete 299
Wax emulsions 349
    effect on compressive strength 157

Wind-speed effect on raindrop impact
    pressure 156
Workability
    control of 32
    effect of
        air-entraining agents 132
        hydroxycarboxylic acids 31–2
        lignosulfonates 31–2
        superplasticizer dosage 91
        water-reducing admixture
            dosage 33
        water reduction 34
Workability loss
    effect of
        high-alkali cement 39, 91
        retarding water-reducing agents 34, 38
        superplasticizers 91–3
        water-reducing agents 34

Xylose 8

Printed in the United States
by Baker & Taylor Publisher Services